DATE DUE

The Future of
the Cognitive Revolution

The Future of
the Cognitive Revolution

Edited by
David Martel Johnson
& Christina E. Erneling

New York   Oxford • Oxford University Press   1997

Oxford University Press

Oxford   New York
Athens   Auckland   Bangkok   Bombay   Buenos Aires
Calcutta   Cape Town   Dar es Salaam   Delhi   Florence   Hong Kong
Istanbul   Karachi   Kuala Lumpur   Madras   Madrid   Melbourne
Mexico City   Nairobi   Paris   Singapore
Taipei   Tokyo   Toronto

and associated companies in
Berlin   Ibadan

Published by Oxford University Press, Inc.
198 Madison Avenue, New York, New York 10016

Oxford is a registered trademark of Oxford University Press

Library of Congress Cataloging-in-Publication Data
The future of the cognitive revolution / edited by David Johnson and Christina Erneling.
        p.   cm.
    Includes bibliographical references and index.
    ISBN 0-19-510333-5; 0-19-510334-3 (paper)
    1. Cognition.   2. Cognitive science.   3. Philosophy and cognitive science.   4. Human
information processing.   5. Artificial intelligence.   I. Johnson, David.   II. Erneling,
Christina E., 1951– .
    BF311.R352   1996
    153—dc20   96-23813

9 8 7 6 5 4 3 2 1

Printed in the United States of America
on acid-free paper

Preface

The chapters in this volume are related to the conference, "Reassessing the Cognitive Revolution," that was held at the Glendon campus of York University, Toronto, in October of 1993. The original idea for the conference came from a debate on this subject between Jerome Bruner and Stuart Shanker, which appeared in the pages of the journal *Language and Communication*. As Johnson explains in more detail at the end of his Introduction, roughly half the chapters included here (the introduction itself, and the contributions of Shanker, Boden, Feldman, Pascual-Leone, Ross, Bialystok, Agassi, Bechtel, Segalowitz and Bernstein, Neisser, Reed, Bruner, Coulter, and Stenlund) are revised versions of either principal papers or of commentaries presented at that conference. The other half (Chomsky's two contributions, and those of Putnam, Green and Vervaeke, Clark, van Gelder, Shotter, Harré, Donald, Johnson, and Erneling's Afterword) were written later and were solicited especially for this volume.

We consistently tried to choose papers that would represent the very wide variety of issues, methods and standpoints present in cognitive science today. Accordingly, we have organized the chapters of the book in a thematic style that reflects not just the particular topics that happened to be discussed at the conference but also the more general situation of cognitive science today. More precisely, our guiding editorial idea has been first to summarize what seemed to us the most serious problems that the relatively new discipline of cognitive science faces at the present moment. Then, in the light of these difficulties, we posed the question of what were the main alternative directions its recent practitioners propose for this interdisciplinary field in the future. Because we intended the book to be accessible to as wide and interdisciplinary an audience as possible, we repeatedly encouraged authors to write in as clear and simple a way as they were able, consistent with considerations of accuracy, truth, and completeness. We also made various minor editorial changes to ensure a still greater degree of uniformity in style and accessibility.

We gratefully acknowledge the help of all those speakers, commentators, moderators, and discussants who participated in the 1993 York conference—including many whose excellent papers were not selected to appear here. We also owe a debt of gratitude to our co-organizers of the conference, Stuart Shanker, Stanley Tweyman, Ellen Bialystok, Andrea Austen, Melinda Hogan, Steven Rice, and Debbie Steele. Ron Schepper spent many long hours getting the book into something approaching publishable form, and we thank him for his dedication and kindness. Similarly, we appreciate the help of Joan Bossart and Joseph Schorn at Oxford University Press, who have been constantly positive and encouraging in guiding us through the often frustrating process of bringing a large-scale project to completion. We want to acknowledge help received from many other colleagues, graduate students, friends, and family members. Here it is especially appropriate to mention the names of Robert Cohen, Joseph Agassi, Jagdish Hattiangadi, and the York Centre for the Philosophy of Science. We acknowledge generous financial support received from many sources, including the York Departments of Philosophy and Psychology, the Graduate Programme in Philosophy, the Social Sciences and Humanities Research Council of Canada, and the Universities of York in Canada and Umeå in Sweden.

Finally, we wish to express our gratitude to our respective spouses, Barbara and Alf, for their unfailing support.

Toronto, Canada   D.M.J.
Umeå, Sweden   C.E.E.
August 1996

Contents

*The Future of
the Cognitive Revolution*

David Martel Johnson

What Is the Purported Discipline of Cognitive Science and Why Does It Need to Be Reassessed at the Present Moment?

The Search for "Cognitive Glue"

Aristotle's theory of soul presupposed that the distinction between plants and animals was clear, obvious, and exhaustive. According to this theory, all living things were possessed of "vegetative souls," which allowed them to grow, reproduce, and be nourished. Only in the case of a few, fairly complicated organisms—that is, animals—were there also "sensitive" souls, which rendered them capable of moving and perceiving. Fewer creatures still— namely, human beings at the top of the pyramid of life—had "rational souls" in addition to souls of the other two sorts, by virtue of which they were able to think and reason. When I was a student of philosophy at Yale in the 1960s, all this struck me as quaint and almost certainly mistaken. But at least it seemed to encapsulate one important and genuine division among living things. Thus, it came as a surprise when I learned that the university administration planned to close the Botany and Zoology departments and transfer all associated staff to a single, combined department to be called something like Life Sciences. One of my roommates at the time was a biologist. When I asked him about this administrative change, he assured me it was not primarily a cost-cutting measure, nor a piece of empire-building on the part of politically ambitious individuals. Instead, he claimed it was a natural, even inevitable development out of certain facts that scientists had discovered in the relatively recent past.

Think, for example, of the lowly mushroom. Naive observers might take organisms of this sort to be plants. But closer inspection shows they reproduce by means of spores rather than seeds, that their metabolism has nothing to do with photosynthesis, and that they possess many other non-plant-like characteristics as well. Such properties—together with similarities between mushrooms and certain other living things like molds—have forced taxonomists to locate them as well as the other organisms just mentioned in a new category ("fungi"), which they now consider just as basic as the more ancient categories of plants and animals, but separate from both.

3

At present, biologists still argue about exactly how many major types of living things they need to recognize, and what exactly the relations are between the categories in question. (See, e.g., Gould, 1980, pp. 204ff, and Wilson, 1992, *passim.*) But, as a result of the empirically inspired, post-Aristotelian reorganization of this field, they now agree almost unanimously that any properties that separate kinds are not as important as the single factor that binds all (or almost all) of them together. To be explicit, most biologists now suppose the vast majority of forms, genera, and species of life so far discovered are expressions of just one molecular structure — deoxyribonucleic acid, or DNA for short.[1] And, because of this, references to the structure of DNA have become the basis for a shared conceptual framework, in terms of which biologists are able to relate and adjudicate the many different branches, aspects, and problems of their discipline. For example, E.O. Wilson refers to this "common language" in the following (1992, p. 75):

> What drives evolution? This is the question that Darwin answered in essence and twentieth-century biologists have refined to produce the synthesis, called neo-Darwinism, with which we now live in uneasy consensus. To answer it *in modern idiom* is to descend below species and subspecies to the genes and chromosomes, and thence to the ultimate sources of biological diversity. (Italics added.)

"Cognitive science," or the proposed combination of artificial intelligence, psychology, philosophy, linguistics, neurophysiology, anthropology, etc., as discussed and explained, for example, in Howard Gardner's book (1987), is apparently parallel. What I mean is that those people who refer to themselves as cognitive scientists hope someday to be in roughly the same position that biologists now occupy. In other words, cognitive scientists speculate that observationally discovered facts eventually will compel them to accept the idea that all the disciplines listed above are basically identical, and differ merely in surface aspects like the methodology each employs, the special, limited part of the unified field each examines, and so on. At present, however, this remains a hope and aspiration, but not a fact.

Are we justified in supposing that there is some one type, sort, or kind of things that is capable of playing approximately the same unifying role in cognitive science that DNA plays in modern biology? If so, then the name it would be plausible to assign to this type is "mind." (Thus, presumably, eliminative materialists would answer "no" to the question just asked, and all other theorists of mind would say "yes.") To provide a background for the main themes discussed in the 25 chapters brought together in this book, I want to review three successive attempts to answer the preceding question positively which philosophers and scientists have proposed over the course of the twentieth century. The three positions of which I speak are "structuralism," influential roughly from 1880 to 1920; "behaviorism," from about 1920 to 1960; and "(philosophical) functionalism," from 1960 to the present.

Structuralists — whom many people also called introspectionists — claimed that conscious experience was the "glue" that bound together all the cognitive sciences. In fact, they also maintained that experience of this sort was the unifying factor for the non-cognitive sciences as well, as shown by their further assertion that, whereas the subject matter of psychology was conscious experience considered as relative to human beings, the proper subject matter of physics was conscious experience relative to nothing at all. (See Heidbreder, 1933, p. 122.)

It will be clarifying to focus on a particular case. Although Edward B. Titchener taught psychology at Cornell University in the United States for many years in the late nineteenth and early twentieth centuries, he continued to think of himself during all that time as a follower of Wilhelm Wundt in Germany, with whom he briefly had studied early in his career. I consider him a useful representative of structuralism, if only because his account of conscious experience as the unifying link among all the sciences was especially simple, direct, and explicit. For instance, one of his intellectual projects was to compile an encyclopedic list of all the basic forms such experiences could take.[2] The reason he thought this exercise was worth doing was that it was theoretically possible to distinguish each science in terms of its special way of "compounding" these various elements. Another example: Titchener supposed nothing else was available to constitute the meaning of the thought a person typically expressed by means of a word—that is, "green," "rabbit," "prestidigitation," etc.—except some specific (type of) conscious mental image. Thus, trained researchers ought to be able to determine—and agree on—the (inner) meaning of any such word, by the simple process of noticing the image(s) that occurred in their respective fields of consciousness whenever they spoke or silently contemplated this word in a meaningful way.

As a matter of historical fact, however, Titchener and those influenced by him never managed to amass the large body of universally accepted experimental results they originally envisaged. Some historians of science say or imply that the crucial factor that undermined the credibility of Wundtian/Tichenerian psychology was attacks on it by behaviorists like J. B. Watson. (See, e.g., Boring, 1957, p. 419 and Gardner, 1987, p. 109.) But it seems to me that the internal dynamics of this program made its ultimate destruction inevitable—a case of suicide rather than murder. For instance (a point behaviorists were later to emphasize), self-examination shows that there are many sorts of mental states that are not regularly associated with corresponding (kinds of) conscious experiences like (e.g.) mental images. Thus, a person cannot distinguish, just by paying attention to his or her inner experiences, the belief, desire, hope, or fear that it is raining, from the different belief, desire, hope, or fear that it is snowing, because mental states of these sorts do not feel (look, sound, etc.) like anything at all.

A still more fundamental difficulty with the structuralists' program is that its recommended method made it "too easy" to arrive at hypotheses that were supposed to be both scientific and confirmable. Thus, imagine structuralist researcher *A*, who concludes from examination of his inner experience that the meaning of the thought usually expressed by word *W* is image *I*; and structuralist *B*, who forms the equally sincere conviction on a similar basis that the meaning of the same thought that also occurs to her is image *J*. The theoretical approach these two people share does not provide a means of adjudicating this disagreement. To state the same point in broader terms, Titchenerians never were able to specify any natural limits, checks, or controls on introspective observation, of a sort that might have assured the scientific legitimacy and usefulness of the method they proposed. For example, they often made the explicit demand that every result of bona fide scientific procedure should be replicable by other investigators. (See, e.g., Heidbreder, 1933, p. 131.) But appeals to introspection did not provide a sufficient means of meeting this demand. This shows it was inevitable from the beginning that one structuralist's findings never would manage to duplicate (and hence also confirm) the findings of other structuralists.

The focus of behaviorism—our century's next widely accepted conception of cognitive science—was the "reflex arc," or the direct, unmediated relation between an organism's inputs or stimuli on one side and its outputs or responses on the other. Those who subscribed to this general view maintained that the mind was not something hidden, but overt; because of this, ordinary, nonintrospective observation ought to be a sufficient means of answering every question one possibly could ask about it. In other words, behaviorists did not consider the actions of humans and other animals (including what humans said) to be a mere set of clues about the identity and properties of certain underlying mental causes. Instead, they took such things as constitutive of mental states themselves. Furthermore, they maintained that the "insides" of mind, or that which presumably mediated experiential inputs with behavioral outputs, was an unknown but practically and theoretically unimportant "black box." Or, to adopt Ludwig Wittgenstein's alternative metaphor from mathematics, they supposed it was something for which a person always could "divide through." They accepted this view because, according to them, the essence of all science lay in the fact that it was empirical and observational. Therefore there seemed no other way in which the study of mind could be legitimately scientific. (See, e.g., Watson, 1930/1990, p. 14.)

However, like structuralism before it, behaviorism's period of ascendency proved relatively brief. For at least three reasons, the great majority of philosophers, psychologists,[3] and other cognitive scientists no longer subscribe to this approach. First, most people now think behaviorists' claim that introspection is not able to tell us anything about mental contents and properties is a perverse denial of something that every individual knows intuitively. For example, ordinary experience shows that a normal person under normal circumstances can discover by inward reflection as opposed to external observation such facts as that he or she now feels a pain behind the left eye, or is now tasting the flavor of strawberries, or has just remembered the whereabouts of his car keys.

A second reason for rejecting behaviorism is this. Imagine a critic who claims it is incoherent to identify mental states with certain sets of behaviors, since a mental state obviously could remain the same, even if all the behaviors with which it once was associated were changed. Behaviorists' standard reply to this objection was first to grant that behavior alone was not a sufficient basis for analyzing a mental state, but then to insist further that this difficulty could be overcome by also taking account of dispositions to behave. For example, no behaviorist can produce a convincing account of Brown's belief that Henry VIII of England married a series of six different queens, just by compiling a list of things Brown did and said. But an analysis of this sort again becomes possible if, in addition, the behaviorist takes account of such things as Brown's proclivity to accept "6" as the number of Henry's queens and reject all other numbers; her disposition to recommend certain published accounts of Henry's life as accurate and condemn other sources as mistaken; and so on and on. But the trouble with this reply is that dispositions also cannot be a key to an adequate answer, because there is an even more intractable problem implicit in the seemingly innocent phrase, "and so on." What I am talking about is the fact that behaviorists have no principled way of preventing their lists of dispositions from becoming not just lengthy, but infinitely long, which means that their recommended accounts are not merely incomplete, but uncompleteable in principle.

To summarize the preceding point, consider a contrast. I claimed before that one of the main problems with Titchenerian structuralism was that it was "too loose." In other words, the theory had no resources to prevent its practitioners from proposing supposedly scientific hypotheses in casual, offhand, merely subjective ways that were not under proper controls. But on the other side, behaviorists apparently were guilty of making the opposite mistake of being "too rigorous." That is, the methodological principles to which these theorists subscribed seemed to have the result of making it impossible for someone to arrive at any scientific hypotheses about mind at all.

Finally, it is not only the case, as behaviorists supposed, that every mental state has an essential connection with certain stimuli (causes) and responses (effects). It also is true that each mental state is essentially connected with other mental states. For instance, all beliefs presuppose (the existence and effectiveness of) certain desires—and, correlatively, all desires presuppose certain beliefs. We see this when we consider cases like the following. Behaviorists (e.g., Gilbert Ryle, 1949) say a crucial component in X's belief that the ice on the lake is thin is X's active—not merely verbal—refusal to stand or walk on it. But this is only plausible if one assumes that the person in question has the desire not to die by drowning. On the different assumption that X is suicidal, and therefore has the opposite desire, it becomes reasonable to expect that his belief that the ice is thin will be expressed precisely by a willingness to walk on it.

The third and last account of cognitive glue I want to discuss here is one that many (and perhaps a majority of) philosophers and scientists accept today—the so-called functionalist theory of mind (in the philosophical rather than psychological sense of this word). Several developments in technology and mathematics played a role in the introduction and elaboration of this view. One was the invention by people like Norbert Wiener of complex "self-reading" or "feedback" control systems for machinery—that is, anti-aircraft guns. Another was the working out of mathematical information theory by Shannon and Weaver, among others. But the single most important technological inspiration for the functionalist account of mind was the creation of the digital computer—an achievement based on the other two just mentioned, as well as theoretical work by researchers like Alan Turing and John von Neumann.

Why did people suppose these achievements had a bearing on the ancient problem of how one should understand mind? The basic reason was that they provided apparent counterexamples to the Watsonian dogma that the only legitimately scientific way of talking about mental states and properties was to specify them exhaustively in terms of observable behavior. To be more concrete, these technological developments held out the possibility of creating detailed models of "the mind's inner workings" on a middle or intermediate level of mental representations, which explained and related the two, more obviously observable levels of sensory inputs on one side and behavioral outputs on the other.

Early supporters of functionalism (e.g., Turing machine functionalists) insisted that human beings were literally nothing more than digital computers. (See Putnam, this volume.) This meant that humans' sensations and actions had to be—respectively—data input to, and computational output from, such a computer. Furthermore (a point that was supposed to distinguish functionalism from behaviorism) inner mental states were identical with abstractly specified and relationally defined computational states of a machine table—or, in other words, the list of software instructions that causally controlled the computer's operations. However, as Putnam points out in his chapter, the definition of a Turing machine specifies that any such device cannot be in more than one state at a time. Accordingly, Turing machine functionalists were forced to suppose that memory, learning, and action were not states of the organism (machine) itself, but instead took the form of information somehow present both in an organism's input from, and also output to, its environment. Yet— to put it mildly—we have no good reason to suppose human minds really are "constructed" along these lines.

Another problem with functionalism arises from the fact that one apparently can assign an indefinite number of equally valid interpretations to the same complex of computational states. For example, we might interpret a given web of internal, abstract causal states in a computer's memory as the political and military structure of the ancient Persian empire at the time of Xerxes, or as a hypothetical hurricane system, or as the Ugandan economy in 1973, or as a spiral galaxy located in the direction of the constellation Leo, or as a whirlpool, and so on. But functionalists want to say that mentality occupies a different, and more basic position than hurricanes, galaxies, and the rest. The reason is that minds are supposed to *be* computers in a way these other things are not; therefore minds can claim to supply the "natural" interpretation for a particular set of computational states in a sense that does not also apply to the others. Yet, as far as I know, no functionalist has succeeded in justifying the idea that minds have this special status, and it is hard to see exactly what form such a justification could take.

People commonly refer to the mid-twentieth-century advent of the functionalist theory of mind as the "cognitive revolution." (For example, we find these words in Howard Gardner's subtitle, 1987.) But I think this name is misleading for the following reason. The functionalist theory only appears to have a revolutionary impact, if, when, and insofar as one contrasts it with the behaviorist theory that immediately preceded it. (For example, the accounts of functionalism given by Ned Block, 1980, 1981, and by William Lycan, 1990 present the matter this way.) But one no longer thinks of functionalism as revolutionary when one locates it in a wider context—for example, when, as I have done here, one contrasts it not just with behaviorism but also with structuralism. It is plausible to suppose that a revolution should involve some sort of revolving, where what before was the top becomes the bottom and what previously was the bottom becomes the top. For instance, behaviorism's replacement of structuralism was a clear case of a revolutionary development because, although the former identified mind with inner as opposed to outer events, the latter claimed the exact opposite. (For this reason, the psychologist Donald Hebb used to refer to the change from structuralism to behaviorism as "the American revolution"; see Edward Reed, this volume.) But functionalism was different. It did not propose to negate or displace either of its predecessors, but instead amounted to a conservative compromise designed to combine—and thereby preserve—elements and aspects of each of these forebears.

Let me spell out the last claim more explicitly. As opposed to behaviorism, functionalists grant that consideration (including introspective examination) of inner mental states can give us knowledge of at least some of the contents and properties of our minds. Furthermore, as opposed to structuralism, they also maintain that ordinary, external observation is a source of (some) knowledge about mental states. This shows that functionalism is essentially a combination.

All this having been said—and in spite of certain enthusiastic claims of its proponents—I do not think functionalists have succeeded in salvaging the best, most defensible points from these two earlier theories, and avoiding all their weaknesses. For example, functionalists have been so preoccupied with avoiding the problems associated with behaviorism that they failed to notice that they again have opened themselves to one of the main difficulties associated with structuralism. The difficulty I am talking about is that of making it "too easy" for a person to propose uninformed, ad hoc hypotheses about the nature of mind, as contrasted with hypotheses that are objective and empirically significant. To restate: I noted reasons in the previous discussion for concluding that introspection was not a sufficient means of distinguishing valid, well-confirmed accounts of mental states from other, more questionable hypotheses that lacked scientific merit. In similar style, I now point out that one also cannot draw such a distinction by appealing—as "cognitive glue"—to computer-program-like, abstract causal complexes of inputs and outputs, combined with mechanisms for selecting, relating, and transforming inner states. For instance, suppose a functionalist like Ray Jackendoff sketches a plausible-sounding, apparently adequate complex of computational states as a proposed explanation for various phenomena associated with consciousness (as he did in 1987). There are no scientific observations, experiments, or theoretical considerations capable of preventing another functionalist (e.g., Chomsky) from proposing some different, incompatible abstract structure with just as strong and valid a claim to be the "right" explanation of these same phenomena.

Accordingly, functionalists and their critics face two important questions today. The first is what, if anything, is one able to salvage (i.e., defend successfully) from this special way of thinking about the mind? (Again, see Putnam, this volume.) Similarly—if there are reasons for thinking that functionalism needs to be replaced by one or several different approaches—in what directions should we expect theorizing about mind to proceed in the future? The chapters in this volume explore these questions from several angles, and propose a correspondingly wide and varied set of answers.

Early versions of roughly half these chapters (Bruner, Shanker, Agassi, Boden, Feldman, Ross, Reed, Neisser, Stenlund, Bechtel, Bialystok, Pascual-Leone, Coulter, and this introduction of my own) were presented at a conference entitled "Reassessing the Cognitive Revolution: An Interdisciplinary Seminar." It was organized by myself, Christina Erneling, Stuart Shanker, Ellen Bialystok, and Stanley Tweyman, and held at Glendon College of York University, Toronto, on October 22–24, 1993. After the conference, it occurred to Christina Erneling and me that several dimensions of the general topic still had not been discussed at sufficient length. Therefore, in putting together this book, we decided to solicit contributions from other people as well. (Thus, it now also includes two chapters by Chomsky, and one each from Putnam, Harré, Clark, Donald, van Gelder, Shotter, Segalowitz and Bernstein, Dror and Dascal, Green and Vervaeke, myself, and Christina Erneling's Afterword.) We consider it a positive quality of the book that, as contrasted with predecessors like Block (1980, 1981) and Lycan (1990), it is not a text in philosophy alone, but one that treats problems in a genuinely interdisciplinary way. Furthermore, the fact that all the chapters with the possible exception of one (van Gelder) are original rather than reprints means that the book has a good claim to be a representative snapshot of an important and changing field, as it exists at the present moment.

Notes

1. The exceptions—if one counts them as living—are the so-called retro-viruses, which are based on RNA rather than DNA.
2. Cf. Fancher (1979/1990, p. 170):

> From such painstaking introspections, Titchener concluded that there exist more than forty-three thousand distinguishable elements of sensory experience, more than thirty thousand being visual and eleven thousand auditory in nature. He found just four specifiable elements involved in taste, and three in the sensations of the alimentary tract.
>
> True to his premise, Titchener believed he found an elemental sensory base for virtually everything [i.e., every mental state] he analyzed introspectively, even including Wundt's key processes of apperception and attention.

3. Psychologists often accept behaviorism as a methodological principle, but not also a description of what exists. Thus, for example, although they are careful to express data, results, tests, criteria, procedures, etc. in terms of behavior in order to bestow properly objective status on their experiments, they do not suppose this commits them to the further doctrine that mental states are identical with behavior.

References

Block, N. 1980. *Readings in Philosophy of Psychology,* Vol. One Cambridge: Harvard University Press.

Block, N. 1981. *Readings in Philosophy of Psychology*, Vol. Two. Cambridge: Harvard University Press.

Boring, E. G. 1957. *A History of Experimental Psychology*, 2nd ed. New York: Appleton-Century-Crofts.

Fancher, R. E. 1979/1990. *Pioneers of Psychology*. 2nd ed. New York and London: W.W. Norton.

Gardner, H. 1987. *The Mind's New Science: A History of the Cognitive Revolution*. New York: Basic Books.

Gould, S. J. 1980. *The Panda's Thumb: More Reflections in Natural History*. New York and London: W.W. Norton.

Heidbreder, E. 1933. *Seven Psychologies*. New York: Appleton-Century-Crofts.

Jackendoff, R. 1987. *Consciousness and the Computational Mind*. Cambridge: MIT Press.

Lycan, W. G. (ed.). 1990. *Mind and Cognition: A Reader*. Cambridge and Oxford: Basil Blackwell.

Putnam, H. (this volume). "Functionalism: Cognitive Science or Science Fiction?"

Reed, E. S. (this volume). "The Cognitive Revolution from an Ecological Point of View."

Ryle, G. 1949. *The Concept of Mind*. London: Hutchinson.

Watson, J.B. 1930/1990. "An Excerpt from 'Talking and Thinking,' Chapter 10 of *Behaviorism*," pp.14–22 in Lycan (1990).

Wilson, E. O. 1992. *The Diversity of Life*. Cambridge: Harvard University Press.

PART ONE

David Martel Johnson

GOOD OLD-FASHIONED COGNITIVE SCIENCE

Does It Have a Future?

What does it mean to speak (paraphrasing John Haugeland, 1989) of "good old-fashioned cognitive science"? We propose to use this phrase as a name for the general view that—as also expressed by Putnam in his chapter in this volume—a living, thinking, intelligent human being is a digital computer made of flesh and blood. Another way of stating the same idea is to say that a computer of this type—or, more correctly, the software the computer runs and instantiates—is a (simplified) model of the human mind, which illustrates the basic principles of organization and functioning that the mind actually employs. Still more narrowly, good old cognitive science is the doctrine that theorists are justified in analyzing every human thought as a series of computations, in the (classical) sense of rule-governed changes. And because of this, one cannot understand the true nature of the mind either by focusing on what people ordinarily do and say, or on the structures, processes, and functions of people's brains. Instead, it is necessary to pay attention to an "intermediate" ("software") level of abstract principles presumably instantiated (at least in a rough and approximate way) by these other two, more concrete levels.

The chapters included in this first, introductory part of the book pose the question of whether we reasonably can expect its defenders to revise, elaborate, and extend this once popular view successfully into the future, or whether the many problems associated with this conception will force people to abandon it and adopt a substantially different approach. The answer Noam Chomsky proposes (or presupposes) in the first chapter is that we have no convincing reasons for suspecting that good old cognitive science is not essentially sound. Therefore it should be possible to solve all its seeming difficulties by the simple method of working it out in a more careful, exhaustive, and empirically justified way than people did in the past. In particular (as his and the other chapters in Part Two show in greater detail), Chomsky considers analysis of language an appropriate instance and guide to confirming the correctness of a computational account of the mind in general, because language is the one part, aspect, or "module" of the human mind, for which we possess the greatest amount of generally accepted information. By contrast, Putnam argues in the chapter immediately following Chomsky's that any such project of revision and repair is bound to fail because the problems associated with this now classical account (also sometimes called "functionalism") cannot be overcome by mere "fiddling," but require a radical shift in viewpoint.

The other chapters in this part are split in the type of answers they endorse. On one hand, in different ways, Boden, Feldman, Pascual-Leone, Ross, and Bialystok all say "yes"—that, despite its problems, good old-fashioned cognitive science is (or has to be) salvagable in one form or another. On the other hand, Shanker takes a more pessimistic view, and, like Chomsky and Putnam, answers "no."

References

Boden, M. 1977. *Artificial Intelligence and Natural Man*. New York: Basic Books.

Chomsky, N. 1972. *Language and Mind*. New York: Harcourt Brace Jovanovich.

Haugeland, J. 1989. *Artificial Intelligence: The Very Idea*. Cambridge: MIT Press.

Johnson-Laird, P.N. 1988. *The Computer and the Mind*. Cambridge: Harvard University Press.

McCorduck, P. 1979. *Machines Who Think*. New York: W.H. Freeman and Co.

Pylyshyn, Z. 1984. *Computation and Cognition: Toward a Foundation for Cognitive Science*. Cambridge: MIT Press.

Noam Chomsky

Language and Cognition

The study of language and the study of cognition are ancient disciplines, extending back thousands of years. Often the inquiries have proceeded in isolation from one another, sometimes as a matter of principle, particularly in the more self-conscious modern period. There also have been attempts to integrate these studies in a more unified discipline.

Throughout this history, a number of fundamental questions have come to the fore that have far-reaching import concerning human nature, questions concerning thought and its expression, knowledge and understanding and their origins, the sources and nature of action, and the bases for human interactions with the material and social world.

In the current period, a number of these questions, which had languished and been largely forgotten, have been revived and addressed within the framework of what has been called "cognitive science," sometimes the "cognitive revolution." George Miller, who was one of the most influential figures in this change of course in the study of mind and behavior, has traced its origins to a conference at MIT in September 1956, where papers were presented on human experimental psychology, computational models of reasoning, and generative grammar, all in their incipient stages.[1]

The time and the place were appropriate for a renewal of these classical concerns. There was a great deal of interest at that time, and in that place, in the project of unified science, a common naturalistic approach to all problems that can be addressed by rational inquiry. The development of quantum theory had explained "most of physics and all of chemistry" so that "physics and chemistry have been fused into complete oneness . . . ," to borrow the words of Dirac and Heisenberg. A dramatic recent step forward was the explanation of the nature of life in biochemical terms. These processes of linking of levels of understanding took markedly different paths; large parts of biology were reduced to known biochemistry, while physics had to be "expanded" and radically modified to incorporate known chemistry. These achievements eliminated mysterious vitalist notions while introducing new and no less puzzling ideas about the nature of the material world. It was also a period of considerable technological euphoria, particularly with digital computers just becoming available, and this mood was enhanced by new intellectual tools of information theory, cybernetics, and the behavioral sciences. In the background, there was an element of triumphalism in the political culture, which doubtless contributed to the sense that barriers to progress were rapidly falling. In this atmosphere, it was natural to confront the next great challenge after the unification of organic and inorganic processes: the human mind and its manifestations.

Some might argue that the terms "cognitive science" and "cognitive revolution" promise more than the substance can bear. The judgment depends on the answers to several questions, among them: Has there been a further unification of psychology and the core natural sciences, either by reduction or expansion or modification of foundations? More narrowly, have unifying ideas and principles been found that truly integrate the various domains of the cognitive sciences: visual perception, motor control, language and its use, thought and planning, questions of intentionality and belief-desire psychology, and so on? Still more specifically, how has our understanding of particular topics advanced over the past 35 years?

I would like to touch on some of these questions, but only in a general way and at a level of superficiality for which I apologize in advance, from a standpoint that reflects personal interests in the study of language from a point of view close to that of my paper on language at the 1956 Cambridge conference. As points of reference, I would also like to consider several approaches to the cognitive sciences, including the conception formulated by those who have sought most carefully to elucidate the nature of this discipline in recent years; and an approach that we can extract, realistically I think, from what might be called "the first cognitive revolution" of the seventeenth and eighteenth centuries—one way of assessing how much progress we have really made.

Consider first the two projects of unification: the broad project of linking psychology with the core natural sciences, and the narrower search for common grounds for the cognitive sciences. These projects differ in character. A general faith in the unity of nature may lead us to expect that the broad project has a target, whether we are capable of reaching it or not; there must, we assume, be links between the true theory of the workings of the mind and other parts of the natural sciences. The narrow project of unification may, however, have no issue because there is nothing much to be discovered. There is no prior reason to suppose that there are overarching principles that unify various cognitive systems and processes beyond the level at which they fall together with many others. It is a traditional view that there are indeed mechanisms of general intelligence and the like that operate across all cognitive systems, and this has also been a guiding intuition in more recent thinking. In domains in which anything much is understood, the assumption appears to be without foundation, and as a guide to research it has hardly been productive. That should occasion no surprise. There is little reason to suppose that alone among the complex biological structures of the world, the human mind/brain is a relatively unstructured and homogeneous system, and the evidence suggests otherwise.

One leading theme of the modern phase of cognitive science—some would say, its defining characteristic—is an interest in computational models of mind, the thesis that perception, learning, judgment, planning, choice of action, and so on, can be illuminated by thinking of them as software problems. Thus, on the receptive side, we consider the mind in a certain state, presented with an external physical array represented as signals to the mind, which lead it to assume a new state and to form some new symbolic representation of the input signals. If we focus attention on the new representation and how it is produced, we are studying perception; if we focus on the state change, we are studying learning or growth, as when we study changes that are internally induced through maturational processes, possibly also triggered by external events. When we focus on the states themselves, we are studying cognitive systems. Insofar as some component of the mind has reached a relatively stable state, we can distinguish more clearly between perception and learning or growth, something that was not done during the first cognitive revolution. As objects of inquiry, then, we have the cognitive states themselves and processes of several kinds, including the mapping of signal to symbolic representation in a given state, and the state transitions induced by external events or internal maturation. This inquiry has been quite fruitful in certain domains, language being one. That is not surprising, since language is an infinite digital system, hence tailor-made for computational approaches.

In the language case, a particular state of the language faculty can be taken to be a computational system of rules and representations that generates a certain class of structured expressions, each with the properties of sound and meaning specified by the language in question. We can seek to determine the initial state of the language faculty and to find out how internal processes and external events induce state changes to a relatively stable final state. And we can investigate the parsing processes that map signals to percepts, related more or less closely to structured expressions of the language being processed. For certain parts of language, parsing is fast and generally accurate, in that it matches the structures of the internalized language. It is also well known that, in general, the parser and the language are not closely matched in this sense, so that parsing is often difficult or fails outright. Contrary to what is sometimes supposed, it is not a criterion of adequacy for grammars that they allow for parsing that is easy and quick, because the real-world objects do not have this property in general, though they do for the parts of language that are readily used, a near tautology.

The second cognitive revolution, if we may call it that, introduced an important change of perspective, from the study of behavior and its products to the inner mechanisms that enter into how behavior is interpreted, understood, and produced. Behavior and its products, such as a corpus of utterances, are no longer the object of inquiry, but are just data, of interest insofar as they provide evidence for what really concerns us, the inner mechanisms of mind and the ways they form and manipulate representations, and use them in executing actions and interpreting experience.

This avowed mentalism should be understood as a relatively uncontroversial realist perspective. It undertakes a study of a real object in the natural world, the brain and its functions. From this point of view, a grammar of a particular language and what is sometimes called "universal grammar" are theories of states of the brain, an attained state of the language faculty in the first case, the initial state prior to external experience in the second. The study of these states of the brain is a normal part of the natural sciences, and is answerable to any kind of evidence, in principle. The states, representations, mappings, principles, and conditions postulated are properties of the brain, described at a certain level of abstraction from mechanisms that we assume to exist, though we do not yet know what they are, a familiar story in the natural sciences. The terms of this account are legitimate, and its ontological assumptions justified, if they provide for adequate description and insightful explanation, and open productive research programs for inquiry.

The shift in perspective was controversial at the time, and remains so today. One leading structuralist, Fred Householder, commented derisively (1966) that "A linguist who could not devise a better grammar than is present in any speaker's brain ought to try another trade." From the naturalistic point of view, this would be like saying that a biologist who could not devise a better circulatory system than humans have ought to try another trade. On the non-naturalist grounds that underlie this critique, the tacit assumption seems to be that there is some object, a language, independent of speakers and their brains; the speaker may have some confused picture of this object, but a qualified linguist ought to be able to find a better one. Or perhaps the assumption is that there are certain functions that should be fulfilled, say, effective communication; the speaker does it one way, but the linguist should be able to find a better way. Whatever is meant exactly, this position contrasts with the one I have just described, which does not assume the existence of any such external object or any range of functions to be fulfilled, and does not suppose that the speaker has a theory or picture of anything. Rather, the speaker's language faculty has reached a certain state, and the linguist tries to find out what it is, the linguist's grammar being a proposal to this effect.

More recent critics argue that it is not only wrong but even immoral, "irresponsible" some allege, to regard one's theory of a language, or of language in general, as a theory of a state of the brain, thus subject to the conditions of any other scientific theory. Since there are no other objects in the natural world that the linguist's theory is about, the demand apparently is that the linguist construct a theory of some non-natural object. Again, the tacit—and sometimes explicit—assumption seems to be that there are entities independent of what people are and what they do, and these objects are what theories of language are about, and further, *must* be about, on pain of irresponsibility. Again, we are left in the dark about these curious entities and how we are to identify their properties. Considerations of communication, the theory of meaning, the theory of knowledge, and folk psychology have also been adduced to argue that there are independent entities, external to the mind/brain, of which each of us has only a partial and partially erroneous grasp—always leaving as a mystery the manner in which they are identified, except by stipulation, and what empirical purpose is served by assuming their existence.

I think there are ample grounds for skepticism about all of these moves, but without exploring that matter here, even if there is some reason to believe that these entities exist and to study them, the claim that this is the only valid endeavor, and that one *must* reject the naturalistic approach of mentalist linguistics, has been supported by no serious argument. The legitimacy of the study of the states of the brain implicated in language use can hardly be denied. We can ask whether particular theories of these states and their functioning are true or false, but the injunction that some strange category of researchers called "linguists" must keep to a different enterprise is not easy to comprehend.

This shift to a mentalist perspective was a step toward scientific naturalism in two major respects: with regard to the prospects for unification, and the nature of evidence. The project of unification seeks an accounting of properties, structures, and functioning of the brain in terms of entities and principles of the core natural sciences, possibly modified to achieve this end; it can progress insofar as we have some grasp of these properties, structures, and processes. As regards data, the standpoint of the natural sciences is instrumental. Data are important only insofar as they provide *evidence*, that is, evidence *for*, a relational notion. The goal is to find evidence that provides insight into a hidden reality. Adopting this approach, we abandon the hopeless search for order in the world of direct experience, and regard what can be observed as a means to gain access to the inner mechanisms of mind.

From this point of view, there is no privileged category of data for the study of language, or any other topic of the cognitive sciences. The contrary position is, however, widely held. Thus, it is common to distinguish between two kinds of data, termed "linguistic" and "psychological," the former including judgments concerning form and meaning, the latter the results of other kinds of experiment. Advocates of what they call "Platonistic linguistics" hold that only linguistic data are relevant to determining the properties of English, an abstract object of which you and I have only a limited grasp. "Linguistic evidence" is relevant to linguistics, "psychological evidence" to psychology. The regular practice of leading figures in linguistics is then fundamentally in error. For example, the efforts of Edward Sapir and Roman Jakobson to find perceptual correlates to phonological properties, and to test the validity of phonological theory in these terms, involve a kind of category mistake. If the study of on-line processing reveals differential priming effects that correspond to, or conflict with, certain assumptions about the position and typology of empty categories, these results have no bearing on the theory of empty categories—contrary to what just about any practicing linguist would assume.

Adopting the same assumption, others argue that while the psychological evidence bears on "psychological reality"—meaning truth, presumably—the linguistic evidence is not of the right *kind* for determining truth or falsity. Quine, for example, has held that standard linguistic arguments for the placement of phrase boundaries leave the matter completely unresolved, though click experiments do bear on where these boundaries are. In this case, the so-called "linguistic evidence" is far richer and more compelling, but it seems not to be properly certified for establishing matters of fact, so it doesn't matter how powerful it is.

Similar attitudes are implicit in the position of Donald Davidson, Michael Dummett, and others, that the theories we construct only specify a body of knowledge that would be required to speak a given language. We should not proceed further to attribute this body of knowledge to the speaker. For such attribution, some other kind of evidence is needed. The kinds of evidence we can amass are defective in character, not just in weight and significance. Or the assumption may be that this property of the speaker lies beyond any possible evidence, or that there is no empirical fact of the matter with regard to the speaker's cognitive state. The narrower pro-posal that we *should*, more modestly, limit our goal to specification of the required body of knowledge has equally curious consequences. Thus, it would demand that we dismiss evidence from Japanese in studying English, a conclusion that is explicit in Quine's influential radical translation paradigm, which restricts inquiry to evi-dence from a specific community, indeed evidence from a particular experiment. All such proposals depart radically from standard linguistic practice. Suppose the study of Japanese leads to the conclusion that the initial state of the language faculty has a certain property. No serious practitioner would hesitate to regard the conclusion as relevant in principle to the choice between alternative proposals for a theory of English, on the well-grounded assumption that the language faculty is shared. This perfectly rational step makes sense only if we attribute a cognitive state to the speaker, contrary to the stipulation that this move is unwarranted or impermissible.

Such attitudes toward evidence are a sharp departure from a naturalistic stand-point, and have no merit, as far as I can see. The arguments have to be taken seri-ously, but do not withstand scrutiny and can be understood, I think, as a lingering form of mind-body dualism that should have been long abandoned, a kind of methodological dualism, replacing the metaphysical dualism of the Cartesians. But while Cartesian dualism was quite rational from a naturalistic standpoint, on factual assumptions soon proven wrong, the new methodological dualism has no redeem-ing virtues.

From the same mentalistic-naturalistic point of view, there is also no privileged external standpoint from which to pass judgment on the legitimacy of entities and processes that are postulated in the course of construction of explanatory theory. The postulated elements are justified in terms of the roles they play in successful explanation, there being no external criterion, no independent foundational condi-tions, that theory construction must meet, apart from considerations so general as to have no particular relevance here.

We thus adopt the position that the natural sciences, including this enterprise, are "first philosophy." As Michael Friedman puts the point in a recent paper, citing Kant's *Prolegomena*, "the philosophers of the modern tradition," from Descartes, "are not best understood as attempting to stand outside the new science so as to show—from some mysterious point outside of science itself—that our scientific knowledge somehow 'mirrors' an independently existing reality. Rather, (they) start from the *fact* of modern scientific knowledge as a fixed point, as it were; and their problem is not so much to justify this knowledge from some 'higher' standpoint as to articulate the new *philosophical* conceptions that are forced upon us by the new science. . . . (In particular,) Kant's inquiry into the a priori conditions of the possi-bility of mathematics and natural science is not intended to provide an otherwise absent justification for these sciences, but rather precisely to establish a reorganiza-tion and reinterpretation of metaphysics itself," in ways imposed by mathematics and the science of nature. As Kant put it, these sciences stand in no need of philo-sophical inquiry for themselves, "but for the sake of another science: metaphysics."[2]

Such ideas are not now considered particularly controversial with regard, say, to quantum theory; it is a rare philosopher who would scoff at its weird and counterin-tuitive principles as contrary to right thinking and therefore untenable. But this standpoint is commonly regarded as inapplicable to cognitive science. Somewhere between, there is a boundary. Within that boundary, science is self-justifying; the critical analyst seeks to learn about the criteria for rationality and justification from the study of scientific success. Beyond that boundary, everything changes; the critic applies independent criteria to sit in judgment over the theories advanced and the entities they postulate. Again, I think we see here residues of a pernicious dualism.

The point is illustrated in Quine's ontological argument for his well-known indeterminacy thesis.[3] The argument is that if we take the distribution of "states and relations over elementary particles" to be fixed, we will still find alternative theories of language consistent with this fixed distribution but inconsistent with one another, so that there is no fact of the matter. The same is true of two theories of organic chemistry, or of the neurophysiology of vision, and so on; therefore, there is no fact of the matter in these cases either, by this argument. And if it turns out that the world really consists of strings vibrating in ten-dimensional space, then there will be no fact of the matter about elementary particles, by the same argument. Fixing one part of science, we will find "indeterminacy," in Quine's sense, in other parts. This will remain true as closer relations are established among domains of empirical inquiry. The entire discussion seems only to reduce to familiar forms of skepticism, understood for centuries to be without essential interest.

The same point is illustrated in the ongoing debate over attribution of rules and rule-following in the cognitive sciences. It has repeatedly been shown that we can explain puzzling and complicated facts about what speakers know by attributing to them mental structure involving rules and representations that are beyond conscious awareness, and are often quite abstract, in that they have no direct point-by-point connection to articulate form or semantic judgments: with empty categories, abstract case, syntactic operations of LF movement that do not feed the instructions to the vocal apparatus, and so on. Some of these assumptions have a fair explanatory range over widely varied languages, and a degree of explanatory depth as well. But according to another influential doctrine of Quine's, whatever the explanatory force achieved, we are not permitted to attribute the postulated structures to the mind/brain. We are entitled only to say that the behavior we observe *fits* the rules, as the planets satisfy Kepler's laws but without an internal representation of these laws. Behavior can also be *guided* by rules that can legitimately be attributed to the mind/brain, but only if the speaker is aware that this is so. Fitting and guiding, so construed, exhaust the possibilities.

The standard analogy to Kepler's laws is misleading. In that case, the doctrine permits us to attribute to the planets the properties required to explain why they fit these laws. By the proposed analogy, then, we should attribute to the person properties which will account for the fact that his or her behavior "fits" the principles of, say, binding theory. Attribution of mass will not suffice. One plausible proposal — in fact, the only one with any explanatory reach — takes these properties to be as formulated in a computational theory of rules and representations. But the doctrine has it that we are not entitled to proceed in this way, attributing the computational system postulated, however powerful the explanatory force. Evidently, there is some external criterion, apart from scientific practice, that tells us what ontological moves are legitimate.

Perhaps again we are at the mysterious divide between linguistic and psychological evidence, so let us keep to the latter. Take the case of blindsight, where a patient denies seeing a stimulus but can point to it, manipulate it, make judgments based on it, and so on. By Quine's criterion, this patient only fits the rules and is not guided by them, unlike the person who acts the same way but with awareness. Scientists dealing with such examples follow a different course, assuming "covert knowledge" with mental representations beyond awareness.[4] But here we are saved from the intolerable consequences of Quine's stipulation. We have *psychological evidence* that the patient is following the rules, so perhaps we may speak of guidance even without awareness—assuming that any sense can be made out of this curious distinction in types of evidence. Or perhaps there is another way. John Searle, who has been the most vigorous advocate of these notions, argues that in the case of blindsight we have only "blockage," not "inaccessibility in principle," which alone bars attribution of rule-following by his Connection Principle. In the linguistic case, however, we have inaccessibility in principle.

This will become a substantive proposal if some principled grounds are given to distinguish mere blockage from inaccessibility in principle. Searle does not address the problem, except to give examples on both sides of the alleged divide. A look at some cases indicates that it would be an error to try to overcome this crucial gap and thus to reformulate Searle's Connection Principle as an empirical thesis that could be evaluated.

Suppose we have extensive evidence that John uses a certain algorithm to extract square roots, and is fully aware of this. Then we are allowed to attribute this algorithm to him. Suppose that evidence as rich as we like shows that Bill uses the same algorithm, not some different one, but he has no ability to gain awareness, say, because of some birth defect. We must take this again to be a case of "blockage," not inaccessibility in principle, or else the doctrinal stipulation is again revealed to be pointless. Suppose that Bill's birth defect made him indistinguishable in the relevant respect from Tom, a member of a different species. We conclude then that Tom is using the same algorithm, suffering only from "blockage." Otherwise the pointlessness of the stipulation is as obvious as before. We therefore reach the conclusion that *X* suffers only from blockage when some possible species could gain awareness of the algorithm in question. Now take Jones, who, according to all the evidence we have, is using a system of linguistic rules inaccessible to him. The entire point of these efforts is to show that here attribution of the computational system is illegitimate, so it must be a case of inaccessibility in principle, not mere "blockage." We therefore are driven to the empirical claim that there is no possible species like humans except that they could gain awareness of these rules and representations. Tracing this absurd conclusion to its source, we reject the idea that the contents of the mind are necessarily accessible to consciousness in some quite mysterious sense.

Note that from a naturalistic perspective, no such pointless problems arise; such notions as "accessible in principle" would simply be dismissed, and no external standpoint is accepted for judging ontological claims, apart from the normal conditions of rational explanation. We need not ask where the boundary lies between string theory and the study of mind. There is none. There are just various ways of studying the world, with results of varying firmness and reach, and a long-term goal of general unification.

Adopting this position, we assess the credibility of assumptions at various levels on the basis of explanatory success. In the case of language, by far the most credible assumptions about what the brain is and what it does are those of the computational theories. We assume, essentially on faith, that there is also an account in terms of quarks, atoms, molecules, and cells, though without expecting the operative principles and structures to be detectable at these levels. And with a much larger leap of faith, we also tend to assume that there is an account in neurological terms, though it may well be that the relevant elements and principles of the brain, at roughly this level, have yet to be discovered. It would come as no great surprise to find that the much more firmly based computational theories of the mind will provide the guidelines for the search for such mechanisms, much as nineteenth-century chemistry, based on such notions as molecule, valence, and the Periodic Table, provided crucial empirical conditions that led to radical revision of fundamental physics. The familiar slogan that the mental is the neurophysiological at a higher level may prove accurate, but, for the present, those committed to scientific naturalism have more reason to feel secure about the mental than about the neurophysiological. The common belief to the contrary lacks substantive grounding, and seems to me, again, to reflect a lingering dualism.

Keeping to a naturalistic perspective, we identify a component of the mind/ brain—"the language faculty"—that is responsible for the ability to speak and understand, including a variety of physical manifestations. Its initial state is apparently common to the species to fine detail, and biologically isolated in striking respects. Under natural social conditions, the faculty changes state in a regular way through childhood, stabilizing in a steady state that undergoes little subsequent change. We can identify the language that a person has with the state of the language faculty. Having a language entails having a range of knowledge, belief, understanding, ability to express and interpret thoughts, and in general a range of cognitive abilities. We therefore may properly call the language faculty a cognitive system, and its various states, cognitive states. It is not an input system or an output system, though it is accessed by such systems, which may be dedicated to this task. This cognitive system and its states do not fall naturally into the typology of modules in Jerry Fodor's sense, nor does the investigation of it fall within the Marr hierarchy of computational, algorithmic, and physical implementation, since it is not a process but rather a cognitive system. If we adopt the classical epistemological program of Hume and others, what is nowadays sometimes called "naturalized epistemology," then we may call the states of this faculty systems of knowledge, though recognizing that the concept of knowledge employed departs in some respects from ordinary usage, and more radically from current philosophical theories of knowledge, matters I have discussed elsewhere and won't pursue here.

We now have a general research program: to identify the shared initial state, the class of attainable states, and the internal and external factors that effect the transition and eventually stabilize the state attained; and to discover how the state attained enters into language use. In all these respects, there has been considerable progress. Beyond there lies the problem of how the language faculty developed in the species.

Richard Lewontin argues that there may have been "no direct natural selection for cognitive ability at all," and that study of the evolution of the language faculty is likely to remain in the realm of what he calls "story telling."[5] Progress beyond this stage may well await much deeper understanding of the physical properties of complex systems and the conditions that determine their variety and character. There also remain the two problems of unification that I mentioned earlier, again, problems for the future.

The language that a person has is a generative system that assigns structural descriptions to expressions over an infinite range, each structural description spelling out properties of sound, meaning, and structure, insofar as these are determined by the language faculty. So conceived, a language is a function regarded in intension, that is, a particular characterization of a function. The approach treats language from a completely internalized and individualist perspective. Capitalizing on idiosyncrasies of English, let us call language in this sense "I-language," that is, a concept of language that is intensional, internal, and individual. This concept of language captures one traditional view, that a language is, in Sapir's terms, a particular way of understanding. Of the various concepts of language that have been proposed, this one seems the least controversial. It is also the only concept that has been formulated with any clarity, and it is the one that is actually used in linguistic inquiry, I believe, whatever its practitioners may profess.

Other notions appear widely in the literature—for example, the concept of "formal language," construed as a designated class of well-formed expressions, or "common language," shared by a community. The notion of formal language, borrowed from the formal sciences, appears to have no status at all. There has been no serious proposal as to what such an object might be, and there is no known gap in linguistic theory that would be filled by answering the question. Correspondingly, the notion of weak generative capacity has no clear formulation, and no known empirical status if defined one or another way. As for the notion of "common shared language," which plays a fundamental role in much inquiry into language and meaning, while it causes no problems in informal usage, it remains obscure and perhaps irremediably so. It is doubtful that it can be clarified for the purposes of theoretical discourse. In any event, if there are notions of language apart from I-language, they have yet to be presented in a useful way.

In formulating a constructive research program we depart in various ways from ordinary usage with its teleological and normative conditions, which have no place in a naturalistic inquiry and merely lead to confusion if adopted uncritically. It is sometimes thought that these notions are significant for the study of folk psychology, but even that is doubtful, since to a large extent they are cultural artifacts, not reflecting any very significant features of natural "common sense." This is a large topic in itself, which I will again have no time to pursue here.

This strictly internalist framework reaches quite far toward accommodating hard questions about the form and meaning of expressions of natural language, and the principles of language design. This design must provide symbolic systems that specify sound and meaning, and a computational system to determine how they are associated. The theory of the initial state of the language faculty must therefore provide systems for phonetic and semantic representation. The phonetic system must satisfy three basic conditions of adequacy. It must be *universal*, in the sense that an expression of any actual or potential human language is representable within it. It must be an *interface*, in that its elements have an interpretation in terms of the motor-perceptual system. And it must be *uniform*, in that this interpretation is language-independent, uniform for all languages, so as to capture all and only the properties of the system of language as such. The same three conditions hold for the representation of meaning, what some call the level of "logical form." If it is to capture what the language faculty determines about the meaning of an expression, it must be universal, in that any thought expressible in a human language is representable in it; an interface, in that these representations have an interpretation in terms of other systems of the mind/brain involved in thought, planning, and so on; and uniform, in just the sense that the phonetic system is. The conditions are more obscure than in the case of the phonetic analogue, because the systems at the interface are much less well understood, but there is nonetheless a wealth of evidence firm enough to allow substantive inquiry.

A good deal has been learned about these questions in the past generation. A plausible conclusion today, I think, is that the computational system is fixed and invariant; in this sense, there is only one human language, much as an intelligent Martian biologist would have expected. Language variation seems to be limited to the lexicon, either general properties of all lexical items or properties of specific ones, variable within a narrow range among languages. Furthermore, specific variation may well be limited to functional or grammatical elements of the lexicon, much as Jerry Fodor has suggested in a slightly different context. Saussurean arbitrariness of course remains; there are no intrinsic connections between sound and meaning for lexical items. Choice among the options left open in the initial state allows for a finite number of essentially different human languages, in the Osherson-Weinstein sense, that is, up to choice of lexical items within the restricted (but unbounded) limits allowed. This research program has opened a wide range of empirical questions and has led to a flood of discoveries about unsuspected features of language, and explanations for some of them that go far beyond anything that could have been anticipated not many years ago.

Certain topics lie beyond the scope of a strictly internalist and individualist approach, for example, questions about physical signals on the phonetic side, and questions of intentionality on the semantic side. As for the latter, there seems to be ample evidence that rich semantic properties of expressions are determined by the language faculty, yielding an elaborate network of intrinsic semantic connections among expressions, including analytic connections as a special case. The concepts associated with elements of the lexicon, in particular, have a surprising and intricate structure, evidently determined by the initial state since it is known without experience. A relation with properties of reference may hold between formal expressions and these concepts, but the relation of conceptual structures to the physical and social world is a different matter; in particular, it is questionable that a notion of direct reference exists for natural language. Recent philosophical literature provides numerous examples to illustrate that our intuitive judgments of what people are talking about rely on presumed properties of the external world, social interactions, and the whole array of normative-teleological considerations that permeate ordinary discourse about language. Slight modification of such examples leads to very different judgments as to what utterances are about. Skepticism is in order about whether these are topics for coherent theoretical discourse, or whether they are too-many-variable problems, too interest-relative and merged too indissolubly with networks of social relations and individual purposes to allow much more than a Wittgensteinian assembly of particulars.

We might compare this research program with those of the first cognitive revolution and its current revival. In the former case, we can construct a composite picture with leading ideas of roughly the following kind. A fundamental contribution of the parts of Cartesian philosophy that we would today call "cognitive science" was the development of a computational-representational theory of mind, particularly in the domain of visual perception. Rejecting the neo-scholastic view that the form of an object is transferred and implanted in the brain of the perceiver, Descartes argued that the sense organs receive a series of stimulations and the resources of the mind produce the representation of some external object. Thus, the eye scans a surface, or a blind man taps it with a stick. The mind then constructs a representation based on this sequence of impressions, using its inner resources, which will interpret it as a regular Euclidean figure, or as men with hats walking in the street. Its mechanisms being innate in fundamentals, the uninstructed mind will interpret a series of stimuli to be, say, a distorted triangle, not an exact instance of the presented spatiotemporal fragment. Once constructed, the percept guides action and understanding. Similar ideas were also applied in the study of language, with a rather modern flair in many ways, involving notions of phrase structure, operations similar to grammatical transformations in the modern sense, a concept of logical form devised to accommodate some inferences involving relational structures, a reasonably clear sense-reference distinction, a focus on explanation rather than mere description and the vernacular rather than a prescriptive model. We might also interpret these as computational-representational theories of mind. The study of vision and language have been central to the second cognitive revolution as well.

Another concern of the first cognitive revolution, also revived in its more recent phase, is the crucial role of innate conditions for the growth of knowledge and understanding, the tendencies and dispositions that Descartes called "innate ideas." It was argued that experience is interpreted not only in terms of ideal forms provided by innate geometry, but also concepts of objects and their relations, of cause and effect, the whole-part relation, notions of symmetry and proportion and other Gestalt properties, a space of functions and characteristic uses, a notion of unity of objects— physical objects, melodies, and others—all determined by the resources of the mind, by virtue of its intrinsic structure, from piecemeal and successive presentations.

The broad problem of unification arose in the context of a particular conception of the nature of the material world, understood in the rather commonsensical terms of contact mechanics. The notion of material object thus had identifiable limits, and it was possible to ask what, if anything, lies beyond them. The Cartesians argued that operations of the mind do transcend these limits, notably, the normal use of language, marked by unbounded innovation, freedom from stimulus control, coherence and appropriateness, and a capacity to awaken thoughts that we might have expressed the same way; linguistic behavior was described as appropriate to situations but not caused by them—a fair description, as far as we know. No automaton could behave in this manner, they plausibly argued. Given their metaphysics, it was therefore necessary to postulate a second substance, a thinking substance, which interacted somehow with body; the unification problem, then, was to determine how the interaction took place. In this context, experimental programs were proposed to determine whether a creature that looks like us has a mind like ours. Insofar as the creature passes tests such as those illustrated by the normal use of language, we may assume that the creature has a mind like ours, it was reasonably argued.

As is well known, this program collapsed with Newton's demonstration of the inadequacies of its theory of the material world. Newton's invocation of the "occult property" of action at a distance transcended the common notion of body or material object, and thus made it impossible to formulate the mind/body distinction in anything like the Cartesian sense. Newton himself was disturbed by this consequence, agreeing with the Cartesians that "It is inconceivable, that inanimate brute Matter should, without the Mediation of something else, which is not material, operate upon and affect other matter without mutual Contact"; the idea of action at a distance through a vacuum is "so great an Absurdity," he wrote, "that I believe no Man who has in philosophical matters a competent Faculty of thinking, can ever fall into it."[6]

The proper reaction, which took some time to crystallize, is that there is no determinate notion of body or matter. The world is what it is, period. Whatever ideas are needed to yield understanding and insight into it are judged valid, part of the presumed truth about the world, insofar as they enter into valid explanatory theory, however offensive they may be to our intuitions. We may, for historical reasons, continue to use the term "mental" for certain levels of description and theory, but nothing metaphysical is entailed thereby.

The obvious conclusion was that thinking must be a property of organized matter. Through the eighteenth century there were attempts to develop Locke's musings on this quandary. Reviewing one strain of the history, John Yolton concludes that the thinking-matter controversy was dramatically transformed in the late eighteenth century by the eminent chemist Joseph Priestley, whose conclusion was "not that all reduces to matter, but rather that the kind of matter on which the two-substance view is based does not exist," and "with the altered concept of matter, the more traditional ways of posing the question of the nature of thought and of its relations to the brain do not fit. We have to think of a complex organized biological system with properties the traditional doctrine would have called mental *and* physical."[7] Matter has active properties, "powers of attraction and repulsion," Priestley wrote, just as it has the property of solidity, and the powers of sensation or perception or thought are properties of "a certain *organized system of matter*," the conclusion drawn by La Mettrie a generation earlier; thought in humans "is a property of the *nervous system*, or rather of the *brain*," Priestley continued, just as much "the necessary result of a particular organization" as "sound is the necessary result of a particular concussion of the air." Another eighteenth-century version was that the brain secretes thought as the liver secretes bile. More cautiously, we may say that *people* think, not their brains, which do not, though their brains provide the mechanisms of thought. Similarly, *people* understand a language, not their brains, if we are using terms in their ordinary sense. It is a great leap, which often arouses pointless questions, to pass from commonsense intentional attributions to people, to such attributions to parts of people, and then to other objects.

Taking this tack, we arrive at the naturalistic approach to language and cognition that I outlined earlier. But the second cognitive revolution pursued a somewhat different course. There has been a revival and often profitable development of computational-representational theories of mind, and an associated elaboration of innate conditions on the growth of knowledge and understanding. Questions of unification, however, have commonly not followed the lines just indicated, a fact that merits some thought.

Note that the Cartesian tests for the existence of other minds have no place in a post-Newtonian naturalistic cognitive science. Something like them has been revived as the Turing test for possession of intelligence, but with a very different cast. For the Cartesians, the experiments are part of normal science, an attempt to discover whether some object has a certain property, like a litmus test for acidity. The modern version is an operational test intended to provide a behavioral criterion for intelligence. It is unclear what the purpose is. Operational tests are a dime a dozen, and have no point unless there is some interest in the concept they characterize. In this case, the possible interest is quite unclear. In any event, whatever the point of the Turing test may be, it is not the kind of idea that we find in normal science, as distinct from the Cartesian analogue, which was normal science, though no longer appropriate with the collapse of the Cartesian notion of the material world.

Much the same is true of simulation generally. As in the current revival, the imagination of the first cognitive revolution was stimulated by the automata of the day. In a recent study, John Marshall points out that Jacques de Vaucanson, the great artificer of the period, was explicitly concerned to understand the animate systems he was modeling; he constructed automata in order to formulate and validate theories of his animate models, not to satisfy some performance criterion. His clockwork duck, for example, was intended to be a model of the actual digestion of a duck, not a facsimile that might fool his audience.[8] That is the purpose of simulation generally in the natural sciences. There is little if any role here for operational tests of one or another sort.

If a test were proposed for "digestion," or "higher organism," or "flying," or "understanding language," or any other notion, the natural scientist would simply disregard it, unless it somehow shed some light on actual existing systems that are under investigation. It makes sense to study possible chemical elements consistent with physical law, possible human languages consistent with the initial state of the human language faculty, possible realizations of John Smith's genetic endowment as the environment varies, or possible people consistent with some shared endowment. But it would make no sense to ask what would count as a chemical element if physical laws were different, or whether communication in bees is a language, or whether some imagined entity sharing properties with humans is a John Smith variant, or a possible person. These are matters of decision, not fact.

Suppose that some variety of Turing test is proposed, and both people and a computer program pass it, by the judgment of an observer. If this result is for some reason taken seriously, the questions for the naturalist are: (1) Why do humans pass the test, that is, what are the processes and structures of the mind/brain that account for the result? (2) Why did the computer program pass the test? And finally, (3) is there anything interesting about the common achievement; more specifically, since the program is of no interest for science in itself, does its success tell us anything interesting about humans? A computer program is, after all, basically a theory written in an odd notation, and we ask the same questions about it that we ask about any other theory, as did Vaucanson.

Take, say, a chess-playing program. Suppose that it does as well as Gary Kasparov, and thus passes a version of the Turing test. For the natural scientist, the appropriate questions are: (1) How does Kasparov play chess? (2) How does the program operate? (3) Does the second question tell us anything about the first, that is, is it a useful theory about Kasparov? If not, the procedure is of no interest, at least for psychology; it is unclear, in fact, what purpose it has, except to take the fun out of chess. Similarly, in the case of programs that simulate translation, or (if this were within the realm of the imaginable) language understanding, and so on; apart from possible applied interest, they merit attention only insofar as they provide insight into how humans do what they do. It is of no interest to devise a criterion for performance matched by some other device.

The matter is viewed differently in contemporary cognitive science. John Haugeland asks "how one might *empirically* defend the claim that a given (strange) object plays chess." We would properly conclude that it does, he argues, when a skeptic is convinced that it meets a performance criterion. He then concludes that no program could meet such a criterion for language understanding.[9] This recalls the Cartesian argument, but with the different cast just indicated. The Cartesians wanted to know whether the strange object has a certain property, a mind like ours. Its ability to meet some performance criterion may provide relevant evidence, but is of no interest in itself, for the natural scientist, for whom the notion "playing chess" has no significance unless it helps us understand how a real organism does what it does. It is not even clear what it means to speak of an empirical defense of the claim that the object plays chess. The question of whether an airplane flies, in the sense in which birds fly, is a matter of decision, not fact; natural languages handle it differently. The question of whether a program plays chess also seems to be a matter of decision, not fact, hence not susceptible to empirical defense. There seems no clear point to the questions that are being raised, or any interest in whatever answers are given to them.

The departure from the natural sciences is sometimes presented as a point of principle. In a lucid exposition of contemporary cognitive science, Ned Block identifies it with the computer model of the mind, a "level of description of the mind that abstracts away from biological realizations of cognitive structures." Note that such abstraction might be viewed in several ways. It might be *contingent*, in that we don't know the biological mechanisms and therefore must abstract from them; or in that they don't happen to bear on some particular problem. Or it might be *exploratory*: we want to see what we can learn by considering only certain features of these mechanisms, as in a branch of mathematics that treats planets as mass points satisfying physical laws. But Block argues that the abstraction is one of principle. The computer model of the mind assumes certain primitive processors (usually though not necessarily with arbitrary Turing machine capability); and for cognitive science, "it does not matter" whether the primitive processors are realized in gray matter, switches, little creatures running around, or whatever. Consequently, psychology, taken as the science of the mind, is not a biological science, if the computer model of the mind is correct. Block adds that he makes "this obvious and trivial point to counter the growing trend toward supposing that the fact that we have brains that have a biological nature shows that psychology is a biological science." Furthermore, given the "anti-biological bias" of this approach, if we can construct automata in "our *computational* image," performing as we do by some criterion, then "we will naturally feel that the most compelling theory of the mind is one that is general enough to apply to both them and us," as distinct from a biological theory of the *human* mind, which will not apply to these devices.[10]

Regarding the automata as theories, programs for some universal Turing machine, the computational model of mind takes them to be of intrinsic interest as theories of some category of systems sharing properties of humans. There are as many such categories and such theories as there are performance criteria.

Cognitive science, so conceived, is driven by performance criteria and is concerned with systems that satisfy them; it is not obvious why humans enter into the procedure at all, or, if so, how. In these respects, cognitive science differs from the natural sciences. First, these do not abstract in principle from the realization of postulated mechanisms, and are open to change of theory if something relevant is learned about them; thus, if the biologist were to find that one theory of empty categories can be readily realized in cellular structures and another not, that would be as relevant to the naturalistic linguist as results about on-line processing or judgments about meaning or well-formedness. Second, the natural sciences do not care about tests and criteria for some category of performance except insofar as they lead toward understanding of the real systems under investigation.

This nonnaturalist approach serves as a kind of puzzle-generating algorithm. Suppose we have some physical system S constructed out of certain elements and exhibiting some range of behavior. Take P to be a class of "primitive processors" with some properties of the elements from which S is constructed, and take B to be a subclass of the behaviors of S specified by some criterion. Thus, taking S to be people or their brains, we might take P to be elements with some properties of neurons and B to be actions that resemble playing chess or understanding Chinese, by some criterion. Consider now the class K of systems that manifest B and are constructed from P. For the natural scientist, the process is interesting if it sheds light on S. Otherwise, it is a waste of time. But for cognitive science, understood nonnaturalistically, we have a puzzle: do members of K share some deep property of S, like mind, or understanding, or intentionality? Such puzzles can be generated freely. Thus, take S to be the class of living organisms, P elements with certain properties shared by cells, and B behaviors that include some of those of physical organisms. Define "abstract biology" as the theory of the category K of systems constructed from P and manifesting B, whether they are constructed of cells or silicon chips, and whatever other differences they may have from real organisms. We now have all sorts of puzzles about elements of K. Are they alive? Do they grow? Do they sense, or breathe, or think? The natural sciences do not proceed in this way. It is unclear why cognitive science should depart from the pattern. We again have a kind of methodological dualism, which does not seem warranted.

The approach to language I mentioned earlier—and, I think, other work on specific topics within the cognitive sciences—adheres more strictly to the naturalistic pattern. We want to discover how actual systems work. Simulation is of value insofar as it contributes to this end, and, that apart, there is no interest in criteria for commonsense notions like intelligence, language, understanding, knowledge, aboutness, etc. There is no reason to expect them to find a place, unchanged, in a principled theory of the mind.

If this is plausible so far, the next step would be to assess the current state of understanding in particular domains and the prospects for narrow or broad unification, and to ask why so many questions still seem to be mysteries that lie beyond our grasp, rather than problems that we can constructively address. That would take us far beyond the scope of these reflections, and largely beyond anything I could say.

Notes

1. Miller (1979). (The paper on generative grammar was mine.)
2. Friedman (1990).
3. See Gibson, and Quine's response, in Hahn and Schilpp (1986).
4. See, for example, Marshall and Halligan (1988).
5. Lewontin (1990).
6. Letter of 1693, cited by Yolton (1983).
7. Yolton (1983).
8. Marshall (1989).
9. Haugeland (1979).
10. Block (1990).

References

Block, N. 1990. The Computer Model of the Mind," in Osherson, D. N. and E. E. Smith (eds.), *An Invitation to Cognitive Science,* Vol. III. Cambridge: MIT Press.

Friedman, M. 1990. "Remarks on the History of Science and the History of Philosophy." Kuhn Conference. Published in Paul Horwich, (ed.), *World Changes: Thomas Kuhn and the Nature of Science.* Cambridge: MIT Press, 1993.

Hahn, E. and Schlipp, P. 1986. *The Philosophy of W. V. Quine.* La Salle: Open Court.

Haugeland, J. 1979. "Understanding Natural Language." *Journal of Philosophy,* 76, 619–32.

Householder, F. 1966. "Phonological Theory: a Brief Comment." *Journal of Linguistics,* 2,1.

Lewontin, R. 1990. "The Evolution of Cognition," in Osherson, D. N. and E. E. Smith (eds.), *An Invitation to Cognitive Science*, vol. 3. Cambridge: MIT Press.

Marshall, J. C. 1989. "On making representations," in C. P. Hagoort Brown and T. Meijering (eds.), *Vensters op de geest.* Utrecht: Stichting Grafiet.

Marshall, J. C. and Halligan, P. W. 1988. "Blindsight and Insight into Visual-spatial Neglect." *Nature,* December 22.

Miller, G. 1979. "A Very Personal History," *Occasional Papers.* MIT Center for Cognitive Science, 1.

Yolton, J. 1983. *Thinking Matter.* Minneapolis: University of Minnesota Press.

2

Hilary Putnam

Functionalism

Cognitive Science or Science Fiction?

There is an ancient form of "functionalism"—so Martha Nussbaum and I have argued—that can be found in Aristotle's *De Anima*.[1] This is the view that our psyches can best be viewed not as material or immaterial organs or things but as capacities and functions and ways we are organized to function. In that wide sense of the term, I am still a functionalist. In this chapter, however, I will be considering a contemporary rather than an ancient way of specifying what it is to be a functionalist, one I myself introduced in a series of papers beginning in 1960.[2]

The leading idea of this more recent view is that a human being is just a computer that happens to be made of flesh and blood, and that the mental states of a human being are its computational states. In this chapter I shall try to explain why I was led to propose functionalism as a hypothesis, and why I no longer think this contemporary sort of functionalism is correct.[3]

Different functionalists have tried to make this leading idea precise in different ways. In my own functionalist writings, I exploited two ideas that I still find very important in the philosophy of science: the idea of *theoretical identification*, and an idea that I introduced to go with it, the idea of *synthetic identity of properties*. The relation between our mental properties and our computational properties (or a subset of them) is, I suggested, just this kind of "synthetic identity" (here some functionalists, notably David Lewis, would disagree, and insist that it must be some sort of conceptual or analytic identity), and functionalism itself, I maintained, should be viewed as an empirical hypothesis, on all four feet with the hypothesis that light is electromagnetic radiation, and not as a piece of conceptual analysis.

The reason that I was never attracted to the idea that functionalism is correct as a matter of analytic truth, or even, in some looser way, as a clarification or "explication" of our ordinary psychological concepts is, in part, that in other areas of science we know that it is wrong to think that statements that make theoretical identifications of phenomena originally described in different vocabularies must be conceptual truths or must follow conceptually from non-question-begging empirical facts in order to be true. Light is electromagnetic radiation; but this is no "conceptual truth."

I introduced the notion of synthetic identity of properties because I wish to be able to say that not only is light passing through an aperture the same *event* as electromagnetic radiation passing through the aperture, but that the *property of being light* is the very same property as *the property of being electromagnetic radiation of such-and-such wavelengths*. In sum, I hold that properties can be synthetically identical, and that the way in which we establish that properties are synthetically identical is by showing that identifying them enables us to explain phenomena we would not otherwise be able to explain.[4]

Applying this idea to the philosophy of mind, I proposed as a hypothesis that just as light has empirically turned out to be identical with electromagnetic radiation, so psychological states are empirically identical with functional states. Here is the hypothesis as I stated it at the time (for simplicity I stated it only for the case of pain, but I made clear that it was intended to hold for psychological states in general)[5]:

1. All organisms capable of feeling pain are Probabilistic Automata.[6]
2. Every organism capable of feeling pain possesses at least one Probabilistic Automaton Description of a certain kind (i.e., being capable of feeling pain is possessing an appropriate kind of functional organization.)[7]
3. No organism capable of feeling pain possesses a decomposition into parts that separately possess Probabilistic Automaton Descriptions of the kind referred to in (2).[8]
4. For every Probabilistic Automaton Description of the kind referred to in (2), there exists a subset of the sensory inputs such that an organism with that Description is in pain when and only when some of its sensory inputs are in that subset.

If this is an empirical hypothesis, however, then the questions one must ask are: How is the hypothesis to be empirically investigated? And what would the *verification* of such a hypothesis look like? My answer at the time was that to investigate the functionalist hypothesis what we have to do is "just to attempt to produce 'mechanical' models of organisms—and isn't this, in a sense, what psychology is all about? The difficult step, of course, will be to pass from models of *specific* organisms to a *normal form* for the psychological description of organisms—for this is what is required to make (2) and (4) precise. But this too seems an inevitable part of the program of psychology."[9]

Notice that no argument was offered for the idea that the task of psychology is to produce "mechanical models of organisms" (in software, not hardware, terms, of course). Indeed, although I soon recognized that Turing machines and Probabilistic Automata could not possibly serve as such models, I held on for a long time to the idea that it is "an inevitable part of the program of psychology" to provide a "normal form" for those "mechanical models." By this I did not simply mean just a normal form for the description of the psychology of human beings, although that now seems to me to be a completely utopian idea, but a normal form for the psychological description of an arbitrary organism.

So that is how we were supposed to investigate empirically the functionalist "hypothesis": We were supposed to construct "mechanical models" of species of organisms, including the human species, or at least the aspects that we take to constitute their psychological functioning; and then, by reflecting on the nature of these models, we were to pass to a normal form in which the "software description" of the psychological functioning of any physically possible organism could be written.

What about the question of empirical verification? This too seemed to me straightforward, at that time. Consider how we verify that light is electromagnetic radiation of such-and-such wavelengths. What we do—here I still subscribe to a classic account—is show that if we identify light with electromagnetic radiation of certain wavelengths, we can deduce the laws of unreduced optics—say, classical geometrical optics or classical wave optics—to the extent that those laws were true (and, incidentally, to explain why they were not perfectly true, as classically stated). In the same way, I thought, if we find a way to identify the properties we speak of in psychology—in ordinary language, or in clinical psychology, or in behaviorist psychology, etc.—with computational properties (properties of our "software"), then we will be able to deduce the laws of these psychological theories, to the extent that they are true (and, incidentally, to show why and how they are not precisely true as presently stated).

However, as I have already mentioned, the normal form for the description of the psychology of an "arbitrary organism" cannot simply be the Turing machine formalism, for a number of reasons.

A "state" of a Turing machine is described in such a way that a Turing machine can be in exactly one state at a time. Moreover, memory and learning are not represented by states, in the Turing machine model, but by information printed on the machine's tape. Thus, if human beings have any states at all that resemble Turing machine states, those states would have (1) to be states a human being can be in at any time, independent of learning and memory; and (2) to be totally instantaneous states of the human being, states that determine together with learning and memory (the contents of the "machine tape") what the next state will be, and thus totally specify the present psychological condition of the human being. Clearly, such states would neither be the familiar propositional attitudes, nor the states postulated by any presently known psychological theory, be it clinical or behaviorist or cognitive or what have you. For other technical reasons, which I won't bore you with, the Probabilistic Automata formalism fares no better. But I claimed that it is not fatally sloppy to apply the notions of a functional description and of a "normal form" for functional descriptions to "systems" for which we have no detailed idea at present what such a description might look like—"systems" like ourselves. Although it is true that we do not now have a "normal form" in which all psychological theories can be written, we know, for example, that "systems" might be models for the same psychological theory without having the same physics and chemistry. A robot, a creature with a chemistry based on something that is not DNA, and a human being might conceivably obey the same or very much the same psychological laws. If they obeyed exactly the same psychological laws they would be (in a terminology that I introduced in the series of papers I have been referring to) *psychologically isomorphic*; and psychologically isomorphic entities can be in the same psychological state without ever being in the same physical states. In short, psychological states are *compositionally plastic*.

So, in the functionalist hypothesis as spelled out by (1)–(4), it became necessary to replace the notion of a "Probabilistic Automaton description" by the notion of a "normal form" description; and what I claimed was that we will know precisely what we mean by a "normal form description" when we know what sort of description would be provided by an ideal psychological theory, where an ideal psychological theory is thought of as providing a "mechanical model" for a species of organism. This is what led me to think that it was an "inevitable part" of the "program of psychology" to provide mechanical models of species of organisms, and a normal form in which such models can be described.

The importance of this issue is this: if we deny that such a normal form can be provided, or that at least a "mechanical model" can be provided in the case of our own species, then the functionalist hypothesis cannot even be stated (not even if it is restricted to the human species). For what the project of functionalism (as opposed to, say, "eliminitivism," that is, the denial that psychology, as opposed to neurology, has a legitimate subject matter) requires is not just that a computational description of the human brain and nervous system should be possible (an "eliminitivist" may well think that that is true), but that such a description "line up" with psychology, in the sense of providing computational properties that can be *identified* with psychological properties.

Psychology and Functionalist Speculation

The original idea of functionalism was that our mental states could be identified with computational states, *where the notion of a computational state had already been made precise by the preexisting formalisms for computation theory, for example, the Turing formalism or the theory of automata.* What I have just outlined is the manner in which the original idea of functionalism quickly became replaced by an appeal to the notion of an ideal "psychological theory." But this "ideal psychological theory" was conceived of as having just the properties that formalisms for computation theory possess.

A formalism for computation theory *implicitly defines* each and every computational state by the totality of its computational relations (e.g., relations of succession, or probabilistic succession) to all the other states of the given system. In other words, the whole set of computational states of a given system are *simultaneously implicitly defined*; and the implicit definition *individuates* each of the states, in the sense of distinguishing it from all other computational states. But no psychological theory individuates or "implicitly defines" its states in this sense. Thus, functionalism conceived of what it called an "ideal psychological theory" in a very strange way. No actual psychological theory has ever pretended to provide a set of laws that distinguish, say, the state of being jealous of Desdemona's fancied regard for Cassio from every other actual or possible propositional attitude. Yet this is precisely what the identification of that propositional attitude with a "computational state" would do. Thus, functionalism brought to the study of the mind strong assumptions about what any truly scientific psychological theory must look like.

There is no reason to think that the idea of such a psychological theory (today such a theory is often referred to as "conceptual role semantics") is anything but utopian. There is no harm in speculating about scientific possibilities that we are not presently able to realize; but is the possibility of an "ideal psychological theory" of this sort anything more than a "we know not what"? Has anyone suggested how one might go about constructing such a theory? Do we have any conception of what such a theory might look like? Even if we had a candidate for such a theory, the question remains: how would we go about verifying that it does implicitly define the unreduced psychological properties? One hears a lot of talk about "cognitive science" nowadays, but one needs to distinguish between the putting forward of a scientific theory, or the flourishing of a scientific discipline with well-defined questions, from the proferring of promissory notes for possible theories that one does not know even in principle how to redeem.

Functionalism and Semantic Externalism

Ever since I wrote "The Meaning of 'Meaning'," I have defended the view that the content of our words depends not just on the state of our brains (be that characterized functionally or neurophysiologically), but on our relations to the world, on the way we are embedded in a culture and in a physical environment. A creature with no culture and no physical environment that it could detect outside its own brain would be a creature that could not think or refer, or at least (to avoid the notorious issue of the possibility of private language) could not think about or refer to anything outside itself. That, given our physiology and our environment, H_2O is the liquid we drink, has everything to do with fixing the meaning of "water," I claim. Au is the substance that experts refer to as "gold"; and the cultural relations of semantic deference between us laypersons and those experts has everything to do with fixing the reference of "gold" in our lay speech, I claim. Mere computational relations between speech events and brain events do not, in and of themselves, bestow any content whatsoever on a word, any more than chemical and physical relations do. But this implies that no mental state that has content (no "propositional attitude") can possibly be identical with a state of the brain, even with a computationally characterized state of the brain. Even if this is true, however, it has been suggested that one might abstract away from the content of a word, in the sense I just used that notion (the sense in which content determines reference, and, in the case of assertions, is what can be assigned the values "true" and "false"), by simply *ignoring* all the external factors. The result of this abstraction is supposed to be a new notion of content, "narrow content" (the original notion being "wide content"); and it has been suggested that this notion of narrow content is the proper notion when our purpose is psychological explanation.

The advantage of the suggestion, from a functionalist point of view, is that "narrow content," by definition, has to do only with factors *inside* the organism; thus, there is at least the hope that narrow contents might be identifiable with computational states of the organism, thus realizing a version of the original functionalist programs. But the suggestion has problems.

The key problem is totally obscured by the habit of tossing the term "narrow content" around in the literature as if the notion were really well defined. The problem is that we possess neither a way of individuating "narrow contents" nor a set of unreduced psychological laws involving "narrow contents" (unless the laws of "folk psychology" are supposed to be about "narrow contents"—a suggestion I find it hard to take seriously). But the very idea of a theoretical identification presupposes that the concepts to be reduced are already under some kind of scientific control (recall the case of optics, or of thermodynamics). To introduce a set of concepts that at present figure in no laws (the "narrow contents" of our familiar propositional attitudes), and then immediately to begin talking of searching for theoretical identifications of these "narrow contents" with computational states of the brain (which, as we noted earlier, also have not been defined, since we have the problem of what *formalism* is being envisaged when one talks of "computational states" here) is to engage in a fantasy of theoretical identification. It is to mistake a piece of science fiction for an outline of a scientific theory that it only remains for future research to fill in.

In "The Meaning of 'Meaning,'" I myself suggested that one might speak of "narrow contents" in the following way: For me to be in a mental state that has the "narrow content" *there is water on the table*, or whatever the proposition *p* in question might be, is just for me to be in a total brain-state such that (1) some person *P* with some language *L* might be thinking a sentence *S* which is *syntactically* the same as the sentence I am thinking; and (2) the sentence in the language *L* on the occasion of use by person *P* in question could be *translated* (preserving ordinary or "wide" content) as "There is water on the table" (or whatever *p* is in question). Thus, if a Twin Earth speaker of Twin English, whose "water" is actually XYZ, and an Earthian whose "water" is H_2O are in the same brain-state when they think the words "Water is on the table," we would say (on this proposal) that their words have the same "narrow content" even though they refer to different liquids as "water," since the Twin Earther is in the same brain-state as someone—say myself—whose words "Water is on the table" have the *wide* content in question. Another proposal, advanced by Jerry Fodor, is to say that two thoughts have the same "narrow content" if the *wide content* of the thoughts in question (or of the corresponding utterances, either in natural language or in "mentalese") vary in the same way with environmental and cultural parameters. Like my proposal, this makes the notion of narrow content parasitic on the ordinary ("wide") notion of content. Neither proposal provides an *independent* notion of narrow content that could be the subject of psychological theorizing.[10] Ned Block has proposed that "narrow contents" might be identified with *conceptual roles*.[11] But we lack an unproblematic conception of what we mean by the "conceptual role" of a sentence—especially if we are skeptical about the analytic/synthetic distinction—and it is also contentious to claim that "conceptual roles," if the notion *can* be made precise, in any way correspond to the contents of propositional attitudes.

Of course, one might decide to drop the notion of "narrow content," and say: "Very well then. If mental states are individuated by contents which are themselves partly determined by the community and the environment, then let us widen the functionalist program, and postulate that mental states are identical with *computational-cum-physical states of organisms plus communities plus environments*. But how useful is it to speak of "computational-cum-physical states" of such vast systems?

The Issue of Utopianism

I have charged functionalism with utopianism, with being "science fiction" rather than the serious empirical hypothesis that, in my early papers, I hoped to provide. I have also suggested that bringing in such concepts as "narrow content" and "conceptual role" only drains the functionalist proposal of its original substance—it turns it into a case of using concepts that stand for we-know-not-what as if they had serious scientific content. An example may help to make it clear what I mean by this charge.[12]

Suppose that we encounter a primitive culture in which people are observed to say (or think) "Sheleg" when it snows. Folk psychology—and we may assume, "the ideal psychological theory" as well, if there is such a thing—tells us that when it snows, people are likely to say and think "It's snowing." So it is certainly compatible with psychological theory that these people are saying (and thinking) that it is snowing when they say and think "Sheleg." But, of course, there are other possibilities. What a characterization of the assertibility conditions and the inferential relations (these are usually what people have in mind when "conceptual role" is spoken of) between these Sheleg-sayings-and-thinkings and the other sayings and thinkings and doings of these people is supposed to accomplish is to provide us with a way of *individuating* the propositional content of "Sheleg" in the sense of distinguishing that content from all other possible contents.

But distinguishing a content from all other *possible* contents on the basis of any finite supply of facts about inferential relations and prima-facie assertibility conditions is a tall order. Let us suppose that the members of this tribe have a religion according to which the one infallible sign of the anger of their gods is the falling of snow, and that what "Sheleg" actually means is that the gods are angry. If this is the case, then, of course, it will make a difference to how these natives talk. But can we say in advance just *what* difference it will make? Can we (or an "ideal psychological theory" that we could envisage being able to construct) survey all the possible differences it *could* make?

Such a theory would have to be able to describe *the beliefs of a believer of any possible religion*. Or, to take a different example, consider the case of a tribe all of whose members are superscientists. They may be saying "quantum state such-and-such" when it snows, or making a comment in a physical theory we don't have yet. A psychological theory that is able to individuate the contents of such sayings and doings would presuppose knowledge of a physical theory that we have not even imagined. In short, it looks as if an ideal psychological theory, a theory that would be able to determine the content of an arbitrary thought would have to be able to describe the content of every belief of every possible kind, or at least every human belief of every possible kind, even of kinds that are not yet invented, or that go with institutions that have not yet come into existence. That is why I say that the idea of such a theory is pure "science fiction."

What finally pushed me over the anti-functionalist edge was a conversation I had one day with Noam Chomsky. Chomsky suggested that the difference between a rational or a well-confirmed belief and a belief that is not rational or not well-confirmed might be determined by rules that are innate in the human brain. It struck me at once that it ought to be fairly easy to show, using the techniques Gödel used to prove his incompleteness theorems, that if Chomsky is right, then we could never discover that he is right; that is, if what it is rational and not-rational to believe is determined by a recursive procedure that is specified in our ideal competence description D, then it could never be rational to believe that D *is* our ideal competence description. And I was able to show that this indeed is the case without too much trouble.[13]

This argument does not apply directly to the present discussion, since what we are talking about here is not determining what is well confirmed but determining what the *contents* of thoughts are. The Gödel incompleteness theorems show that any description of our logical capacities that we are able to formalize is a description of a set of capacities that we are able to go beyond. But if this is true of our deductive logical and inductive logical capacities, why should it not be true of our interpretative capacities as well? Interpretation involves decisions that, interwoven as they are with our understanding of all the topics we can think about, are unlikely to be susceptible of *complete* formalization. It is true that we can't *prove* that functionalism can't be made less vague than it presently is; but neither do we have any reason to believe that it can. In short, it seems to me that if there *is* an "ideal psychological theory," that is, a theory does everything that the functionalist wants a "description of human functional organization" to do (*let alone* a "normal form for the description of the functional organization of an arbitrary organism"), then there is no reason to believe that it would be within the capacity of *human beings* to discover it. But it is worth pausing to notice that the *notion* of a "complete description of human functional organization" is itself a tremendously unclear notion. The idea that there *is* such a "complete description," even if a recognition procedure for it does not exist, surely goes beyond the bounds of sense.

Of course, those who are sympathetic to functionalism have not given up as a result of my recantation; there are a number of what we might call "post-functionalist" programs on the market. One kind of post-functionalist program seeks to avoid the difficulties inherent in the "implicit definition by a theory" idea that was at the heart of classic functionalism,[14] by relying *entirely* on external factors to fix the contents of thoughts. Dretske[15] and Stalnaker,[16] for example, try to define the content of thoughts as well as of expressions in a language by simply looking for *probabilistic relations* between the occurrences of thoughts and expressions and external states of affairs, bypassing entirely the question of the functional organization of the speaker, which would presumably come in at a later stage in the account. But both Loewer[17] and myself[18] have argued that the information-theoretic concepts on which Dretske and Stalnaker rely cannot individuate contents finely enough. In response to this problem, Fodor proposes to rely not on information-theoretic notions, but instead on the notion of causality. In *Renewing Philosophy* I argue (1) that the notion of causality Fodor employs itself presupposes intentional notions,[19] and (2) that the assignments of contents that result if we look only at the causes of utterances are the wrong ones.

The Relevance of Interpretative Practice

Before I close, there is one objection to my whole line of argument that I need to consider. It can be stated as follows: I have assumed that our ordinary practice of interpretation—what is often called "translation practice," although what is involved in interpretation is certainly much more than *translation*[20]—is the appropriate criterion for identifying the content of propositional attitudes. But is this not too "unscientific" a criterion for the purposes of a scientific psychology?

Let me begin by considering just one aspect of this large question. In "The Meaning of 'Meaning'," I argued that it would be in accordance with standard interpretative practice to say that the term "water" on Earth has the extension H_2O (give or take various impurities) and the term "water" on Twin Earth has the extension XYZ (give or take various impurities).[21] Since these are names of different liquids (much in the way "molybdenum" and "aluminum" are names of different metals), interpretative practice would conclude that believing that a lake is full of "twater" (Twin Earth water) is a different propositional attitude than believing that the lake is full of (Earth) water. And this is perfectly reasonable if the Earthians and the Twin Earthians in question are scientifically sophisticated. But suppose we are dealing with speakers who do not yet know about the difference between Earth water and Twin Earth "water"? Surely no psychologist would regard the fact(s) that water is H_2O/twater is XYZ as relevant to the subject in *this* case? *For psychological purposes,* shouldn't we say that the Earth subject and the Twin Earth subject who both assent to the sentence "The lake is full of water" have the *same* "belief"? Most "cognitive scientists" would certainly answer "yes."

But what exactly is the force of "for psychological purposes" here? If a psychologist finds that one of her subjects believes that molybdenum is a light metal and that another subject believes that aluminum is a light metal, mustn't the psychologist find out whether either subject has any beliefs that differentiate aluminum from molybdenum? Otherwise, hasn't she failed to determine whether the first subject's belief is "the same belief" as the second subject's "for psychological purposes"? Or is the fact that the each subject believes that molybdenum is *called* "molybdenum" and aluminum is *called* "aluminum" *enough* to make these "different beliefs for psychological purposes"? Would the Earthian belief and the Twin Earther's belief become "different beliefs for psychological purposes" if Twin Earth English had a different *word* for the liquid? So if Twin Earth English has the word "twater" rather than "water," the belief becomes different *even if the liquid is the same.*

Compare the situation in evolutionary biology. Ernst Mayr has long urged that the evolutionary biologist employs what is essentially the *lay* notion of a "species," and that the idea of replacing that notion by a more "scientific" notion (or defining it "precisely") is misguided. The nature of the theory, with its anti-essentialism and its emphasis on variety, explains why there is not and cannot be a sharp line between a "species" and a "variety" (many populations are classified as "species" by one classification and as "varieties" by another). Perhaps it might seem more precise to replace talk of species by talk of "reproductively isolated populations"—but then exactly what would count as a significant "population" (three rabbits in a cage?), and what as significant "reproductive isolation" (would purebred Golden Retrievers count as an object of evolutionary theory?). If evolutionary biology, a science with far better credentials than "cognitive science," better serves the task of accounting for what we are interested in by sticking to ordinary imprecise ways of speaking, is it really necessary or desirable for cognitive psychology to depart from ordinary ways of individuating beliefs and other propositional attitudes?

But supposing that it sometimes is, we are still not driven to the fantastic supposition that we are in possession of or can usefully imagine a way of individuating propositional attitudes in utter independence from our normal standards and practices. If we want, for certain purposes, to ignore all facts about water and twater that are not known to the subjects in a particular experiment, we can decide to do so. But even if we identify the content of Earth English "water" (respectively, Twin Earth English "water") with some salient set of beliefs about the liquid(s) that are known to the subject(s) (as Akil Bilgrami proposes in his recent *Belief and Meaning*), *those* beliefs must ultimately be identified by the unformalized (and probably unformalizable) standards implicit in ordinary interpretative practice. We do not have to engage in science fiction, by imaging that "cognitive science" suddenly delivers a way of individuating propositional attitudes wholly independent of the "whirl" of our spontaneous interests and reactions.[22] As Bilgrami argues, the result is not a notion of "content" that ignores the role of the environment in individuating beliefs; it is simply a notion that employs facts about the external referents of various terms *only when those facts are known to the subject*. This is still an "externalist" notion of content, even if it departs, in motivated ways, from the way content is (sometimes) individuated in translation.

In fact, this is not the first time that one of the Moral Sciences has been seduced (or at least tempted) by the dream of laws and concepts as rigorous as those of physics. Even before he hit on the term "sociology" for the subject he was proposing, Auguste Comte proposed that our social theories and explanations would reach such a state, and he confidently proposed that before long all our social problems would be solved by "savants" acquainted with the new Social Physics.[23] That the greatest sociologist of the subsequent century, Max Weber, would employ in his work such imprecise terms as "Protestantism," "mass party," and so on—terms that require interpretative practice for their application—would have seemed to Comte a betrayal rather than a fulfillment of his hopes for the subject. Twentieth-century positivists extended Comte's dream from sociology to history, insisting that eventually history would be systematized and made explanatory by the application of "sociological laws." Yet today we have evolutionary biologists such as Ernst Mayr and Stephen Gould arguing that evolutionary biology should not be expected to look like physics *because* evolutionary explanations are historical explanations, and, they say, we cannot expect history to resemble physics! Clearly, no one—or almost no one—still hopes that history will be able to dispense with notions that depend on what I have called "interpretative practice"—notions like the Renaissance, or the Enlightenment, or, for that matter, the nation state, or ethnic conflict, etc. I am convinced that the dream of a Psychological Physics that seems to be thinly disguised under many of the programs currently announced for "cognitive science" will sooner or later be realized to be as illusory as Comte's dream of a Social Physics.[24]

Notes

I would like to thank Ned Block and Jim Conant for valuable comments on earlier drafts of this chapter.

1. Cf. Putnam (1992a).

2. "Minds and Machines," in Sidney Hook (ed.), *Dimensions of Mind* (New York: New York University Press, 1960); "Robots: Machines or Artificially Created Life," *The Journal of Philosophy* LXI (November 1964), pp. 668–91; "The Nature of Mental States," published as "Psychological Predicates" in W. Capitan and R. Merill (eds.), *Art, Mind and Religion* (Pittsburgh, University of Pittsburgh Press, 1967); "The Mental Life of Some Machines," in H. Castañeda (ed.), *Intentionality, Minds and Perception* (Detroit: Wayne State University Press, 1967); all of these are reprinted in Putnam (1975).

3. I explained at length why I gave up this view in *Representation and Reality* (1988).

4. For a detailed discussion, see "On Properties" (Putnam, 1975a).

5. Cf. "The Nature of Mental States," in Putnam (1975b, p. 434).

6. A Probabilistic Automaton is a device similar to a Turing machine, except that (1) its memory capacity has a fixed finite limit, whereas a Turing machine has a potentially infinite external memory; and (2) state transitions may be probabilistic rather than deterministic. In this chapter (see n. 5) I assumed that the Probabilistic Automata in question were equipped with motor organs and with sensory organs.

7. A Description of a Probabilistic Automaton specifies the functional states of the Automaton and the transition probabilities between them.

8. Note that this rules out Searle's "Chinese Room."

9. Putnam (1975b, p. 435).

10. Fodor, however, thinks that "wide content" can be defined in *causal* terms. For a criticism of his proposal, see Chapter III of my *Renewing Philosophy* (1992b).

11. Cf. Block (1985). (Cf. my *Representation and Reality*, 1988, pp. 46–56, for a discussion.)

12. I take this example from my "Putnam, Hilary," in *Companion to the Philosophy of Mind* (Cambridge: Cambridge University Press, 1994), pp. 507–13, Samuel Guttenplan, (ed.) I have borrowed a few sentences from that paper in this chapter.

13. Cf. my "Reflexive Reflections" (1985, pp. 143–53).

14. This idea was central not only to my version of functionalism but also to the somewhat different version proposed by David Lewis (1972) and part II of Lewis (1983). In Lewis's view, we already *have* the ideal psychological theory required to implicitly define the content of an arbitrary thought: it is just folk psychology!

15. Dretske (1981, 1986).

16. Stalnaker (1984).

17. Loewer (1987).

18. Cf. my "Computational Psychology and Interpretation Theory" in Born (1987).

19. In particular, Fodor needs to assume a distinction between contributory causal factors and *the* cause of an event. I regard this as an intentional notion because what is *the* cause of an event depends on the interests we have in the context; it is not something that is inscribed in the phenomena themselves.

20. More is involved in interpretation than just translation because, for one thing, in our own language we may be able to describe the extension of a term that we cannot translate into it (e.g., I can say in English "In Choctal, *wakai* is the name for a kind of snail"—I may not know if there is a word in English for that kind of snail). Moreover, paraphrase and even commentary are part of interpretation, and are relevant to our identification of propositional attitudes.

21. Actually the practice is more complex than this: for a discussion cf. *Representation and Reality* (1988, pp. 30-33), and my *Realism with a Human Face* (1990, p. 282).

22. In speaking of the "whirl" of our interests and reactions here I am thinking of the following description of Wittgenstein's view: "We learn and teach words in certain contexts, and we are expected, and expect others, to be able to project them into further contexts. Nothing insures that this projection will take place (in particular, not the grasping of universals nor the grasping of books of rules), just as nothing insures that we will make, and understand, the same projections. That on the whole we do is a matter of our sharing routes of interest and feeling, sense of humour and of significance and of fulfilment, of what is outrageous, of what is similar to what else, what a rebuke, what forgiveness, of when an utterance is an assertion, when an appeal, when an explanation—all the whirl of organisms Wittgenstein calls 'forms of life.' Human speech and activity, sanity and community, rest upon nothing more, but nothing less than this. It is a vision as simple as it is difficult, and as difficult as it is (and because it is) terrifying." (Cavell, 1969, p. 52).

23. Cf. "Sciences and Savants" in Fletcher (1974).

24. I do not, however, claim that the scientism I am criticizing comes directly from the idea of imitating physics. Two more direct sources are (1) the idea that our propositional attitudes will someday be explicated by such things as "belief boxes" and "desire boxes" in the brain (and their contents will be individuated by formulas in "mentalese"); and (2) more fundamentally, the idea that psychology will be absorbed in computer science—the very idea that was behind my own functionalist utopianism. With respect to (1), let me remark that the whole idea of "mentalese" depends on the idea that there could be a language (the "language of thought") with the property that the contents of its sentences are *completely insensitive to context of use*. (If "mentalese" lacks this property, then "sentences in a belief box" cannot serve as a record of beliefs by themselves; two subjects may have the same sentences in their "belief box," if that metaphor makes any sense, and have *different* beliefs. On this, cf. my "Computational Psychology and Interpretation Theory" (1987). For an argument that no language has this kind of context-insensitivity (and, I would add, we do not have any idea what a language with context-insensitivity would *be*—yet another example of the constant appeal to concepts that look "scientific" but refer to We Know Not What in "cognitive science")—see Travis (1981).

References

Bilgrami, A. 1992. *Belief and Meaning*. Oxford: Basil Blackwell.

Block, N. 1985. "Advertisement for a Semantics for Psychology." *Midwest Studies in Philosophy*, 10.

Born, R. (ed.). 1987. *Artificial Intelligence: The Case Against*. London: Croom Helm.

Cavell, S. 1969. *Must We Mean What We Say?* New York: Charles Scribner and Sons.

Dretske, F. 1981. *Knowledge and the Flow of Information*. Cambridge: MIT Press.

Fletcher, R. (ed.) 1974. *The Crisis of Civilization: The Early Essays of Auguste Comte*. London: Heinemann.

Lewis, D. K. 1983. *Philosophical Papers*, vol. 1. Oxford, Oxford University Press.

Lewis, D.K. 1972. "Psychophysical and Theoretical Identifications." *Australasian Journal of Philosophy*, 50, no. 3, pp. 249–58.

Loewer, B. 1987. "From Information to Intentionality." *Synthese*, 70, no. 2, pp. 287–316.

Putnam, H. 1994. "To Functionalism and Back Again," in S. Guttenplan, and Putnam, Hilary, (eds.), *Companion to the Philosophy of Mind*. Oxford: Basil Blackwell.

Putnam, H. 1992a. "Changing Aristotle's Mind," in M. Nussbaum, and E. Rorty, (eds.), *Essays on Aristotle's "De Anima."* Oxford: Oxford University Press.

Putnam, H. 1992b. *Renewing Philosophy.* Cambridge: Harvard University Press.

Putnam, H. 1990. *Realism with a Human Face.* Cambridge: Harvard University Press.

Putnam, H. 1988. *Representation and Reality.* Cambridge: MIT Press.

Putnam, H. 1987. "Computational Psychology and Interpretation Theory," in R. Born, (ed.), *Artificial Intelligence: The Case Against.* London: Croom Helm.

Putnam, H. 1985. "Reflexive Reflections." *Erkenntnis, 22.*

Putnam, H. 1975a. *Philosophical Papers*, Vol. 1. *Mathematics, Matter and Method.* Cambridge: Cambridge University Press.

Putnam, H. 1975b. *Philosophical Papers,* Vol. 2. *Mind, Language and Reality.* Cambridge: Cambridge University Press.

Putnam, H. 1974. "Sciences and Savants," in R. Fletcher,(ed.), *The Crisis of Industrial Civilization. The Early Essays of August Comte.* London: Heinemann.

Putnam, H. 1970. "On Properties." Reprinted in Putnam (1975a).

Putnam, H.1967a. "The Nature of Mental States." Reprinted in Putnam (1975b).

Putnam, H. 1967b. "The Mental Life of Some Machine." Reprinted in Putnam (1975b).

Putnam, H. 1964. "Robots: Machines or Artificially Created Life." Reprinted in Putnam (1975b).

Putnam, H. 1960. "Mind and Machines." Reprinted in Putnam (1975b).

Stalnaker, R. 1984. *Inquiry.* Cambridge: MIT Press.

Travis, C. 1981. *The Uses of Sense.* Oxford: Oxford University Press.

3

Stuart Shanker

Reassessing the Cognitive Revolution

The inspiration for this chapter—and in many ways, for the conference "Reassessing the Cognitive Revolution" that we held at York—was Jerome Bruner's *Acts of Meaning. Acts of Meaning* is very much an act full of meaning, and, some will say, of hubris. In it Bruner denounces materialism, reductionism, physicalism, computationalism, and information-processing: in short, he repudiates artificial intelligence (AI) (see Bruner, 1990, pp. ixff., 4ff.). So concerned is he to combat what he sees as the "trivializing" effect that AI has had on psychology that he calls for a "renewing" and a "reorientation" of the cognitive revolution. To find one of the architects of cognitive science disavowing what so many of us had come to regard as the foundation of the discipline is an event worthy of more than just an article, or even a conference. It demands—as Bruner clearly intended—that we undertake a full review and reassessment of the cognitive revolution, and the dominating role AI so quickly came to play in it (see Shanker, 1992, 1993).

With the benefit of hindsight it is now clear that there was not one but two major and independently occurring revolutions in the sciences of the mind in the 1950s. Each came from a different tradition. The cognitive revolution drew heavily on the work of Piaget and Vygotsky, it owed much to Gestalt psychology, and its roots can be traced back to *Denkpsychologie*. AI grew out of automata theory, cybernetics, and recursive function theory, with its roots in behaviorism and nineteenth-century mechanism. How could two such disparate movements not only join together, but indeed, give the impression that they were one? Or perhaps a better way to put this is: How could the cognitive revolution have been so quickly usurped by AI? Far more is involved here than a sociological matter, or an exercise in the history of ideas; we need to understand the deep conceptual reasons why this occurred. For unless we get clear on the reasons the cognitive revolution so swiftly succumbed to the spell cast by AI, we run the risk of sliding back into computational, or some newer version of mechanist reductionism.

There are really two separate questions here: (1) How did AI get involved in the business of doing psychology, and (2) Why did cognitive psychologists embrace AI? The obvious place to begin to answer the first question is with Turing, since he did so much to set the tone of AI. Turing was working on two related questions during the 1940s in his studies of the mechanist thesis: "Can machines think?" and "Do thinkers compute?" Roughly speaking, the difference between these two questions amounts to the difference between strong and weak AI. But Turing saw no real distinction here; as far as he was concerned, the two questions are internally related: in fact, they serve to answer one another.

Turing's starting point was that Turing machines are devoid of intelligence. In his words, they follow "brute force" routines: which, interestingly enough, suggests that they are on the same level as the brutes (with all the Cartesian overtones that this carries). Intelligence as such emerges from self-modifying—that is, *learning*—programs. There are all sorts of substantial philosophical issues raised by Turing's interpretation of his mechanical version of Church's thesis (see Shanker, 1987), but, for our present concerns, we must focus on his attempt to equate *self-modifying program* with *learning program*.

To do this, Turing needed a psychological theory. We know from Hodges that Turing began to read fairly seriously in psychology at this time, and his writings make it clear just who it was that he was reading. In *Intelligent Machinery: A Heretical View*, Turing insists that:

> Just as we can say of the student exposed to teachers who have been intentionally trying to modify his behaviour that, at the end of the period a large number of standard routines will have been superimposed on the original pattern of his brain, so too, by applying appropriate interference, mimicking education, we should hope to modify the machine until it could be relied on to produce definite reactions to certain commands. (Turing, 1959)

This argument comes straight out of Pavlov and Lashley: What we call "learning" is the result of new neural pathways, brought about by conditioning. According to Turing, "the cortex of the infant is an unorganized machine, which can be organized by suitable interfering training" (Turing, 1947). So we get the following, orthodox behaviorist argument:

> The training of the human child depends largely on a system of rewards and punishments. . . . Pleasure interference has a tendency to fix the character, i.e., towards preventing it changing, whereas pain stimuli tend to disrupt the character, causing features which had become fixed to change, or to become again subject to random variation." Accordingly, "It is intended that pain stimuli occur when the machine's behaviour is wrong, pleasure stimuli when it is particularly right. With appropriate stimuli on these lines, judiciously operated by the 'teacher', one may hope that the 'character' will converge towards the one desired, i.e., that wrong behaviour will tend to become rare." (Ibid., p. 121).

The concepts of extinction and positive reinforcement on which Turing placed so much emphasis in his "learning"-based version of the mechanist thesis were thus directly culled from behaviorist writings; it was by employing "analogues" of pleasure and pain stimuli that he hoped "to give the desired modification" to a machine's "character" (ibid., p. 124). As he put it in "Intelligent Machinery, A Heretical Theory":

> Without some . . . idea, corresponding to the "pleasure principle" of the psychologists, it is very difficult to see how to proceed. Certainly it would be most natural to introduce some such thing into the machine. I suggest that there should be two keys which can be manipulated by the schoolmaster, and which can represent the ideas of pleasure and pain. At later stages in education the machine would recognize certain other conditions as desirable owing to their having been constantly associated in the past with pleasure, and likewise certain others as undesirable (Turing, 1959, p. 132).

Turing's post-computational version of the mechanist thesis is based on the behaviorist picture of a *learning continuum*, "with simple negative adaptation (habituation, or accommodation, and tropisms, which are orientating responses and are known to be mediated by fairly simple physico-chemical means) at one end, and maze-learning, puzzle-box learning . . . and ape-learning . . . in stages of increasing complexity, leading to human learning at the other end" (George, 1962, p. 180). Basically, all Turing proposed was to begin this continuum with Turing machines: but, of course, when one says "all," one is referring to introduction of an entirely new paradigm on which to base the continuum (viz., the mechanically calculable functions that constitute the "atomic tasks" underpinning AI's view of intelligence).

This same behaviorist orientation can be seen in Newell and Simon's early work, and was explicitly declared in Miller, Galanter, and Pribram's *Plans and the Structure of Behavior*. But then, this leaves us with a striking problem: How was it that, if AI was originally presented within the framework of behaviorism, the founders of the cognitive revolution, who above all else were supposedly intent on subverting behaviorism, could have embraced AI? The answer to this question is that the pioneers of the cognitive revolution did not see AI as an advanced version of behaviorism but, quite the opposite, viewed AI as providing a valuable vehicle for modeling cognitive processes. Once again, it is helpful to go back to Turing.

Turing's psychological theory amounted to an attempt to forge a union between *Denkpsychologie* and behaviorism. The picture he co-opted was that of the mind, when solving a problem, proceeding via an unbroken chain of mechanical steps from α to ω, even though the subject may only be aware of α, δ, ξ, and ω. In Adriaan de Groot's highly evocative phrase, α, δ, ξ, and ω could be said to occur *above*, and all the intervening steps *below* the "threshold of introspection" (see de Groot, 1965). Thus, by mapping the subject's thought-processes onto a program designed to solve the same problem, we can fill in the intervening—*the preconscious*—steps that led to his resolution of the problem. This is very much the picture underlying Turing's observation in "Can Digital Computers Think?" that "The whole thinking process is still rather mysterious to us, but I believe that the attempt to make a thinking machine will help us greatly in finding out how we think ourselves" (Hodges, 1983, p. 442).

Turing's post-computational version of the mechanist thesis meshed quite closely with an idea that Otto Selz had broached in the 1920s. Selz argued that:

> The individual analysis of task-conditioned thought processes always shows an uninterrupted chain of both general and specific partial operations which at times cumula-

tively (A + B + C) and at times in a stepwise fashion (B after failure of A) impel the solution of the task. These operations are continued until a solution is found or up to a momentary or lasting renunciation of the solution (Simon, 1982, p. 153).

Although Selz' writings were relatively unknown among English-speaking psychologists, the work of de Groot had a profound impact on the evolution of AI. De Groot sought to implement Selz' ideas in an exhaustive investigation of how a broad spectrum of chess-players set about solving board problems. His primary result was that, as Selz had outlined, such problem-solving processes *must* be based on a linear chain of operations: a point that, as de Groot noted in the Epilogue to the later English translation of *Thought and Choice in Chess*, rendered his findings highly compatible with the requirements of chess programming.

The picture of thinking that guides de Groot is the same as what we find in Turing. It postulates that we cannot—where the "cannot" is thought to be empirical—hope to capture the full range of our thoughts in the net of language, either because so much of the thinking-process is subliminal, too rapid or too far removed for our powers of introspection, or it is simply of a nature that outstrips the present possibilities of linguistic expression. With training it might be possible to ameliorate some of these deficiencies; but no amount of laboratory experience can enable a subject to discern the "elementary information processes" out of which human problem-solving is compounded. Computer models thus provide the cognitive psychologist not just with a valuable, but, in fact, an essential adjunct to thinking-aloud experiments. Without them we could never hope to overcome the inherent limitations of introspection. Moreover, it is difficult to see how we could otherwise hope to explain such phenomena as "moments of illumination" or the mind's ability to solve problems of enormous computational complexity in what, from the vantage-point of current technology, seems like an astonishingly small amount of time.

It is important to remember that de Groot presented these ideas long before he became familiar with computer models of chess-thinking. At the outset of his argument he approvingly cites Selz' dictum that the psychologist's goal must be to deliver "a complete (literally: 'gapless') description of the causal connections that govern the total course of intellectual and/or motor processes" in problem solving (de Groot, 1965, p. 13). De Groot's major task was then to explain the phenomenon of pauses in a player's reports followed by a new approach; for "It is often during these very pauses that the most important problem transformations appear: the subject takes a 'fresh look' at the entire problem" (ibid., p. 184). Given that a "subject's thinking is considered one continuous activity that can be described as a linear chain of operations" (ibid., p. 54), "transitional phases have to be assumed in order to understand the progress of the thought process even though the written protocol gives no indication at all" (ibid., p. 113). Thus, we can hypothesize from the program what *must* have been going on during the pauses in the player's mind. For there can be no such lacunae in the computer model: a program with such "gaps" would literally grind to a halt. And the same must hold true for thinking itself, given that it too must be an effective procedure.

Newell and Simon were quick to pick up on on this thesis. For example, as they explained in "GPS: A Program Which Simulates Human Thought":

We may then conceive of an intelligent program that manipulates symbols in the same way that our subject does—by taking as inputs the symbolic logic expressions, and producing as outputs a sequence of rule applications that coincides with the subject's. If we observed this program in operation, it would be considering various rules and evaluating various expressions, the same sorts of things we see expressed in the protocol of the subject. If the fit of such a program were close enough to the overt behaviour of our human subject—i.e., to the protocol—then it would constitute a good theory of the subject's problem-solving. (Newell & Simon, 1961)

Newell and Simon were not looking for a simple match between protocols and programs. Their idea was the same as we find in de Groot: if an agent's brain operates as some sort of computational information-processing device, then the "fragmentary evidence" presented in a verbal protocol could be treated as the "conscious elements" of a mental program whose underlying operations are *pre-conscious*: inaccessible to introspection, but inferrable from the corresponding steps in a computer program that is mapped onto the protocol.[1] This premise—the correspondence between mechanical procedures and pre-conscious mental processes—is precisely what the founders of the cognitive revolution found so promising in AI.

We can begin to discern here one key reason why the burgeoning cognitive revolution was so attracted to AI: it is because the union that Turing forged between recursive function theory, *Denkpsychologie*, and behaviorism was grounded in an epistemological picture that was common to both AI and cognitive psychology. Thus, we find Bruner himself writing in the late 1950s that "Good correspondence between a formal model and a process—between theory and observables, for that matter—presupposes that the model will, by appropriate manipulation, yield descriptions (or predictions) of how behavior will occur and will even suggest forms of behavior to look for that have not yet been observed—that are merely possible" (Bruner, 1959, p. 368). The big appeal of AI lay in the flexibility of the formal models it afforded the cognitive psychologist. That is, AI seemed to offer the possibility of advancing beyond Piaget's "relatively static, unnecessarily static" (group-theoretic) concepts. Thus, the cognitive revolution turned to AI in the hope that "perhaps the new science of programming will help free us from our tendency to force nature to imitate the models we have constructed for her" (Bruner, 1960, p. 23).

Significantly, this common epistemological ground resulted in the resurgence of an issue that analytic philosophy had supposedly laid to rest. In his entry on the "Laws of Thought" in his *Oxford Companion to the Mind*, Richard Gregory explains how:

A deep question is how far laws of thought are innate . . . and how far they are learned, and are products of particular cultures, societies, and technologies. . . . If laws of thought are innate—inherited—it now appears that they can be modified: there are few, if any, limits to how or what we may think. At present, it seems that we shall not be able to specify effective laws of thought in detail before there are adequate computer programs for solving problems—including problem-solving that is not strictly and explicitly logical (Gregory, 1987, p. 430).

This amounts to a direct challenge to one of the most basic tenets of analytic philosophy—and, it must be stressed, a deliberate challenge. Thus, Boden, in her

"Real-World Reasoning," goes so far as to ridicule the idea that "the logician who falls prey to psychologism has been seduced by a temptress of disreputably easy virtue" (Boden, 1981). But the temptress here isn't just some nineteenth-century model of cognition: it is Kantianism.

The very fact that Boole is so commonly cited by AI-theorists as one of the field's most important precursors is proof enough of this fact, for Boole saw his argument as a defense of Kant's transcendental psychologism. Boole starts out *The Laws of Thought* with the psychologistic claim that the "science of Logic" should be founded on the "fundamental laws of those operations of the mind by which reasoning is performed." Gregory draws from this the conclusion that "For Boole, appropriate laws of logic are rules by which the Mind works. It is thus possible to discover how the Mind works by looking at the most effective rules of logic (Gregory, 1981, p. 229). The key themes here, as Gregory's reading suggests, are the Kantian ideas that logic is the science of conception (categorization and judgment) and reasoning, and that the hidden rules of the mind's operations can be inferred from a formal calculus.[2]

The AI version of psychologism—what might be called *cognitive* psychologism—is thus committed to the view that Boole had entirely the right idea about inferring the "mind's hidden processes" from a formal model; he simply used the wrong kind of model. The crux of cognitive psychologism is the premise that the mind is not a logical machine. Hence the problem with classical psychologism is that it assumed a deductive rigor that oversimplified and might even be foreign to mental processes. But that does not preclude the possibility of using more sophisticated mathematical tools to model the "step-by-step procedures of actual thinking."

The resurgence of psychologism—the belief that, by instituting programs in place of the logical calculus, psychologists would be able to construct more successful models of the mind's hidden information-processing operations (as measured in terms of the model's predictive capacity)—was only part of the Kantian framework that joined AI and the cognitive revolution. The other key factor lay in the so-called "functional" definition of concepts: the idea that concepts enable an organism to

1. Identify and classify input information
2. Reduce the complexity of sensory input to manageable proportions
3. Provide the basis for an organism's adaptability: that is, to anticipate the future, to go beyond the information given.

On this picture, the concepts that an organism forms as a result of its interactions with its environment capture the similarities that it perceives. Every time an organism sees something new, it compares this with its stored repertoire of concepts, and if there is a match, it knows right away what it is seeing and what to expect—to the extent, that is, of its previous encounters with that phenomenon. Experience constantly causes an organism to revise its stored concepts. But the fact that concept-formation is a virtually endless mental process does not mitigate the usefulness of even the most primitive of concepts, for "if we had no concepts, we would have to refer to each individual entity by its own name, and the mental lexicon required would be so enormous that communication as we know it might collapse." Above all else, therefore, concepts "decrease the amount of information that we must perceive, learn, remember, communicate, and reason about" (Smith, 1989, p. 501).

As Rumelhart points out, this picture of concepts as "*the building blocks of cognition*" can be traced directly back to Kant. For Kant assumed that the mind is bombarded by a flux of sensations that must be synthesized and organized via concepts. On this view, concepts (schemata) "are the fundamental elements upon which all information processing depends. Schemata are employed in the process of interpreting sensory data (both linguistic and nonlinguistic), in retrieving information from memory, in organizing actions, in determining goals and subgoals, in allocating resources, and, generally, in guiding the flow of processing in the system" (Rumelhart, 1980, pp. 33-34). Given this Kantian picture of the mind forced to impose order on the intrinsically chaotic information it receives if it is to make sense of reality, it follows that "the concept repertoire serves to carve up or 'chunk' the organism's entire world into functional units, yielding a complete mapping of responses into stimulus sets for the length and breadth of the psychological environment" (Flavell, 1970, p. 984).

There are, of course, marked differences between Kant's views and those of the cognitive revolution—most notably, in the latter's repudiation of the idea that the concepts that the mind imposes on reality are fixed. Rather, they are said to emerge from and evolve through a process of adapting to the environment. But the crux of this Kantian picture is that, contra empiricism, concept-acquisition and formation must be an active process. Concepts emerge out of a process of abstraction and analysis: in Piaget's terms, of assimilation and accommodation to incoming stimuli (i.e., modifying information to conform to pre-existing mental structures and modifying mental structures so as to assimilate a broader range of incoming information). What Boden calls the "equilibratory heart" of this process consists in some set of procedures that "assess the match or mismatch between the input (whether example or counterexample) and the current state of the developing concept" (Boden, 1979, p. 137). And, as Boden points out, it is precisely here that AI entered the scene for the cognitive revolution, for Piaget had little to say about the detailed mechanics of this process.

What could be a more compelling vindication of the functional definition of concepts than to build a system that actually demonstrates how, proceeding from a few schematic rules, it would be possible to acquire certain concepts? The point of such an exercise is not to defend Turing's views on strong AI; rather, it is to provide a model of the "complex structure in terms of which [the mind makes] logically subtle comparisons and effect[s] appropriate transformations to the concept it is constructing" (ibid., p. 137). With all the time and attention that was devoted to the problem of getting programs up and running, it is tempting to see AI as nothing more than an attempt at bottom-up engineering whose sole purpose, in Minsky's famous words, was to create machines capable of performing tasks that would require intelligence if done by humans (see Minsky, 1968, p. v). That is, it is easy to lose sight of the fact that, as far as the cognitive revolution was concerned, all of the work on semantic nets and production systems and frames and scripts was seen as different exercises in *concept-analysis*. Recall Harvey, Hunt, and Schroder's point (which they made in 1961, before the full impact of AI had been felt) that "a concept may be viewed as a categorical schema, an intervening medium, or program through which impinging stimuli are coded, passed, or evaluated on their way to response evocation." In other words, what Newell and Simon (1972), Minsky (1975), Winston (1975), and Schank and Abelson (1977) were all doing was *constructing concepts*— programs are just concepts!

The picture that emerges from this brief historical overview is that the union between AI and the cognitive revolution was grounded in a Kantian picture of the nature of cognition. To be sure, establishing this Kantian foundation does not in itself amount to an indictment of the cognitive revolution. Certainly, Piaget was not disturbed by the Kantian overtones of his epistemological theory. And Kant himself welcomed the "confirmation or refutation by experiment" of his transcendental deductions. So why should the cognitive revolution not claim for itself the honor of fulfilling Kant's vision of advancing the "science of metaphysics"? The problem is, however, that the cognitive revolution embraced AI precisely because of this shared epistemological foundation.

What Bruner has in mind when he talks about "renewing" the cognitive revolution is the recognition that psychology must "seek out the rules that human beings bring to bear in creating meanings in cultural contexts. These contexts are always *contexts of practice*: it is always necessary to ask what people are *doing* or *trying* to do in that context" (Bruner, 1990, p. 118). But there is no reason to limit this "reorientation" of the cognitive revolution to a restored awareness of the significance of intentional concepts for the explanation of human actions and emotions. It is surely not the case that the cognitive revolution can only succeed in its stated ambitions if it abandons all interest in explaining, for example, the nature of perception, or reasoning and problem-solving, or concept- and language-acquisition —all those problems that first inspired the cognitive revolution. But we need an entirely new outlook to approach these topics if we are to heed Bruner's warning and avoid the inexorable descent into materialism, reductionism, physicalism, computationalism, or information-processing. If there is any lesson to be learned from the foregoing survey of the origins of the cognitive revolution, it is that our starting-point should be to liberate psychology from the epistemological framework that for three centuries has determined the very questions that psychology is trying to solve and the type of explanations that are deemed appropriate. It turns out, then, that far more is called for than a mere *reassessment* of the cognitive revolution; a full-scale investigation of its Cartesian and Kantian foundations is clearly needed.

Notes

1. We know from Simon that, while this argument may not have been on his and Newell's mind when they first began to work on LT, they quickly recognized the cognitive implications of their program and its compatibility with Selz' thesis. Indeed, Simon taught himself Dutch expressly so as to be able to read *Thought and Choice in Chess*, which had not yet been translated into English (see Simon, 1982).

2. Cf. what Kant says at the beginning of the *Logic*. All of nature, he tells us, is rule-

governed, which includes the operations of the mind. But we are "not conscious" of the rules that we follow in thinking (or in speaking a language). The understanding is the faculty of thinking: it is "bound in its acts to rules we can investigate." How do we discover the "general, necessary rules" whereby Understanding constructs the "contingent rules" that we use in specific discipines? Boole and, following him, the cognitive psychologicians would answer: through the use of formal models. Whether these should be logical, group-theoretic, or computational is seen by all psychologicians as a purely empirical matter.

References

Boden, M. 1979. *Piaget*. Brighton: The Harvester Press.

Boden, M. 1980. "Real-World Reasoning," in *Minds and Mechanisms*. Ithaca, NY: Cornell University Press, 1981.

Bruner, J. 1959. "Inhelder and Piaget's The Growth of Logical Thinking," *General Psychology*, vol.50.

Bruner, J. 1960. "Individual and Collective Problems in the Study of Thinking," *Annals of the New York Academy of Science*, vol.91.

Bruner, J. 1990. *Acts of Meaning*. Cambridge: Harvard University Press.

De Groot, A. D. 1965. *Thought and Choice in Chess*. The Hague: Mouton.

Flavell, J. 1970. "Concept Development," in P.H. Mussen (ed.), *Carmichael's Manual of Child Psychology*, vol.1. New York: Wiley.

George, F.H. 1962. *Cognition*. London: Methuen.

Gregory, R. L. 1981. *Mind in Science*. Harmondsworth: Penguin Books.

Gregory, R. L. 1987. "Laws of Thought," in *The Oxford Companion to the Mind*, Richard L. Gregory (ed.). Oxford: Oxford University Press.

Harvey, O.J., Hunt, D.E., and Schroder, H.M. 1961. *Conceptual Systems and Personality Organization*. New York: Wiley.

Hodges, A. 1983. *Alan Turing*. London: Burnett Books.

Minsky, M.L. (ed.). 1968. *Semantic Information Processing*. Cambridge: MIT Press.

Minsky, M.L. 1975. "Frame-system theory," in P.N. Johnson-Laird and P.C. Wason, *Thinking: Readings in Cognitive Science*. Cambridge: Cambridge University Press, 1977.

Newell, A. and Simon, H. A. 1961. "GPS, A Program that Simulates Human Thought," in E. A. Feigenbaum & J. Feldman (eds.), *Computers and Thought*. New York: McGraw-Hill, 1963.

Newell, A. and Simon, H. A. 1972. *Human Problem Solving*. Englewood Cliffs, NJ: Prentice-Hall.

Rumelhart, D. E. 1980. "Schemata: The Building Blocks of Cognition," in R.J. Spiro, B.C. Bruce, and W.F. Brewer (eds.), *Theoretical Issues in Reading Comprehension*. Hillsdale, NJ: Erlbaum.

Schank, R.C. and Abelson, R.P. 1977. *Scripts, Plans, Goals and Understanding*. Hillsdale, NJ: Erlbaum.

Shanker, S. 1987. "Wittgenstein versus Turing on the Nature of Church's Thesis." *Notre Dame Journal of Formal Logic*, vol. 28, no. 4.

Shanker, S. 1992. "In Search of Bruner." *Language & Communication*, vol.12.

Shanker, S. 1993. "Locating Bruner." *Language & Communication*, vol.13.

Simon, H. A. 1982. "Otto Selz and Information-Processing Psychology," in N. H. Frijda and A. D. De Groot (eds.), *Otto Selz: His Contribution to Psychology*. The Hague: Mouton.

Smith, E. E. 1989. "Concepts and Induction," in M.I. Posner (ed.), *Foundations of Cognitive Science*. Cambridge: MIT Press.

Turing, A. 1947. "Lecture to the London Mathematical Society on 20 February 1947," in D.C. Ince (ed.), *Collected Works*. Amsterdam: North-Holland, 1992.

Turing, A. 1959. "Intelligent Machinery: A Heretical View," in S. Turing, *Alan Matheson Turing*, Cambridge: Heffers.

Winston, P.H. (ed.). 1975. *The Psychology of Computer Vision*. New York: McGraw-Hill.

4

Margaret Boden

Promise and Achievement in Cognitive Science

Just over thirty years ago, only a few days after arriving in the United States, I was rummaging through secondhand books in a bookshop in Cambridge, Massachusetts. One of the volumes had an unpleasantly rough-textured binding, made of a hideous rust-brown fabric (coffee-stained to boot). Heavy underlining defaced almost every page. It was a most unattractive object. Why I picked it up, I shall never know. But it changed my life.

Despite its battered appearance, it was only one year old. (Perhaps that's what "secondhand" means in the U.S.?) It was *Plans and the Structure of Behavior*, by George Miller, Eugene Galanter, and Karl Pribram (1960).

I had never heard of it, or of them. A few days later, I was to shake hands with George Miller. And some weeks later, I was to find his book featured on the reading lists of a couple of the graduate courses I took at Harvard—including Jerry Bruner's seminar at the newly opened Center for Cognitive Studies. But that was for the future. Meanwhile, in the bookshop, it offered an exhilarating, liberating promise. Specifically, it outlined a way to solve problems about the mind, and its relation to the body, which had troubled me since I was a schoolgirl.

Hysterical paralysis, for instance, had intrigued me for years—for here the mind clearly seemed to be influencing the body, even overriding the anatomical realities of nerve-muscle distribution. How could it possibly happen? This hideous brown book enabled me to see that someone's *concept* of an arm, used as a criterion in a program controlling motor movement, could be reflected in the person's movement, or lack of it—irrespective of the underlying anatomy. Indeed, a robot with wires and levers reflecting human anatomy could show a similar paralysis, if it also had a person's "concept" of *arm* that it could use in its motor-control (Boden, 1970).

Two more psychological phenomena that had long puzzled me were multiple personality and fugue. In the bookshop, I read the authors' remarks on personality and personality disorders and was already half-persuaded. Very soon, I was to write a term-paper (Boden, 1965), and then a book (my Ph.D. thesis) (Boden, 1972), extending their ideas to the general psychology and personality theory of William McDougall.

McDougall—who, I later found out, had taught Jerry Bruner at Duke, and warned him of the behaviorist temptations of Harvard—was a vociferous advocate of purposive explanation in human and animal psychology. He depicted the mind as a hierarchical colony of purposive units (or "monads"), integrated in humans by the chief monad, or self. He explained clinical disorders such as multiple personality and fugue in terms of various types of dissociation, due to differing degrees of breakdown in communication between the monads. The resonance with Miller, Galanter, and Pribram was clear. But (unlike Freud, for example) McDougall was philosophically opposed to any form of mechanism. If his anti-reductionist psychology could be interpreted in computational terms, I felt, any could. My second book (Boden, 1977)—an account of artificial intelligence, and its relation to human psychology—was, in my own mind, merely an extended footnote to the first. And my third book, on creativity (Boden, 1990), was prefigured by a chapter on creativity in that 500-page footnote.

My own intellectual life, then, was profoundly influenced by Miller, Galanter, and Pribram. The same could doubtless be said by many others. (Remember that ubiquitous underlining!) Even in the early 1960s, it was obvious that their book had glaring faults and inadequacies. (A popular canard proclaimed that: "Miller thought of it, Galanter wrote it, and Pribram believed it.") Now, it's even more clear that it was always simplistic, often careless, and sometimes mistaken. Nevertheless, it was a book of vision, whose promise was twofold.

On the one hand, it claimed that the way to approach psychology is to use computational concepts, to think of the mind as some sort of program for the body. This offered not only truth, but testing: a computer model enables us not only to state and clarify our theoretical ideas, but to find out whether they have the inferential consequences we believe them to have. On the other hand, it promised to illuminate the whole of psychology: human and animal, individual and social, normal and abnormal, cognitive and motivational, perceptual and motor—not forgetting personality, madness, and hypnosis. Heady stuff, indeed.

Were the authors wrong? Was their promise an honest deception, their enthusiasm an unwitting betrayal? Were they mistaken to claim that their admittedly "fuzzy" and "oversimplified" ideas "may be wrong, [but] they are likely to be a good deal less wrong than the metaphors many psychologists have used heretofore" (op. cit., p. 209)?—Let us take this question in two parts, concerning method and content, respectively.

As for methodology, their vocabulary of TOTE-units (Test-Operate-Test-Exit) was simple in the extreme. And the handful of computer programs they discussed—*GPS, Pandemonium, Argus* (Newell & Simon, 1961; Selfridge, 1959; Reitman, 1963, 1965; Reitman, Grove, & Shoup, 1964)—were only the first crude steps in computational psychology. More subtle and far-ranging footsteps were later taken in AI. Many of these were followed and extended by computational psychologists (Boden, 1988). But even today, there are still many questions left unanswered, and many others whose putative answers are hotly contested—not least by fellow cognitive scientists.

I am not thinking here of disputes over details, although there are many of those. Rather, I have in mind fundamental disagreements over styles of thinking, over ways of doing cognitive science. In short, GOFAI (good old-fashioned AI, Haugeland, 1985) is no longer the only game in town: there are methods now, not dreamt of in Miller's philosophy.

The running battle between symbolic AI and connectionism—a largely misconceived battle, by the way (Boden, 1991)—is one example. Certain types of thinking are still best modeled by GOFAI. Indeed, some leading connectionists allow that a basically connectionist system may have to simulate a von Neumann machine in order to achieve these types of (hierarchical and/or sequential) thought (Hinton, McClelland, & Rumelhart, 1990). And many of the theoretical insights of GOFAI will have to be (some already have been) incorporated in connectionist models of the relevant phenomena. It is already clear, however, that the conventional notion of a computer program (a list of instructions for a von Neumann machine) is not a good way in which to think of all psychological phenomena. Some require us, rather, to be concerned with parallel distributed processing (PDP)—not least for the properties of graceful degradation and content-addressable memory that this virtual architecture brings "for free" (Hinton, et al., 1990). (I say "virtual" architecture because, of course, most current PDP-systems are implemented in sequential machines.)

There are more fundamental reasons, besides its inappropriateness for certain types of information-processing, why GOFAI is out of favor with many people. Such people challenge its account of thinking as computation over semantically (conceptually) interpretable representations, and the associated "physical symbol system" (Newell, 1980; Newell & Simon, 1972) and "language of thought" (Fodor, 1975) hypotheses.

One such attack comes from the connectionists, or more accurately from certain *aficionados* of PDP. They argue that what classicists think of as symbolically interpretable representations are in fact holistic equilibrium-states, grounded in processing-units whose semantics are sub-symbolic (Smolensky, 1987, 1988; Clark, 1989). Indeed, they take it as a virtue of their own approach that it can explain the grounding of concepts and objective thought in non-conceptual content (Clark, 1993; Cussins, 1990).

Another, even more radical, attack comes from three closely related recent movements: situated robotics, cognitive dynamics, and neo-Heideggerian epistemology. Each of these stresses the situated, embodied nature of thought and action, and rejects the commitments to representation and computation of GOFAI and connectionism alike.

Workers in situated robotics do not try to provide their robots with internal world-models or top-down planning abilities, as most AI-roboticists do. Instead, they design the robots bottom-up, resting on simple "reflex" actions directly triggered by environmental cues. Higher levels of reflexes may be provided also, but these can only inhibit some of the lower reflexes, rather than being able to direct low-level activity in detail (by altering the code). The behavior of relatively simple, insect-like, creatures may be captured by this method (Brooks, 1991, 1992). Certainly, it can produce the appearance of simple purposive, goal-directed activity (Maes, 1991). And some human "intelligence" may be suited to this type of modeling. Whether all human (or even non-human) intelligence can be so modeled is quite another matter. To say that GOFAI neglected the evolutionary basis of human minds is true. But that's not to say that it had nothing useful to contribute in respect to those types of thinking that are specifically human. Moreover, we will see that the contrast between GOFAI and the "situated" approach may be more apparent than real.

Proponents of cognitive dynamics claim that we should not posit mental representations, whether symbolic or connectionist, nor discrete computations either (Port & van Gelder, 1995; van Gelder, 1992). Instead, we should conceive of the organism/environment as two mutually coupled dynamic systems. Since neither is independent of the other, the environment is not objectively given: rather, it is brought into being by the cognitive organism—or, better still, mind and environment constitute each other, by continuous reciprocal interaction. Representations are denied, because they imply reference to some objective reality. Computations are denied because they assume internal ("in-the-head") mechanisms functioning largely independent of the environment. Cognition, on this view, is seen as state-space evolution in (non-computational) dynamic systems. The exemplary icon is the steam-engine governor, rather than the digital computer.

This challenge to the Cartesian subject-object dichotomy taken for granted by both GOFAI and connectionism is raised also by the neo-Heideggerians (Varela, Thompson, & Rosch, 1992; Wheeler, 1996). Representations, traditionally understood, are semantically interpretable internal states that encode objective, observer-independent, states of the external world. Neo-Heideggerians, by contrast, argue that observer-dependent worlds are brought forth through situated action and dynamical coupling. A somewhat similar view was taken in the 1920s by the ethologist Jacob von Uexkull, in his description of "a stroll through the worlds of animals and men" (von Uexkull, 1957). A Cartesian realist would insist that to say that different animal species live "in different worlds" is merely to say that they can sense and influence different (objectively present) environmental features. The neo-Heideggerians want to go further than this, by denying that any sense can be made of a purely objective, subject-independent environment.

Their approach raises many controversial philosophical questions. For instance, what are we to say about the existence of the material universe prior to the evolution of sentient creatures? (The conditional answer—*if* we had been around, *then* we would have perceived these things—does not convince me.) Must we not, at least, reject Heidegger's view that world-making is essentially linguistic, for fear of denying that non-human animals inhabit experienced worlds? And how are we to express the "being" of non-sentient things, like rocks, which can certainly have no "being-in-the-world," or *Dasein*? This is not the place for a critique of Heidegger, nor am I competent to offer one. But we should note the philosophical challenge to Cartesianism offered by Heideggerian phenomenology. Were it to succeed, the subject/object distinction as we currently understand it would fall. Given this change in the basic epistemological landscape, some fundamental assumptions of current cognitive science would have to be rejected, and its findings re-interpreted accordingly.

The proponents of GOFAI are not lying down quietly. In a recent number of *Cognitive Science* devoted to a debate on "situated action," GOFAI's high-priest Herbert Simon combats the charges of the situated school (Vera & Simon, 1993). He argues that their insights are not incompatible with his, and not even wholly new. For instance, he cites his well-known example of the ant, whose "complex" pathway is continuously determined by the immediate environmental context (Simon, 1969), and his accompanying claim that "the intellective component of man may be relatively simple" since "most of the complexity of his behavior may be drawn from his environment" (op. cit., p. 83). His longstanding emphasis on the importance of external, as well as internal, representations in problem-solving has been backed up by detailed experimental studies, some of which concern temporally exacting tasks embedded in complex environments—just the challenge thrown down to GOFAI by the defenders of situated action (Card, Moran, & Newell, 1983; Newell & Simon, 1972). Furthermore, he argues that many "situated" models are in fact thoroughly symbolic (and representational), and that those that are not probably cannot be extended to cope with complex tasks. (How can the notion—and the mathematics— of dynamically coupled systems, for example, help us understand cryptarithmetic, or vacation-planning?)

Other researchers also retain the notion of representation—even though they are deeply critical of GOFAI on other grounds. For example, recent experimental work shows that perception is intrinsically dynamic, arguably because it has co-evolved with dynamic processes in the environment, such as the movements of predator or prey (Freyd, 1987, 1992; Miller & Freyd, 1993). These authors criticize GOFAI, and cognitive psychology in general, for ignoring both time and evolution. But they do not reject the concept of representation. Nor do they reserve it only for relatively high-level human thought. To say that perception is dynamic, in their view, is to say that it involves intrinsically dynamic representations.

The second part of our question about whether Miller, Galanter, and Pribram's enthusiasm was misplaced concerned the psychological content they expected to encompass. Human cognitive psychology has been enriched, and advanced, by the computational approach in many ways—far too many to discuss here (Boden, 1988). So too has developmental psychology, which "MGP" hardly mentioned (Karmiloff-Smith, 1992; Klahr, 1992). But what of animal psychology, motivation and emotion, personality, creativity, social behavior, and abnormal psychology? Can all these, as they claimed, be made grist for the computational mill?

Animal psychology was not much addressed by computationalists until relatively recently. Suggestions that ethologists use notions such as planning and production systems to compare the behavior of various species were taken up by some, but roundly opposed by others (Boden, 1988; Montefiore & Noble,1989). The opposition in ethology to using AI-work on planning is largely based on suspicion of the notion of representation (at least where non-human animals are concerned). Many ethologists—and, as noted above, increasing numbers of computational researchers—look for relatively direct organism-environment couplings, as opposed to central direction by internal representations. Nevertheless, recent work on (for example) animals' "theory of mind" was partly prompted by computationally influenced work in human psychology (Premack, 1988).

Some current computational work on animal behavior is done within evolutionary robotics. Here, genetic algorithms are used to evolve the "sensorimotor" morphology of artifactual creatures. For instance, a neural net linking visual sensors with motor effectors may be evolved by "natural selection" within a simple task-environment (Beer & Gallagher, 1992; Cliff, Husbands, & Harvey, 1993; Cliff, Harvey, & Husbands, 1993). Specific linkage-patterns are initially defined by (randomly chosen) rules. These rules can then be randomly mutated for many successive generations. At each generation, the least successful creatures are automatically rejected; only the more successful ones are available for further breeding.

A closely related form of computer modeling, which also involves work on animal behavior, is Artificial Life (A-Life). Much as AI aims to find the principles of intelligence in general, so A-Life seeks the principles of life in general—including reproduction and evolution (Boden, 1996; Langton, 1989; Langton et al., 1992; Meyer & Wilson, 1991; Meyer, Roitblat, & Wilson, 1993; Varela & Bourgine, 1992). Since intelligence is an emergent feature of living systems, AI and cognitive science can be thought of as included within A-Life, or (ahistorically) as offshoots of it.

Insects figure largely in this new form of computational work (Lestel, 1992). (Indeed, some wags say that "AI" now stands for "artificial insects.") Insect behavior is relatively simple (compared with language-understanding or theorem-proving, so beloved of classical AI), and relatively dependent on environmental triggers. In some cases, insect neuroanatomy is fairly well understood. And many insects, such as ants and bees, show "social" organization, wherein the entire colony exhibits emergent behavior of a type that cannot be attributed to any individual animal. Accordingly, examples of computational ethology include work on the swarming of bees and the path-following of ants (as well as the flocking of birds, fish, or buffalo; Reynolds, 1987). And computational neuroethology relates animal behavior to detailed hypotheses about the neural processing involved, for instance, in the sensorimotor system of the hoverfly (Beer, 1990; Cliff, 1991).

Other animals, however, are also studied. And, as computational ethology develops, we may expect to see more models of the behavior of the higher animals. One of the many unsolved problems is the level at which the theoretical concept of "representation" becomes unavoidable. As we have seen, Simon—and others—claim that (especially if we allow for *external* representations) this level is considerably lower down the phylogenetic "scale" than most A-Lifers appear to believe.

As yet, relatively little computational work has been done on motivation and emotion. Purposive behavior has been a concern of computational psychology from the early days (Boden, 1972; Newell & Simon, 1961; Reitman, 1965; Reitman, Grove, & Shoup, 1964). However, most computer models still have only one or two goals (plus subgoals). Truly multiple-goal systems, when we achieve them, will show aspects of emotional behavior. Any intelligent system capable of scheduling multiple conflicting goals, and of reacting to emergencies and opportunities, would need the sorts of goal-scheduling mechanisms characteristic of emotions (Beaudoin & Sloman, 1993; Frijda, 1986; Ortony, Clore, & Collins, 1991; Sloman, 1987, 1992). To some extent, we can think of what is needed in terms of interrupts, opportunism, and priority-setting. But there are many unanswered questions concerning the general architecture that would be needed for scheduling the activities of a highly complex creature.

These questions would need to be answered in an adequate account of personality, for personality is largely a matter of idiosyncratic values, motivations, and affective tendencies, associated with distinctive patterns of belief (what McDougall termed "sentiments," organized under the "master sentiment of self-regard" (Boden, 1972, chs. 6 & 7; McDougall, 1908, 1923). Some of the earliest computer models were attempts to capture aspects of personality (Tomkins & Messick, 1963), but these were unavoidably crude. Recent computationally grounded philosophical work has depicted the personality as a complex structure of beliefs and desires, and the self as a reflexively constructed entity: a "center of narrative gravity" (Dennett, 1991).

Creativity, too, can be understood in computational terms (Boden, 1990). Some creative ideas are surprising novel combinations of familiar ideas—brought about by association, analogy, or inductive classification. Connectionism has given us some promising ideas about how this sort of (improbabilist) creativity can happen. Other creative ideas are even more deeply surprising, for they seem to involve ideas that simply could not have occurred before. This type of (impossibilist) creativity can be explained in terms of the mapping, exploration, and transformation of conceptual spaces. If a dimension (an aspect of a style of thinking) is transformed in some way, then ideas can be generated that previously would have been impossible. They are related to the untransformed space, and so are potentially intelligible; at the same time, however, they lie outside it. Computational ideas about the definition, exploration, and transformation of conceptual spaces (generative systems) can help us to understand this sort of creativity.

Some of the earliest work in computational psychology was done by a distinguished social psychologist, Robert Abelson (1973, 1981). He used this approach to study attitude-change, and interpersonal relations of many different (but systematically related) kinds. Members of the Yale AI-group took up and developed some of his ideas (Dyer, 1983). With those exceptions, very little computational work has focused on social psychology. But there is no reason, in principle, why Abelson's good example should not be followed. Cognitive dynamicists would argue that GOFAI models (Abelson's included) tend to prioritize the individual mind, with social facts being included as influences on minds. They would see their approach as better suited to conceptualize the reciprocal influences of individual and group. On this approach, two dynamic systems may be mutually related; changes in the one lead to changes in the other, and vice versa. Moreover, the dual-system can be seen as a single system, which itself may be closely coupled with another. However, given Simon's strictures noted above, it is questionable whether this approach is needed if we are to model the complex relations within and between social groups and institutions of varying sizes.

Abnormal psychology, if neurosis is "abnormal," was a concern of some of the very earliest work in computer modeling. Simulations of Freudian defense mechanisms, and of the ways in which they could change (and affect the expressibility of) beliefs, were developed in the early 1960s (Boden, 1977, chs. 2 & 3; Colby & Gilbert, 1964). Simplistic though they were, these models suggested that the complex mental processes involved in neurosis could in principle be understood in computational terms. (One of their many crudities was an over-simple model of analogy-finding; recent AI-work offers much richer accounts of analogy (Holyoak & Thagard, 1989a, 1989b; Mitchell, 1993).) Later, a computer program was written to model the effects of neurotic anxiety and defensive thinking on speech—not only its content, but also its pauses, restarts, and self-corrections (Clippinger, 1977). Paranoia, too, has been modeled in AI-terms (Colby, 1981). And the potential relevance of computational ideas to phenomena such as fugue, hysterical paralysis, and multiple personality disorder has already been mentioned (see also Boden, 1994).

Several groups have explored the possible applications of AI-methods to psychiatry (Hand, 1985). Most clinical applications concern ELIZA-like interviewing systems, which can be used to elicit patients' foci of anxiety in an impersonal, and therefore non-threatening, manner. But research-oriented modeling may go deeper. Ideas about mental structure have passed both ways between computational and clinical psychology. Some concern the mind's global functional architecture (Shallice, 1988). Others provide detailed computational models of specific types of performance. One persuasive example is a connectionist model of deep dyslexia: the "neural" processing within this system appears to explain the puzzling pattern of varied symptoms characterizing this type of dyslexia (Hinton & Shallice, 1991; Plaut & Shallice, 1993).

Although Miller and his co-authors did not dwell on it, it was implicit in their work (and in that of McCulloch and Pitts before them; McCulloch & Pitts, 1943; Pitts & McCulloch, 1947) that computer models of actual neural circuits might eventually be achieved. Today, there is increasing interest in neurocomputational modeling (Schwartz, 1990; Hanson & Olson, 1990). Various specific areas of the brain are being simulated, sometimes in great detail. Examples include the synapses between the parallel fibers and Purkinje cells of the cerebellum (Miall, Weir, Wolpert, & Stein, in press), the neurons of the posterior parietal cortex (Zipser & Anderson, 1988), and the cortical processing of visual invariants (Rolls, 1992). Biophysicists are using similar techniques: phenomena now being simulated include enzyme-actions at individual synapses, Hodgkin-Huxley channels, synaptic diffusion and inhibition, and the details of specific dendritic trees.

Another topic not dwelt on in *Plans and the Structure of Behavior* is consciousness. Whether the existence of consciousness *as such* could be explained in computational terms is highly controversial. Most psychologists, computational or otherwise (and including Miller and friends), simply ignore this question. Consciousness is typically taken for granted; no scientific or philosophical explanation for its existence is sought. Recently, several extended discussions of consciousness have appeared, both within (Dennett, 1991; Jackendoff, 1987) and outside (Bock & Marsh, 1993; Marcel & Bisiach, 1988) the computational paradigm. In addition, a new journal (*Consciousness and Cognition*) was launched in 1992. Even so, many of these discussions avoid the philosophical problems: they assume the existence of consciousness, and ask what conditions determine its occurrence, or favor one type rather than another.

Dennett (1991) is an exception. But whether one is convinced by his broadly functionalist "disqualifications" of qualia is another matter. I think it fair to say that, if we take for granted the occurrence of subjective consciousness, then computational accounts can help us to understand why *this* sort, rather than *that* sort, happens in certain circumstances. Dennett, for instance, argues persuasively that if we assume that there is indeed "something which it is like" to be a bat (Nagel, 1974), then we can get some (admittedly limited) idea of just what it is like, by studying the bat's behavior and neuroanatomy. The more we know about the information-processing of which bats are capable, the less vague our notion of what it might be like to be one. It does not follow, however, that a computational account—or a neurophysiological one, either—could explain consciousness *as such*. This is still an unsolved philosophical problem (Boden, in press).

With hindsight, then, most of the promises made by Miller, Galanter, and Pribram have been fulfilled, or appear likely to be fulfilled, by later achievements. This is so with respect not only to methodology, but also to content.

They were right to suggest that psychology should be computational. But "computational," here, must include methods going beyond the bounds of GOFAI—not to mention TOTE-units: much as physics uses many types of mathematics, so cognitive science will need many formalisms besides those envisaged in 1960. They were right to ask that psychological theories be computer-testable. And they were right in being prepared to posit internal representations of some sort. Perhaps, where inner plans are concerned, they overdid it—although their discussion of animal "instinct," and of cultural artifacts, explicitly allowed for the Plan's being largely in the world, not in the head. But whether, and where, internal representations exist is not a question to be settled by (behaviorist or dynamicist) *fiat*.

They were right, too, in suggesting that a computational approach might illuminate many different content-areas of psychology. Even phenomena like motivation, emotion, and creativity, which most people assume to be beyond the reach of computational theory, can helpfully be thought of in these terms.

In sum, the trio's Plans have not all been realized. But they helped us to learn a great deal about the Structure of Behavior. (A pity, though, about that execrable binding.)

References

Abelson, R. P. 1973. "The Structure of Belief Systems," in R. C. Schank and K. M. Colby (eds.), *Computer Models of Thought and Language*, pp. 287–340. San Francisco: Freeman.

Abelson, R. P. 1981. "The Psychological Status of the Script Concept." *American Psychologist*, 36, pp. 715–729.

Beaudoin, L. P., and Sloman, A. 1993. "A Study of Motive Processing and Attention," in A. Sloman, D. Hogg, G. Humphreys, D. Partridge, & A. Ramsay (eds), *Prospects for Artificial Intelligence*, pp. 229–38. Amsterdam: IOS Press.

Beer, R. D. 1990. *Intelligence as Adaptive Behavior: An Experiment in Computational Neuroethology*. New York: Academic Press.

Beer, R. D., and Gallagher, J. C. 1992. "Evolving Dynamical Neural Networks for Adaptive Behavior."*Adaptive Behavior,* 1, pp. 91–122.

Bock, G. R., and Marsh, J. (eds.). 1993. *Experimental and Theoretical Studies of Consciousness*. Chichester, England: Wiley.

Boden, M. A. 1965. "McDougall Revisited." *Journal of Personality*, 33, pp. 1–19.

Boden, M. A. 1970. "Intentionality and Physical Systems." *Philosophy of Science,* 37, pp. 200–14.

Boden, M. A. 1972. *Purposive Explanation in Psychology*. Cambridge: Harvard University Press.

Boden, M. A. 1977. *Artificial Intelligence and Natural Man*. New York: Basic Books.

Boden, M. A. 1988. *Computer Models of Mind: Computational Approaches in Theoretical Psychology*. Cambridge: Cambridge University Press.

Boden, M. A. 1990. *The Creative Mind: Myths and Mechanisms*. London: Weidenfeld & Nicolson. (Paperback ed., expanded, London: Abacus, 1991.)

Boden, M. A. 1991. "Horses of a Different Color?," in W. Ramsey, S. P. Stich, and D. E. Rumelhart (eds.), *Philosophy and Connectionist Theory*, pp. 3–20. Hillsdale, NJ: Erlbaum.

Boden, M. A. 1994. "Multiple Personality and Computational Models." In A. Phillips-Griffiths (ed.), *Philosophy, Psychiatry, and Philosophy*, pp. 103–114. Cambridge: Cambridge University Press. 1994.

Boden, M. A. (ed.). 1996. *The Philosophy of Artificial Life*. Oxford: Oxford University Press.

Boden, M. A. (in press) "Can Consciousness Be Scientifically Explained?" In J. Cornwell (ed.) *Consciousness and Human Identity*. Oxford: University Press.

Brooks, R. A. 1991. "Intelligence Without Representation." *Artificial Intelligence,* 47, pp. 139–159.

Brooks, R. A. 1992. "Artificial Life and Real Robots," in F. J. Varela and P. Bourgine (eds.), *Toward a Practice of Autonomous Systems: Proceedings of the First European Conference on Artificial Life*, pp. 3–10. Cambridge: MIT Press.

Card, S., Moran, T. P. & Newell, A. 1983. *The Psychology of Human-Computer Interaction*. Hillsdale, NJ: Erlbaum.

Clark, A. 1989. *Microcognition: Philosophy, Cognitive Science, and Parallel Distributed Processing*. Cambridge: MIT Press.

Clark, A. 1993. *Associative Engines: Connectionism, Concepts, and Representational Change*. Cambridge: MIT Press.

Cliff, D. 1991. "The Computational Hoverfly: A Study in Computational Neuroethology," in J.A. Meyer and Wilson, S. W. (eds.), *From Animals to Animats: Proceedings of the First International Conference on Simulation of Adaptive Behaviour*, pp. 87–96. Cambridge: MIT Press.

Cliff, D., Husbands, P. and Harvey, I. 1993. "Evolving Visually Guided Robots," in J.-A. Meyer, H. Roitblat, & S. Wilson (eds.), *From Animals to Animats 2: Proceedings of the Second International Conference on Simulation of Adaptive Behaviour*, pp. 374–83. Cambridge: MIT Press.

Cliff, D., Harvey, I. and Husbands, P. 1993. "Explorations in Evolutionary Robotics," *Adaptive Behavior*, 2, pp. 71–108.

Clippinger, J. H. 1977. *Meaning and Discourse: A Computer Model of Psychoanalytic Discourse and Cognition*. Baltimore: Johns Hopkins.

Colby, K. M. 1981. "Modeling a Paranoid Mind." *Behavioral and Brain Sciences,* 4, pp. 515–33.

Colby, K. M., and Gilbert, J. P. 1964. "Programming a Computer Model of Neurosis." *Journal of Mathematical Psychology,* 1, pp. 405–17.

Cussins, A. 1990. "The Connectionist Construction of Concepts," in M. A. Boden (ed.), *The Philosophy of Artificial Intelligence*, pp. 368–440. Oxford: Oxford University Press.

Dennett, D. C. 1991. *Consciousness Explained*. Boston: Little, Brown.

Dyer, M. G. 1983. *In-Depth Understanding: A Computer Model of Integrated Processing for Narrative Comprehension*. Cambridge: MIT Press.

Fodor, J. A. 1975. *The Language of Thought*. Hassocks, Sussex: Harvester Press.

Freyd, J. J. 1987. "Dynamic Mental Representations." *Psychological Review,* 94, pp. 427–38.

Freyd, J. J. 1992. "Dynamic Representations Guiding Adaptive Behavior," in F. Macar, V. Pouthas, and W. J. Friedman (eds.), *Time, Action, and Cognition: Towards Bridging the Gap*, pp. 309–23. Dordrecht: Kluwer Academic.

Frijda, N. 1986. *The Emotions*. Cambridge: Cambridge University Press.

Hand, D. J. 1985. *Artificial Intelligence and Psychiatry*. Cambridge: Cambridge University Press.

Hanson, S. J., and Olson, C. R. (eds.). 1990. *Connectionist Modeling and Brain Function: The Developing Interface*. Cambridge: MIT Press.

Haugeland, J. 1985. *Artificial Intelligence: The Very Idea*. Cambridge: MIT Press/Bradford Books.

Hinton, G. E., McClelland, J. L. and Rumelhart, D. E. 1990. "Distributed Representations," in M. A. Boden (ed.), *The Philosophy of Artificial Intelligence*. Oxford: Oxford University Press. (First published as Chapter 3 of D. E. McClelland and J. E. Rumelhart (eds.), *Parallel Distributed Processing: Explorations in the Microstructure of Cognition, Vol. I: Foundations*, pp. 77–109. Cambridge: MIT Press.)

Hinton, G. E., and Shallice, T. 1991. "Lesioning an Attractor Nework: Investigations of Acquired Dyslexia." *Psychological Review*, 98, pp. 74–95.

Holyoak, K. J., and Thagard, P. R. 1989a. "Analogical Mapping by Constraint Satisfaction." *Cognitive Science* 13, pp. 295–356.

Holyoak, K. J., and Thagard, P. R. 1989b. "A Computational Model of Analogical Problem Solving," in S. Vosniadou and A. Ortony (eds.), *Similarity and Analogical Reasoning*, pp. 242–266. Cambridge: Cambridge University Press.

Jackendoff, R. 1987. *Consciousness and the Computational Mind*. Cambridge: MIT Press.

Karmiloff-Smith, A. 1992. *Beyond Modularity: A Developmental Perspective on Cognitive Science*. Cambridge: MIT Press.

Klahr, D. 1992. "Information-Processing Approaches to Cognitive Development," in M. H. Bornstein and M. E. D. Lamb (eds.), *Developmental Psychology: An Advanced Textbook*, 3rd ed. Hillsdale, NJ: Erlbaum.

Langton, C. G. (ed.). 1989. *Artificial Life*. Reading, Mass.: Addison-Wesley.

Langton, C. G., Taylor, C., Farmer, J. D. and Rasmussen, S. (eds.). *Artificial Life II*. Reading, Mass.: Addison-Wesley.

Lestel, D. 1992. "Fourmis Cybernetiques et Robots-Insectes: Socialite et Cognition a l'Interface de la Robotique et de l'Ethologie Experimentale." *Information Sur Les Sciences Sociales*, 31 (2), pp. 179–211.

McCulloch, W. S., and Pitts, W. H. 1943. "A Logical Calculus of the Ideas Immanent in Nervous Activity." *Bulletin of Mathematical Biophysics*, 5, 115–133. (Reprinted in W. S. McCulloch, *Embodiments of Mind*, pp. 19–39. Cambridge: MIT Press, 1965.)

McDougall, W. 1908. *An Introduction to Social Psychology*. London: Methuen. (23rd ed., enlarged, 1936.)

McDougall, W. 1923. *An Outline of Psychology*. London: Methuen.

Maes, P. (ed.). 1991. *Designing Autonomous Agents*. Cambridge: MIT Press.

Marcel, A. J., and Bisiach, E. (eds.). 1988. *Consciousness in Contemporary Science*. Oxford: Oxford University Press.

Meyer, J.-A., and Wilson, S. W. (eds.). 1991. *From Animals to Animats: Proceedings of the First International Conference on Simulation of Adaptive Behaviour*. Cambridge: MIT Press.

Meyer, J.-A., Roitblat, H. and Wilson, S. (eds.). 1993. *From Animals to Animats 2: Proceedings of the Second International Conference on Simulation of Adaptive Behaviour*. Cambridge: MIT Press.

Miall, R. C., Weir, D. J., Wolpert, D. M., and Stein, J. F. In press. "Is the Cerebellum a Smith Predictor?" *Journal of Motor Behaviour*.

Miller, G. A., Galanter, E., and Pribram, K. H. 1960. *Plans and the Structure of Behavior*. New York: Holt.

Miller, G. F., and Freyd, J. J. 1993. *Dynamic Mental Representations of Animate Motion: The Interplay Among Evolutionary, Cognitive, and Behavioral Dynamics*. Cognitive Sciences Research Paper CSRP-290, University of Sussex, 1993. (Submitted to *Brain and Behavioral Sciences*.)

Mitchell, M. 1993. *Analogy-making as Perception*. Cambridge: MIT Press.

Montefiore, A., and Noble, D. (eds.). 1989. *Goals, No-Goals, and Own Goals: A Debate on Goal-directed and Intentional Behaviour*. London: Unwin Hyman.

Nagel, T. 1974. "What Is It Like to be a Bat?" *Philosophical Review*, 83, pp. 435–51.

Newell, A. 1980. "Physical Symbol Systems." *Cognitive Science* 4, pp. 135–183.

Newell, A., and Simon, H. A. 1961. "GPS—A Program that Simulates Human Thought," in H. Billing, ed., *Lernende Automaten*, pp. 109–124. Munich: Oldenbourg. (Reprinted in E. A. Feigenbaum and J. Feldman (eds.) *Computers and Thought*, pp. 279–96. New York: McGraw-Hill, 1963.)

Newell, A., and Simon, H. A. 1972. *Human Problem Solving*. Englewood Cliffs, NJ: Prentice-Hall.

Ortony, A., Clore, G. L. and Collins, A. 1991. *The Cognitive Structure of Emotions*. Cambridge: Cambridge University Press.

Pitts, W. H., and McCulloch, W. S. 1947. "How We Know Universals: The Perception of Auditory and Visual Forms." *Bulletin of Mathematical Biophysics*, 9, pp. 127–47. (Reprinted in W. S McCulloch, *Embodiments of Mind*. Cambridge: MIT Press, 1965.)

Plaut, D. C., and Shallice, T. 1993. "Deep Dyslexia: A Case Study of Connectionist Neuropsychology." *Cognitive Neuropsychology*, 10 (5), pp. 377–500.

Port, R. F., and van Gelder, T. (eds.). 1995. *Mind as Motion: Dynamics, Behavior, and Cognition*. Cambridge: MIT Press.

Premack, D. 1988. "'Does the Chimpanzee Have a Theory of Mind?' Revisited,' in R. M. J. Byrne and A. Whiten (eds.), *Machiavellian Intelligence: Social Expertise and the Evolution of Intellect in Monkeys, Apes, and Humans*. Oxford: Clarendon Press.

Reitman, W. R. 1963. "Personality as a Problem-Solving Coalition," in S. S. Tomkins and S. Messick (eds.), *Computer Simulation of Personality: Frontier of Psychological Theory*, pp. 69–100. New York: Wiley.

Reitman, W. R. 1965. *Cognition and Thought: An Information-Processing Approach*. New York: Wiley.

Reitman, W. R., Grove, R. B., and Shoup, R. G. (1964) "Argus: An Information-Processing Model of Thinking." *Behavioral Science*, 9, pp. 270–81.

Reynolds, C. W. 1987. "Flocks, Herds, and Schools: A Distributed Behavioral Model." *Computer Graphics*, 21 (4), pp. 25–34.

Rolls, E. 1992. "Neurophysiological Mechanisms Underlying Face Processing Within and Beyond the Temporal Cortical Areas." *Philosophical Transactions of the Royal Society, London (Series B)*, 335, pp. 11–21.

Schwartz, E. L. (ed.). 1990. *Computational Neuroscience*. Cambridge: MIT Press.

Selfridge, O. G. 1959. "Pandemonium: A Paradigm for Learning," in D. V. Blake and A. M. Uttley (eds.), *Proceedings of the Symposium on Mechanisation of Thought Processes*, pp. 511–29. London: H. M. Stationery Office.

Shallice, T. 1988. *From Neuropsychology to Mental Structure*. Cambridge: Cambridge University Press.

Simon, H. A. 1969. *The Sciences of the Artificial*. Cambridge: MIT Press.

Sloman, A. 1987. "Motives, Mechanisms, and Emotions," *Cognition and Emotion* 1, pp. 217–33. (Reprinted in M. A. Boden (ed.), *The Philosophy of Artificial Intelligence*, pp. 231–47. Oxford: Oxford University Press, 1990.)

Sloman, A. 1992. "Prolegomena to a Theory of Communication and Affect," in A. Ortony, J.

Slack, and O. Stock (eds.), *Communication from an Artificial Intelligence Perspective: Theoretical and Applied Issues*, pp. 229–260. Heidelberg: Springer.

Smolensky, P. 1987. "Connectionist AI, Symbolic AI, and the Brain." *AI Review,* 1, pp. 95–10.

Smolensky, P. 1988. "On the Proper Treatment of Connectionism" (with peer-commentary and author's reply). *Behavioral and Brain Sciences,* 11, pp. 1–74.

Tomkins, S. S., and Messick, S. (eds.). 1963. *Computer Simulation of Personality: Frontier of Psychological Theory.* New York: Wiley.

van Gelder, T. 1992. *What Might Cognition be if not Computation?* Indiana University Cognitive Science Technical Report 75.

Varela, F. J., and Bourgine, P. (eds.). 1992. *Toward a Practice of Autonomous Systems: Proceedings of the First European Conference on Artificial Life.* Cambridge: MIT Press.

Varela, F. J., Thompson, E., and Rosch, E. 1991. *The Embodied Mind: Cognitive Science and Human Experience.* Cambridge: MIT Press.

Vera, A. H., and Simon, H. A. 1993. "Situated Action: A Symbolic Interpretation." *Cognitive Science*, 17, pp. 7–48.

von Uexkull, J. 1957. "A Stroll Through the Worlds of Animals and Men," in C. H. Schiller (ed.), *Instinctive Behavior*, pp. 5–82. New York: International Universities Press.

Wheeler, M. (1996). "From Robots to Rothko." In M. A. Boden (ed.) *The Philosophy of Artificial Life*, pp. 209–236. Oxford: Oxford University Press.

Zipser, D., and Anderson, R. A. 1988. "A Back-propagation Programmed Network that Simulates Response Properties of a Subset of Posterior Parietal Neurons." *Nature,* 331, pp. 679–684.

5

Carol Fleisher Feldman

Boden's Middle Way

Viable or Not?

Boden's paper, and its important predecessors—including the books *Artificial Intelligence and Natural Man*, 1977 (*AINM* hereafter), *Computer Models of Mind*, 1988 (*CMM*), and *The Creative Mind*, 1990 (*CM*)—have staked a claim to a distinctive territory in an embattled field. She calls for a psychology that is at once computational and humanistic. Many AI hardliners share her view that a scientific psychology must be computational. And they believe that they can model human minds *insofar as they are real* within such a science. But, in many ways, the hardliners make strange bedfellows for Boden, who cares principally about such essentially human (and most unmachinelike) expressions of mind as creativity. She is surely the most original thinker in cognitive science today.

Indeed, her natural allies would seem to be the AI critics. But she brushes critics of both main camps aside. To objections that computers are irrelevant to scientific psychology because they cannot model the essentially human, on the one side, and to objections that essentially human cognition is described with variables of a kind that can have no place in a scientific psychology on the other, her response is to look for a bridge between mind and machine. Critics of the first kind (e.g., Searle, 1984) claim that computational models of mind tell us little about important, and real, features of the human mind. Critics of the second kind (e.g., Fodor, 1980) claim that science can never explain the human mind as it is commonly understood in the folk culture, and should not try. According to Boden, members of both groups of critics believe that there is an unbridgeable gulf between computer models and human experience as it is commonly understood, and both groups explain the gap in the same way—that while human experience is based in meaning, especially meaning related to context, computational systems have no level of meaning, and especially of context-dependent meaning. What underlies this, as Boden sees it, is that critics of both groups subscribe to the erroneous view that computational systems are all syntax and no semantics. It is on this central point that Boden disagrees with them.

For Boden, computer programs have a semantics that is close enough to human meaning to bridge the gap, once we take a sufficiently functionalist view of meaning. Computer programs have an "intrinsic causal-semantic aspect" (*CMM*, p. 238). "Although such a program cannot use the symbol 'restaurant' to mean *restaurant*" . . . its "inherent procedural consequences" . . . "give it a toehold in semantics" of a causal, if not a denotative, kind (ibid., p. 250, italics in original). Thus, for Boden, it is possible in principle, and within limits she is willing to live with, for computers to model meaning and, therefore, to serve as a scientific psychology of the essentially human functions of mind—of creativity, understanding, consciousness, intentionality.

The gap between the two worlds is only partly bridged in this way, for a causal notion of meaning is not going to satisfy many of the critics, nor will a functionalist toehold in semantics. Nor will a system behaving in a way that is instrumentally, rather than really, intentional. Nor a system taking context into account in the limited ways computer programs can. Nor, finally, a system doing apparently creative things. This general line of argument is always presented in a carefully qualified way, and leads Boden to the honest conclusion that it may not ever be possible to capture the essential common core of these divergences between man and machine—the "what it's like," or *qualia*, of human experience. Rather, the approximation could, in principle, be close enough.

One way to come closer would be a new way of writing programs, and of talking about everyday psychology, that shared a new common set of definitions, preferably procedural, that worked as a neutral ground between them—a kind of man/machine *Esperanto*. "(A) unified theory of computation could illuminate how *reflective* knowledge is possible. For, given such a theory, a system's representations of data and of processes . . . would be essentially comparable" (*CMM*, p. 249, italics in original). Indeed, a thoroughgoing commitment to process in both AI and psychology is, for Boden, a fundamental tenet. And it creates a context, free of troublesome representational aspects of mind, where AI models could adequately capture everything about (functional aspects of) mind—a matter we will return to. The point for now is only to note the depth of Boden's commitment to find a bridge between machine and mind that makes them formally equivalent, so that one can serve as model of the other. And finally there is a third possible solution, in the new PDP technology with its "tantalizing, human-like capacity" (*CM*, p. 120).

For Boden, the chief thing is to bridge the gap and create a unified scientific psychology that is both computational and life-like. One cannot help admiring Boden's desire for such a program, however quixotic it may seem, for Boden's program is one that, in a more generalized form, is widely shared in the scientific psychological community today. To put the general case crudely, many right-thinking psychologists today want to end an era of looking under the methodologically illuminating light, so to speak, with all the loss of naturalism that attended it, and start finding a reasonable way to do science about the real workings of mind as we see them in life. It is partly because "situated cognition" seems so much smarter and more complicated than the laboratory varieties of it, partly because it seems to include a host of interesting interpretive mental processes that are quite unlike the logical (and illogical) forms of thought that constitute the bulk of the literature in experimental cognition so far. In general, there is a lot of method finding going on in this field at the moment.

In this respect, if Boden differs from this general group, it is only because she already has found her method of choice. The specific basis of her affection for the computer model, as the method of choice for all psychology, is in the fact that it runs. The computer lets one see whether a particular solution would be brought about by means of a hypothesized process; it is a general-purpose theory testing device that embodies the TOTE unit in actual fact. Thus, for Boden, at the very least, computer models are appropriate tools in the service of a humanistic psychology. Here she would find good support in such thoughtful PDP research as that of Thomas Shultz, who has done a number of interesting studies of psychological phenomena to see how the machine would run when certain assumptions (usually derived from the experimental literature) were made, and how it would develop over run time, when certain developmental assumptions were made. Shultz et al. (1994) say, "Connectionist models are not, by themselves, psychological theories. Rather, they are powerful tools that may, along with more conventional empirical and theoretical work, help us to develop more coherent psychological theories" (p. 51).

But unlike Shultz, for example, Boden also claims that in principle computer models can accurately enough match human minds to expose how the mind works in its myriad and even mysterious ways. Nevertheless, far from reducing man to a mere machine, Boden hopes to show (in *AINM*) that "properly understood, artificial intelligence can *counteract* the dehumanizing influences of natural science that have been complained of by critics of the role of 'scientism' in urban-industrial cultures" (ibid., p.5; Boden's italics).

In summary, then, Boden's position takes the hardliners' stance that models of mind must be given in computational terms, and claims, along with the hardliners, and against the AI critics, that computers have meanings that are close enough to life to model human minds. But, unlike the hardliners, she does this in a non-reductive spirit, and tries to keep the most complex forms of human cognition at the center of the cognitive science enterprise. It seems not unfair to say that in terms of the canons of the field, her position takes a perspective orthogonal to canonical perspectives of hardliner and critic alike, and more optimistic than either. To appreciate her view, it may be best to step back from the canonical views, into her own, and try to see what animates her.

The first, and unquestionably most important step, is to note that Boden is a thoroughgoing functionalist in a representationalist world. Boden's computational humanism, if I can call it that, is greatly facilitated by a particular, functionalist view of both the computer and of the human mind, the latter derived from McDougall's purposivism (Boden, 1972). As we have noted, Boden emphasizes, as her defining trait of the computer, that it runs. The content of prior states as representations is not essential, nor even the form or content of the procedures for operating on them as, for example, whether they are algorithmic or stochastic. What *is* essential is that any claims about what the human mind can do can be expressed as actual real time doings. It is this emphasis on process in Miller et al.'s TOTE unit that captured her, and it is what makes her a computationalist today. As for the mind, what creativity, for example, amounts to is creative functioning—*not* a special static, representational state.

Moreover, the computer facilitates the study of human mental process by having the (contrastive) virtue that *its* procedures of "thought" are done inside a box that is transparent rather than, as is the human's, black to the scientific eye. Thus, whereas the actual steps in the processing of any rule or model in the human mind are inaccessible to third person, or scientific, observation, not just the outcome, but also the steps along the way of machine problem-solving are done in the scientific light of day. This is, of course, true of GOFAI; but it is much less true of PDP models.

Boden's rather distinctive emphasis on process protects her to some extent from the accusation that she has done an injustice to *qualia*. In a sense, she sidesteps them altogether, for she does not imply that the computer's representations are like our first-person experience. She simply doesn't make any claims about representations in art or in life. The discrepancy between first- and third-person experience does not arise as so sharp a contrast in strictly functional descriptions. Whether this is because, with their focus on process, they have no central variables that are experiential, or because they cede experiential ground to the third-person point of view, I cannot tell. The two versions could have interestingly different implications for Boden's work, but, for present purposes, it may suffice to note that both equally permit evasion, without commitment, about *qualia*.

But even without representations, there are problems with Boden's strong isomorphism. Why should the steps a computer goes through in any process shed any light on how humans do it? In her earlier book, *AINM*, Boden dismissed worries that artificial models might not, eventually, simulate human process. Her dismissal was not very convincing, and many of her colleagues who had also cheerfully made that assumption earlier on became discouraged with it later. Boden perceived connectionism as the perfect solution, for here she saw a model that not only revealed human process by stipulation but also by virtue of such real matters as similar function and similar structure. As she observes in *CM*, connectionist models have such human design features as flexible pattern matching, pattern completion, analogue pattern matching, contextual memory, weak constraints, and the capacity to learn to do better. "In short, connectionist systems have 'associative memory,' grounded in both meaning and context" (*CM*, p.119). The price we pay is that in the sense that GOFAI is a transparent tool, PDP is not, for we can no longer see inside.

With connectionist models, even though we cannot go so far as to say that computers themselves are necessarily creative, we at least can understand how human creativity is possible. So with the advent of connectionism, computers have definitely graduated for Boden from being, if I may so put it, a "mere" model, to being in many essential respects a "right" model. She says, "The key point is . . . that brains are, to some extent, *like* . . . connectionist systems" (ibid., p.114). Whereas preconnectionist models were driven by programs and so seemingly had all and only the intelligence their programmer put into them, connectionist models "do these things 'naturally,' without being specifically programmed to do them. . . . Rather, their associative memory and tantalizing, human-like, capacities are inevitable results of their basic design" (ibid., p.120).

Boden likes the promise of PDP, likes it so much that when she points to its defects as a model of mind, she fails to mention the most serious of its limitations. She notes that the brain has only one-way signals, that units in the brain are connected to many more others than in connectionist models, and that there is no analogue to the wide diffusion of neurochemicals. But these all speak only to its limitations as a brain model. In PDP the real problems arise in modeling mind. She seems to believe that it is not just brains, but minds too, that PDP models. What I may have missed is any mention of the fact that PDP is almost entirely a reactive system, that its mental events are largely associative, and that even its adherents (see Rumelhart et al.) think it has no elements that can be used to model the mind, but is simply a device for testing the possibility of certain outputs coming from certain inputs by means of *its own* (even if brain-like) procedures: "There are central issues, however, which remain difficult to describe within the PDP framework. We have particularly in mind here, the process of thinking, the contents of consciousness . . . the nature of mental models, the reason for mental simulations, and the important synergistic role of language in thinking and in shaping our thought (Rumelhart et al., 1986, p. 38). This is not the kind of stuff that can equate machine with mind.

Boden herself realizes this. For she notes that "many people—including some leading connectionists—argue that to do conscious reasoning . . . the brain may have to function *as though* it were a digital computer. That is, it may have to follow strict, hierarchical and even sequential rules" (*CM*, p.131)—even though it is really a connectionist machine. And elsewhere she notes that unlike PDP, the GOFAI model has at least one guarantee of, in some sense, embodying mental function in that it always depends on a program written by a (conscious, thinking) human.

And so we return full circle to the unsatisfactory earlier model that would not justify strong AI.

My own conclusion is that we are running in circles in an effort to get from *seems* (as in this computer process *seems like* a human mental process) to *is* (as in this computer process *is* a human mental process) without doing any real work to get there. And perhaps we are even trying to affirm the claim that {this computer is like a human mind}, in the more static language of representation and structure. It is a mystery to me why a functionalist like Boden would reach so hard for equivalence— or at least for functional isomorphism of these two distinctive functional systems with their different derivational histories, and even, it seems at times, for the structural[1] isomorphism necessary to assert the identity of the two systems. I can shed no further light on it here. So let me turn now to what I think would be the real work necessary to make such claims stick.

There have been some instructive recent attempts to point to exactly what it is that makes the computer not really a human mind—including those by Goldman (1993), Searle (1992), Penrose (1990), and others. They point variously to understanding, to meaning, to consciousness, and to qualia—that is, the "what it's like to be" of first-person reports. Plainly there are some differences of emphasis here; and there may turn out to be important substantive differences among elements of this list—that is, there may, in fact, be several different facets of human mental life that a computer does not have, and even traits that it lacks for different reasons.

For present purposes, let me roughly group these objections around the common view that there are aspects of human experience that are fundamentally different for the person having them and for the scientific observer (Nagel, 1974). To make the case for strong AI, then, one has to dismiss differences between first- and third-person accounts by denying that they are important, or meaningful, or real, or scientifically manageable. Indeed, first-person experience itself must be a mere epiphenomenon; or, if real, it can be nothing but its computerized version.

But for Boden, with her emphasis on the operational aspects of mind, such a denial of human experience is not necessary nor with her humanistic emphasis is it attractive. For the big differences between first- and third-person experience do not perhaps appear when the program is running, so to speak, but rather in how, and on what, and to what end, it runs. And, in fact, Boden is an avowed agnostic about the computability of qualia.

But that, in a sense, is not enough. For what she needs to do to make the claim that any computer model, even a functional model, is a right model of any construal of mind is to show us not just that there's a model there and that it runs, but to defend the claim of isomorphism with evidence. This chore is not one belonging to Boden alone. But because her strong humanism leads to her having unusually high ambitions for computationalism in the human domain, she particularly must take on the job of justifying claims of equivalence, if she wants to make them. This important task remains undone.

If at the start of the cognitive revolution there was a fundamental ambiguity that was torn apart to define the essential stance of today's warring camps, Boden can be seen as adhering to the original ambiguity as a third way that tries to hold commitments to human minds and computers together. This is because, for Boden at her best, as for many at the start of the cognitive revolution, the computer was simply a tool, a neutral instrument for testing theories. One wanted to study the human mind, and this powerful tool came along that offered to help, a tool that made memory drums look like the nineteenth-century equipment they were. Why not use it? But somehow the tool became a model, and after 30 years, many early computer optimists are pessimists, and many original pessimists find psychology trivialized by seductive machines. But it was perhaps innocent playing with a new and terrific piece of laboratory machinery at the start. At least that is how I remember it. And fun.

Boden carries some of this original enthusiasm into her work to the present day and we cannot but salute her for it. In a sense, her work is all of a piece with the *aha* experience she describes having in that bookshop many years ago. And because it is so much of a piece, perhaps understanding it can help us to better understand the origins of the standoffs of today.

Notes

I thank the Spencer Foundation for support of this work through its grant, "Studies in Cultural Psychology," to Jerome Bruner.
   1. David Kalmar suggested the useful language of function and structure to describe the two parts of Boden's claims.

References

Boden, M.A. 1972. *Purposive Explanation in Psychology*. Cambridge: Harvard University Press.

Boden, M.A. 1977. *Artificial Intelligence and Natural Man*. New York: Basic Books.

Boden, M.A. 1988. *Computer Models of Mind: Computational Approaches in Theoretical Psychology*. Cambridge: Cambridge University Press.

Boden, M.A. 1990. *The Creative Mind: Myths & Mechanisms*. New York: Basic Books.

Fodor, J.A. 1980. "Methodological Solipsism Considered as a Research Strategy in Cognitive Psychology." *Behavioral and Brain Sciences,* 3, pp. 63–110.

Goldman, A.I. 1993. "The Psychology of Folk Psychology." *Behavioral and Brain Sciences,* 16 (1), pp. 15–28.

Miller, G.A., Galenter, E. and Pribram, K.H. 1960. *Plans and the Structure of Behavior*. New York: Holt.

Nagel, T. 1974. "What Is It Like to be a Bat?" *Philosophical Review,* 83, pp. 435–457.

Penrose, R. 1990. *Precis of* "The Emperor's New Mind: Concerning Computers, Minds, and the Laws of Physics." *Behavioral and Brain Sciences,* 13, pp. 643–705.

Rumelhart, D.E., Smolensky, P., McClelland, J.L., and Hinton, G.E. 1986. "Schemata and Sequential Thought Processes in PDP Models," in J.L. McClelland, D. E. Rumelhart, and the PDP Research Group (eds.), *Parallel Distributed Processing: Explorations in the Microstructure of Cognition. Vol. 2: Psychological and Biological Models,* pp. 7–57. Cambridge: MIT Press.

Shultz, T.R., Schmidt, W.C., Buckingham, D., and Mareschal, D. 1994. "Modeling Cognitive Development with a Generative Connectionist Algorithm." Prepared for T. Simon and G. Halford (eds.), *Developing Cognitive Competence: New Approaches to Process Modeling*. Hillsdale, NJ: Erlbaum.

Searle, J.R. 1984. *Minds, Brains and Science*. Cambridge: MIT Press/Bradford.

Searle, J.R. 1992. *The Rediscovery of the Mind*. Cambridge: MIT Press.

6

Juan Pascual-Leone

Metasubjective Processes

The Missing *Lingua Franca* of Cognitive Science

The force of the real is made palpable by the foundering of thought.
(Jaspers, 1971, p. 71)

Thus man would be, and in various senses of the word, a fantastic
animal (Ortega y Gasset, 1957/1980a, p. 251)[1]

Cognitive science has brought with it a keen awareness of the difference between process and performance in humans, and forced recognition that high cognitive functions in humans (thinking, problem solving, intelligence, creativity, etc.) are not reducible to perceptual/motor or learning processes. Human performance, objective as much as subjective, is often construed by cognitive science as caused by brain processes that can be properly modeled by logico-mathematical means (computer programs, neuronal-connectionist instantiations, etc.); or, in a more idealist Galilean epistemology, as processes that literally are logico-mathematical and rule-governed (i.e., formally equivalent to computer programs, connectionist networks, or complex electronic machines). This latter claim, often called *the hard cognitive science* (computer-reductionist) assumption, does literally construe humans as biological computing machines; this is the position against which I argue and for which I offer a methodological remedy.

Many scientists question this hard cognitive science position. A common objection of neuroscientists is that the brain is a biodynamic complex organization whose patterns of processing, outputs (bodily actions), and inputs (recurrent patterns from "reality"—sensorial, motor, and other reactions from the body) are most unlike those of digital computers. Whereas the brain is subject to a long history of learning within and across situations, computers can have at most a limited amount of prepackaged "experimental" learning.

Developmental psychologists object to hard cognitive science because its models are usually paradigmatic (i.e., restricted to a type of situation or paradigm) instead of being organismic (i.e., modeling the psychological organism as a whole); and maturational or developmental processes are not easily modeled in it. Moreover, individual differences often are ignored, and the problem of a psychological unit (a unit of theoretical analysis for psychological processes, applicable across specialties such as learning, development and neuropsychology) is not addressed. The unit of analysis chosen in cognitive science is often arbitrary: determined by the paradigm, formalism, or computer language adopted. Furthermore organismic constraints that are important across paradigms to explain change in performance over time (such as growth of mental attentional capacity and working memory, etc.) are often ignored.

Cognitive psychologists with phenomenological orientation along with philosophers object that hard cognitive science has not been able to explain, or even describe, consciousness and its role in human functioning, acts of meaning, intersubjective processes, the Self, the Other, intentions, values, among other things—all descriptive notions about human minds often found in folk psychology and sociocultural and clinical psychology. Since society operates at this molar level of description, this level must be real and should be accounted for. Thus, hard cognitive science appears excessively (because exclusively) molecular and mechanistic.

At the same time, from an epistemological perspective, hard cognitive science seems to be uncritically realist vis-à-vis the nature of reality. It is apparent to (post-) modern thinkers that reality is construed by the mind of the beholder, albeit construed in obedience to highly structured and interdependent *resistances* from reality (i.e., external constraints: whether positive—Gibson's affordances—or negative— hindrances to praxis). In other words, there are no substantial chairs, or people, or institutions, or public affairs, or sociocultural events, etc., *as such* out there. The linguistic terms and concepts we use to refer to them stand for complexes of *schemes* (i.e., semantic-pragmatic functional systems of information-processing; or, if you will, packages of relations and interrelations found to hold among resistances, or resistance packages, or packages of packages). "Information" here is not substantive, but rather must be understood in terms of relations as the packaged result of *in-forming*—the process of injecting form. External resistances in-form the body's own processes (sensorial, motor, or mental) making them epistemologically reflect (*epireflect*) the structure of interrelations, and of dependency relations, holding among encountered reality resistances—resistances that emerge as a consequence of praxis.[2] These schemes and scheme complexes are utilized in acts of meaning to categorize experience; and these acts are often enriching, because they not only assign schemes to the input, but they synthesize dynamically new relational configurations to epireflect truly new aspects of in-formation that reality imposes on us. These enriching, information-generating, *dynamic syntheses* of our reflective abstraction are not well modeled by cognitive science, in particular by the old AI (propositional or symbolic as contrasted with connectionist computer simulations).

Unlike humans, computers receive with their programming an already concretized, relevant ontology of data. It is the programmer who engages in the uniquely human activity of inventing and constructing a suitably epireflective ontology (objects, predicates, statements, commands, transformations, etc.) who helps the computer deal with the problems at hand[3].

The Mental Touring Test and Searle's Major Objections to Computer-Cognitivism

A major problem of hard (reductionistic) cognitive science is the assumption that humans as intelligent organizations are just computing machines, which implies that truly novel performances are not possible in humans, although the history of human invention and culture speaks to the contrary. I use the words *truly novel* (Pascual-Leone, 1980, 1987, 1991) to mean performances not prefigured in the combinatorial of a subject's stored prior knowledge, performances that are not solely the result of learning, but are produced by creative (enriching) dynamic syntheses. Among his objections to hard cognitive science ("computer-cognitivism") Searle (1992) makes three criticisms that I propose to discuss within my own perspective. The three criticisms (*C*) are:

(*C1*) The human brain is not organized as an information-processing machine like a computer: the brain structures its own experience, constructs its own "programs," and sets its own executive plans, goals, and values. In contrast, computers must receive plans, goals, and values from a programmer—a truly novel performance (caused by dynamic syntheses) is not accessible to any computer.

(*C2*) Syntax, the logical form of computation used in actual (particularly old AI) computers, is not intrinsic to physics. I understand this claim of Searle's as meaning that syntax, logic, and mathematics are (from a physicalist perspective) purely generic abstractions of physical relations found among possible unspecified objects—"objects whatsoever" (Gonseth, 1936/1974). This is why the same programs can be embodied by many different physical organizations. But by the same token the program—disembodied from its physical embodiment—lacks causal power; it cannot be regarded as a sufficient explanation for the functioning of its working embodiment. Before claiming that a system of computer programs can explain the brain's performance, we must show that the brain does not have any capability that a computer embodiment of this system does not have. Two cases in point are consciousness and affect. Computers do not require either of these. And comments already made show that the brain exhibits different, unique performances in many other respects as well.

(*C3*) "Computational states are not *discovered within* physics, they are *assigned to* physics" (Searle 1992, p. 210). They are assigned to the physical computer, or to the brain, by a scientist's or user's *observer's analysis* that interprets input/output relations using his or her own mental-processing semantics. This is fine in the case of physical computers, because they have programmers and users who provide a semantics to interpret the computer's work as computations. But the brain has neither a human programmer nor a user: despite its learning history and social culture, it is its own user and programmer. We thus cannot claim that a program (or "unitized" programming system—Newell, 1991) is a proper theory of the human brain's intelligence. Before making that claim, we must show that the program (or its unitized system) explains all relevant organismic processes (and output/ input relations) of programmer(s) and user(s) who ensured the computer's successful functioning.

This last condition to C3 is my own, but I presume Searle would agree with it. I developed this idea 10 years ago, under the nickname of Mental Touring Test, to contrast it with the famous Turing test—famous albeit inadequate to evaluate the thesis of hard cognitive science.

The Mental Touring Test takes this form: if a scientist wishes to propose a given computer program (or unitized system) as a valid theory of human processing in domain X (in the psychological organism), he or she first must establish, by theoretical analysis or simulation, that this program in fact explains (reconstructs) all the X-processes (psychological-organismic processes) that programmers who created the program actually have used in this creation. In other words, the scientist must make a "mental tour" into the mind of the original programmer(s)—and user(s)—to see whether the program can serve as computational theory for the mind processes that achieved the computer run.

To evaluate this prescription more closely, consider the form of inference implicit in the hard cognitive science thesis: If the program's process (PP) leads to or implies (=>) the computer's behavior output B, and B actually occurs; and the organismic process (OP) leads to or implies (=>) the organism's behavior B', and B' actually occurs; and B and B' are shown or taken to be equal (or equivalent, =) then, based on the reductionist assumption of computer cognitivism, we should conclude that $PP = OP$, that is, the Organism's Process is the same as the Program Process. We can represent this inference more compactly as follows:

$$\frac{PP => B,\ B,\ OP => B',\ B',\ B = B'}{PP = OP} \tag{1}$$

Where the upper terms are premises and the equation below them is their consequence.

It is apparent that this is at most a scheme of plausible inferences (Polya, 1954), since the consequence does not necessarily follow from the premises. There are many possible ways leading to or producing (=>) B and/or B'; and to make this inference more plausible one should first verify that in *all* important respects the relevant context where $PP => B$ is equal to the context where $OP => B'$. The Mental Touring Test is intended to test a major differential element of this context—that is, the mental processes of the programmer and the user. Points (C1), (C2), (C3), as well as other remarks made above, bring up other important differences between the contexts of PP and OP.

For researchers who would pursue logico-mathematical formalizations, it is worth noticing an implication of theorems about the limitations of such formalisms: no computer program can improve on or explain (i.e., derive, evaluate the truth of) all its own program processes (Kleene, 1967, p. 245, Theorem II)—something that computer programmers could do vis-à-vis programs. As Kleene (1967, p. 246) puts it, referring to Turing machines: "To improve the procedure or machine must take ingenuity, something that cannot be built into the machine." This semantic limitation of programs implies that programmers cannot be modeled by the programs they design.

Thus, the research program of hard or reductionistic cognitive science must be modified. A mediating "science" must be created that relates computer-program models of strong cognitive science or artificial intelligence (old or new) to relevant psychological and neuropsychological processes on the one hand, and to subjective, phenomenological, or naive-psychological descriptions on the other. This mediating "science" or "lingua franca" might be the metasubjective process analysis I proposed, programmatically, some time ago (Pascual-Leone, 1978a).

This analysis would be an explicit, epistemologically formal (i.e., epiformal) way of theorizing, but would not be computationally formal theorizing. This concept of epistemological formality needs to be emphasized. I propose that we engage in empirically based (dialectical constructivist) epiformal metatheorizing; a metatheorizing that is metasubjective. Why metasubjective ? It has to be a metatheory, a theory of the observer, but of an observer carefully modeling subjective and objective processes of the "psychological" organism, not of the computer. We must provide a metatheory that makes possible cogent discussion of the subject's processes, and permits interaction among neuroscience, cognitive development, psychology at large (including phenomenological and naive psychology), and other human sciences. Now, what would this metasubjective theory look like? Of course I cannot say how a metasubjective theory will look from your perspective. You have to discover it. But I may suggest some conditions with the aid of assumptions about human organisms made in my neoPiagetian theory. I hope readers will recognize that the arguments presented in this chapter are not meant to support this particular metasubjective theory, but to support metasubjective analysis in general.

Metasubjective Analysis and its Four Dialectical "Moments"

Science, in particular biological science and psychology, often must take the perspective of an ideal observer who describes the animal or subject within its circumstances (immediate environment or situation) and does so from a "detached" third-party perspective. Auxiliary sciences like physics (e.g., measurement) and chemistry often are used to ensure that this "detached" observer's description of the subject and its situation is objectively accurate. This is the usual scientific methodology.

But in psychology (and other human sciences) a new perspective—that of the subject with its subjective language and subjective analysis—also must be adopted. As Searle (1992) has recently re-emphasized, it is not possible to grasp consciousness without adopting a subjective language and analysis. From this new perspective, subjective descriptions (at least those consensually validated) ought to constrain and be explained by the final causal-process model. As Edelman (1989), puts it: "We must, nevertheless, proceed to examine evidence from all possible sources *provided it does not conflict* with known scientific laws of physics, chemistry, and biology" (p. 19).

Let us call a theory metasubjective (*MS*) if it is capable of describing processes of human performance of various kinds from the perspective of the subject's organism. And in addition it can: (*MS1*) model these processes temporally, so that sequences of steps prescribed by the analysis can be shown to reflect sequences of real processes in the organism. (*MS2*) It also can model causally the critical processes of organismic change, that is, development and learning. I say both because neither of these is reducible to the other. I include in development the endogenous change throughout the life span of content-free organismic capacities or resources—i.e., "hardware" operators if they exist; and in my opinion they do. (*MS3*) It can model the causal formulas of organismic individual differences, including differences in content-free organismic capacities of the brain's hidden hardware. (*MS4*) It can model types of situations in terms of the sorts of processes they would elicit in the human, given his or her praxis (goal, purpose) and his or her strategy of achievement. Here we are talking about task analysis, but task analysis conducted from a metasubjective perspective, and not from a simple observer's perspective. (*MS5*) It has a proper interpretation, which is experimentally testable in terms of causal predictions, within neuropsychological theory. I am not talking about reductionism. I do not believe that psychology can be reduced to the brain; but I do believe it can be put into correspondence with it (Sperry, 1977/1992). And this should happen, because study of the brain can help to clarify the postulates that our psychological theories must have.

This definition shows that metasubjective analysis is a method of process or task-analysis guided by a metasubjective theory. It is a collection of rational heuristics for assigning to tasks a specific model of the mental processes in the subject that intervene in production of the task performance. These models are always relative to a chosen strategy and a stipulated situation.

In order to conduct this process or task analysis, one has to make definite theoretical commitments with regard to functional characteristics of the (psychological) organism. The theory I will assume is neo-Piagetian (Pascual-Leone, 1987; Pascual-Leone & Baillargeon, 1994; Pascual-Leone & Goodman, 1979), but other users of metasubjective analysis could utilize alternative theories.

To do metasubjective analysis, one must complete four dialectical "moments" or phase-methods that are cumulative, the last one being metasubjective analysis proper. These moments or phases are: objective analysis, subjective analysis, ultrasubjective or real-time mental-process analysis, and metasubjective or organismic-process analysis.

Objective analysis is known to all scientists. It adopts the perspective of an ideal observer who looks at the subject from outside, to describe (with the help of appropriate sciences, measurement, etc.) the situation and performance of the subject. If the analysis is predictive, one also needs a theory of the organism that can help to generate predictions in combination with situational analysis. It is in objective analysis where the contribution of J. J. Gibson (1979), and that of other ecological psychologists, is important.

In this phase of analysis, the choice of the psychological unit is arbitrary, and researchers often choose descriptively obvious units. Methods of task-analysis in cognitive science may just be instances of objective analysis, or may go beyond it.

Subjective analysis formulates or describes performance from the perspective of the subject's organism. Ontological commitment regarding organismic processes are necessary in this phase, along with a tacit or explicit choice of psychological unit. But the assumed units no longer can be arbitrary: they must reflect functional processing invariants that capture the essence (infrastructure) of performance at the level of analysis adopted—only in this way is it possible for us to find a correspondence with brain processes, so that we can predict the subject's performance.

In addition to objective analysis, introspection and phenomenology are commonly used in subjective analysis. In introspection one asks: "If I were this subject, in this particular task or situation, what would I do?" Phenomenology asks a generalized version of this question: "If I were *any subject whatsoever*—or any subject of type *j* (an ideal subject of type *j*) in *this type i* of situation, what would I do?"

The mental attitudes that allow analysts to move from introspection to phenomenology involve mental techniques like those that Husserl (1970) called phenomenological reduction (and other kinds of reduction). That is, the analyst must intentionally disregard ("bracket") his or her own conceptual preconceptions, and "knowledge" of the external or internal experience, and instead, mindful of the intended situation, imagine and describe the ideal subject's intended mental operations. Contrasting the descriptions across different "ideal observers" helps to reduce an observer's bias. Philosophical descendants of phenomenology, such as existential phenomenology, hermeneutics, or deconstruction analysis; and psychological techniques such as those of folk psychology, narrative analysis, or empirical theories of mind, add new perspectives to the basic method, helping to ground it empirically.

From the perspective of psychology, a second level of subjective analysis is found in *structuralism*—provided we use the term for any method that aims to generate explicit generic models of the experienced objective-and-subjective (e.g., phenomenological) realities, as well as their generating processes; but disregards real-time, step-by-step temporal aspects of the evolving realities being modeled. In developmental psychology, classic structuralism is illustrated by Piaget's and by most neo-Piagetian theories, as well as by much of psycholinguistics.

A basic method of structuralist analysis is the constructive abstraction of functional-invariant categories (structures) that correspond to sequences of idealized performance patterns. Phenomenology is tacitly or explicitly included in structuralism as a moment; but structuralism adds to it the explicit formulation of process models, a terminology, and a notation for representing psychological units in their functional interrelations.

Structuralism and structural analysis are important methods, but one should not assume that only structure is important: substantive "content" (whether external—i.e., sociocultural—or internal—in the form of brain biodynamic properties) must not be disregarded (e.g., Edelman, 1989; Searle, 1992).

The temporal representation of changes in time—modeling in real time—is not done well with classic structuralist methods. As a result, the obtained psychological units, albeit idealized descriptions of performance patterns, are not properly defined organismic schemes suitable for "real time" analysis. An obvious consequence of this limitation is that structuralist models cannot show the process of transition from one developmental stage to the next, even though developmental stages were discovered with structuralist methods.

Ultrasubjective (or mental-process) analysis adds, to the objective and subjective methods, techniques for modeling process in real time (or in some idealized molar version of the real-time evolution of processes). Examples of these methods are found in computer-simulation approaches to information-processing psychology, but these approaches tend to choose their units arbitrarily; and chosen units are often excessively constrained by the computer and programs that "run" the models in question.

In addition to requiring suitable choices of units for its modeling, the method of ultrasubjective analysis (because it is done in "real time") brings to the fore issues like intentionality, purpose, motives, values, and teleology; and thus forces awareness of processing distinctions such as those between executive processes (which embody plans for action) versus action processes (which implement actions, representations, etc.—performance of any sort). And within each of these, ultrasubjective analysis must distinguish between operative or procedural units (which embody local transformations or procedures) versus figurative or declarative units (which embody mental "objects" or states, signs, conditions, to which mental commands or local transformations can apply). Experiential intentional meanings are concrete (particular, individual entities), and correspond to the purely experiential schemes that Piaget or neoPiagetians call infralogical or mereological.[4] However, in the intellective-representational level of processing the units used, whether figurative or operative, are often generic, describing kinds of knowledge rather than concrete and particular experiences. These generic units may be called concepts by psychologists (e.g., Case, 1985, 1991; Mandler, 1992) and neuroscientists (Edelman, 1989). Generic schemes, concepts in this broad sense of the word, are never found in children before 12 months of age. But from a subjective or ultrasubjective perspective, "generic scheme" may not be the proper term to describe developments, since a baby in fact does not distinguish between the "concept" and its instances (babies have only "practical" concepts). Yet they constitute the infrastructure of meaning in early language (Bloom, 1993) and the initial logical schemes of the child ("logical" is Piaget's name for them). To emphasize more than Piaget did their close connection with the onset of language and their generic character, I have called these schemes logological and distinguished them sharply from properly linguistic schemes (Johnson, 1991; Pascual-Leone & Irwin, 1994). This sort of logological processing appears before language proper, and might constitute its foundation (cf. Bates, Benigni, Brethelton, Camaioni, & Volterra, 1979; Benson & Pascual-Leone, submitted; Bloom, 1993, 1994; Edelman, 1989; Greenfield & Smith, 1976; Mandler, 1992). These logological processes may justify the claim of Ortega y Gasset, already mentioned, that a baby has things *to tell* before he or she can talk, and it is this emotional and cognitive need that sets the conditions for language learning.

Operative units that pertain to *executive processing* take the form of standard plans for action or executive systems; the corresponding figurative or declarative schemes are core knowledge structures—that is, complex temporally organized systems of structures, such as "frames" or "scripts," or "narratives" (as Bruner uses the term). These knowledge structures, often called by developmentalists "operational structures" (Piaget, 1985; Vuyk, 1981) or "central conceptual structures" (Case, 1991), are semantic-pragmatic systems of generic representations; they constitute complex distal-object "shells" stipulating sorts of operative processes that can apply to the "object" in question. A car (from the point of view of the driver) or a restaurant, or an active hockey game (for hockey players) or even a Quick Basic "screen environment" for a programmer, are all represented in the mind by this sort of core knowledge structure.

Another important modality of processing for "real-time" modeling is affective processing; and modes[5] of processing that coordinate affective with cognitive processing are the interpersonal (or intersubjective) mode and its derivative: the personal (personality) mode. Emotions are the affective experiences connected to these two modes. Affects can be formulated psychologically as schemes (Pascual-Leone, 1991). Affects are important in ultrasubjective analysis because cognitive goals are ultimately affective, and affects activate executive schemes and knowledge structures relevant for them. Intersubjective processes are mediated by structures that combine affective and cognitive schemes; they represent interpersonal situations in terms of the person-types, or concrete persons, related by symbols of their typified exchanges (roles, and interpersonal relations of all sorts). This intersubjective mode of structuring a unit explicates the existence of social "collectives" and other social structures (e.g., Mead's "generalized other"). Complex interpersonal phenomena such as moral judgments can be represented in this manner (Stewart & Pascual-Leone, 1992), as I show below.

These basically different modes of processing, necessary in ultrasubjective analysis, seem to be distinctly localized in the human brain—suggesting that these different modes are indeed natural kinds of processing. Operative schemes are always located in the frontal lobe (operative executives in the prefrontal cortex); figurative schemes are located in the parietal, occipital, or temporal lobes, depending on their processing modality (i.e., somatosensory, visual, auditory, etc.); core or central knowledge structures seem to be located in the intersection of the three figurative lobes, the parieto-occipito-temporal region (high tertiary or quaternary associative centers). It is well established that affects are connected with the limbic system—possibly regulated by the prefrontal lobes, which also may intervene in moral judgment and other non-automatized intersubjective processing.[6]

Ultrasubjective considerations—that is, "real time" analysis of process—make explicit the need to distinguish among different modes of processing. The distinction between executive and action schemes is a case in point: because real-time intentional processes are goal-directed, executives and core knowledge structures must be assumed that explain this goal-directedness.

Metasubjective or organismic-process analysis adds, to the methods already mentioned, organismic constraints that have been inferred from psychological research across specialties, including neuropsychology. The mark of metasubjective analysis is the reference to deep purely organismic constructs ("silent" capacities or resources of the brain's "hardware"), which constrain the subjects' performance and thus are needed in deeper analysis.

Metasubjective Processes and the Brain's Content-Free Organismic Capacities

Humans have consciousness, and their consciousness has causal power. They can, by exerting "mental effort," focus on chosen psychological schemes to hyperactivate them, and can (via active central inhibition) disregard schemes, perhaps activated by the situation, that are not interesting to the executives and affective goals that are now dominant (Case, 1992; Fuster, 1989; Pascual-Leone, 1987, 1989; Stuss, 1992). This flexibility of human consciousness for chosing or changing its current "contents" is made possible by an attentional mechanism that, independent of the schemes (the "contents") to which it can apply, modulates the "stream" of consciousness (James, 1892/1961). I have formulated this model of mental attention (*MA*) elsewhere (Pascual-Leone, 1987, 1989, 1990a; Pascual-Leone & Baillargeon, 1994), and described it as constituted by four organismic resources or factors that work in dialectical interaction. Namely:

(*MA1*) The currently dominant executive schemes driven by affective goals. These executives have the power to mobilize and allocate to schemes resources (MA2) and (MA3). In Figure 6.1, these executives are symbolized by the letter *E*.

(*MA2*) A mental "energy" or scheme-booster capacity *M*, also called *M*-operator—a limited resource that grows in power (i.e., the number of schemes it can boost simultaneously) throughout childhood until adolescence. The hyperactivation of schemes by this resource constitutes the "beam of focal attention." Figure 6.1 denotes *H* as the schemes on which mental attention applies at the moment; and it denotes *H'* as the other schemes in the repertoire of the subject that are not now attended to. This mental "energy" or *M*-capacity increases endogenously with human development up to adolescence. Its "hidden" (ideal) measure—the number of schemes it can boost simultaneously—is a quantitative stage-characteristic of Piagetian and neoPiagetian qualitative stages of human development (for empirical validation of this claim see, for instance, Case, 1985, 1991, 1992; de Ribaupierre & Hitch, 1994; Halford, 1993; Johnson, Fabian, & Pascual-Leone, 1989; Johnson & Pascual-Leone, 1989; Pascual-Leone, 1970, 1980, 1987, 1993; Stewart & Pascual-Leone, 1992).

(*MA3*) A capacity for active central inhibition of schemes, or Interruption *I* (also called *I*-operator), that the executive can direct to inhibit schemes that are not currently relevant for the task at hand (Pascual-Leone, 1989; Pascual-Leone, Goodman, Ammon, & Subelman, 1978; Pascual-Leone, Johnson, Goodman, Hameluck, & Theodor, 1981). The famous "searchlight" analogy for mental attention (Crick, 1984; James, (1892/1961) is easily explicated in this model; it results from automatic interruption (by the *I*-operator) of the schemes that are not attended to at the moment (call these schemes *H'*). The automatic interruption is symbolized in Figure 1 by the expression *I* (*H'*). The developmental growth of this interruption mechanism is not well studied, but it is believed to be interlocked to the *M*-operator growth.

(*MA4*) The fourth major constituent of mental attention determines closure of its "beam of attention." It is the subject's endogenous capacity to produce as performance a single integrated whole, even though often many different schemes co-determine the performance in question. This fourth capacity is what Gestalt psychologists and Piaget have called internal or autocthonous "field" processes (e.g., "minimum principle" of perception, "S-R compatibility" principle of performance, etc.). These field processes I attribute to a not-well-understood performance-closure dynamism called *F*-operator, which makes performance (perception, imagery, thinking, language, motor activity, etc.) to be minimally complex while remaining maxi-

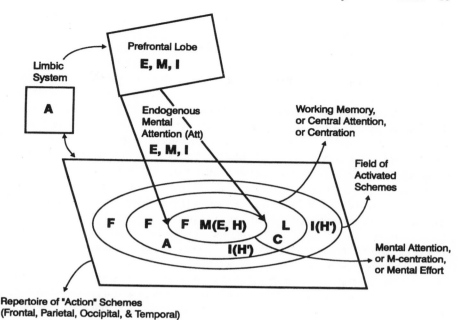

Figure 6.1. Model of Endogenous Mental Attention.

mally adaptive (Pascual-Leone, 1989; Pascual-Leone & Johnson, 1991; Pascual-Leone & Morra, 1991).

The *F*-operator works in coordination with an important silent principle of the brain's hardware. This is the principle of *schematic overdetermination of performance* (SOP). This SOP principle says that performance, at any time, is synthesized by the dominant (most activated) cluster of compatible schemes available in the brain's field of activation at the time of responding (Pascual-Leone, 1969, 1989, 1995; Pascual-Leone & Goodman, 1979) Pascual-Leone & Johnson, 1991; Pascual-Leone & Morra, 1991).

Together, the *F*-operator and the SOP principle constitute a mechanism that produces dynamic syntheses of performance out of the currently dominant cluster of compatible schemes (cf. Pascual-Leone, 1983, 1984, 1987, 1989; Pascual-Leone & Goodman, 1979; Pascual-Leone & Johnson, 1991). This conception of a subconscious dynamic "choice" (caused by *F-SOP*) that generates performance is congenial to modern computational theories of induction (e.g., Holland, Holyoak, Nisbett, & Thagard, 1986; Thagard, 1989), and evokes epistemologically the "relaxation" algorithms used in connectionist and neural programming to resolve conflicts among competing processes (e.g., Smolensky's Harmony function, 1988).

In Figure 6.1, I symbolized this dynamic model of mental attention. The three key constituents of mental attention—the *M-operator* (mental energy *M*), the *I-operator* (central inhibition or interrupt *I*), and the currently dominant set of executive schemes (i.e., *E*)—are symbolized by a rectangular "flashlight" controlled, at least in part, from the prefrontal lobes.

This "flashlight" of mental attention "illuminates," that is, boosts with activation, a region (the inner ellipse in Figure 6.1) of the repertoire of action schemes; this region is the *M*-space (*M*-centration) or focus of endogenous mental attention (Att). Action schemes are schemes other than executives that cause the subjects to act (this performance including representations, inferences, problem solving, etc.). Mental energy is exerted on *E* and on the "chosen" action schemes *H* to empower them to produce performance. This is symbolized in Figure 6.1 by the expression *M (E, H)*. Figure 6.1 assumes that the task the subject is dealing with is a *misleading* situation: schemes that are not relevant and not inside the *M*-space must be interrupted to reduce interference. An example of this sort of situation would be, for instance, an item of the Embedded Figures Test (Pascual-Leone & Goodman, 1979). This interruption is indicated in Figure 6.1 by the expression *I (H')* found in both the outer ellipse (i.e., the field of activation) and the middle ellipse—the working memory or central attention region. Working memory is therefore the set of hyperactivated schemes, whether they are activated by mental attention or by factors such as *A* (affective schemes), *C* (automatized content schemes cued by the input), and *L* (overlearned logical-structural schemes retrieved from memory). Whenever these factors of working memory are boosting misleading schemes, the executive *E* must block them by applying the Interrupt operator, thus reducing the current working memory to the *M*-space. In this sort of misleading situation, the area within the "beam" of attention tends to be the *M*-space (area inside the inner ellipse); the "edge" of this "beam" (the ellipse) results from interruption caused by the *I*-operator; the phenomenological closure of the field of mental attention is done by the *F-SOP* mechanism. Pascual-Leone and Baillargeon (1994) discuss this model of mental attention in greater detail, and use it to model a task mathematically.

This mental attention mechanism is not an infallible *deus ex machina*. If schemes involved are excessively contradictory, mental attention might not generate an adaptive representation of the current circumstances. In this case mental attention could provide instead the experience of "foundering of thought" in the face of uninterpreted reality—an experience that might elicit some anxiety. This experience is Jaspers' "force of the real" (see the epigraph to this chapter). Computers cannot have this experience, for they would get jammed instead.

Vital Reason and the Emergence of Mind

Ortega y Gasset, the great Spanish philosopher of human existence, argued that the essence of human reality is vital reason, that is, insightful, practical rationality about one's life (*vita*) and one's living in all its aspects (Ortega y Gasset, 1957/1980a, 1944/1980b). Vital reason is concerned with the person as a concrete, evolving totality—a totality that extends into the future and the possibilities that a person might realize. David Bakan (personal communication, August 1988) illustrates this projection into the future with the example of *danger*: the assessment of an impending possible (i.e., potentially real) harm, damage, or loss. The subjective experience of danger, as with so many aspects of vital reason, involves two of the human (animal) capabilities that computers do not seem to have: consciousness and affect. Other aspects of vital reason can be found in modern computer programs, such as foresight, planning, and memory.

The original experience from which vital reason begins is the conscious I (the person, the self) facing his or her circumstances—the world in the here and now with objects and others (subjects) in a situation (Pascual-Leone, 1990a, 1990b). For Ortega the world as conscious experience emerges together with personhood, contrasted with the self (the I), both developing to serve the subject's praxis. What made humans different from other animals, in the opinion of Ortega y Gassett, was their abnormal initial power of "fantasy" (thought), that is, a non-programmed superior imagination. This *fantastic fantasy* of humans allows them to evoke internally, and mentally investigate, past or future or possible events or problems of life. Such rich "mental life" shapes and differentiates new emotions (Pascual-Leone, 1990a) and leads to primal (affectively loaded) knowledge of the Other as "alter ego"; a different self who is often a *stranger* endowed with the same power of mental life as I have: an alien self that can easily lead me into altercation, alienation, outrage—all words that share their etymological Sanskrit root with the word other. This emphasis on the other as different, as a potentially dangerous stranger, which Ortega shares with Sartre (1966), reinforces his conviction, contrary to Husserl (1970) but in agreement with Scheler (1973) and many psychologists (e.g., Baldwin, Bowlby, Trevarthen, etc.), that the other (this alien alter-ego) is *not* derivative from the self, but is primary. The other is somewhat prefigured in the innate affects (instinctual drives?) or dispositions related to Love (e.g., attachment, affiliation, anger, hate, etc.) and power (e.g., mastery, dependency, etc.) (Pascual-Leone, 1991). With naive folk psychology, with Baldwin (1908, 1894/1968), Mead (1934), Bruner (1990), and many others, the vital reason of Ortega prescribes that self is constructed, within this affectively driven "intersubjectivity," from the dialectical (mutuality, reciprocity, etc.) relations with others.

As Piaget would say, animals and humans are independent, and at times unpredictable, sources of causality. Controlled interactions with animals or humans (or human-programmed computers) requires each "player" to abstract the *dialectic* of his or her mutual, reciprocal interactions. This is the true beginning of social relations. Humans (the others) emerge in praxis as the animals that are "presumed" (Ortega, 1980a, p.102) to have a subjectivity and a mind like ours. Thus, humans are experienced not only as sources of independent causality, but as sources of premeditated (mentally mediated, thoughtful, planful, willful) causation. The initial mechanism for this intersubjective experience of the other and his or her power of premeditated causation is affective, and is largely innate. As Ortega (1980a, p. 112) says, "Before each of us develops self-awareness, we have had the basic experience that there are those who are not me, the others." Modern infant research in psychology is consistent with this view (e.g., Bloom, 1993; Bruner, 1975, 1977; Stern, 1977, 1985; Trevarthen, 1990).

The emergence of language is, from the perspective of vital reason, not so much the cause as the consequence of a fantastically growing mental processing (imagination, mind) that "must" be communicated to others. The infant invents as much as learns language, adopts it from the culture, without needing a programmer; and this is because he or she needs—an affective need—to communicate (Bloom, 1993). As Ortega puts it, the baby *has to tell*; and this need to tell creates what Piaget would have called a disequilibration: a problem. And language is acquired in the attempt to solve it—initiating emergence of the mind that Bruner (1990), Bloom (1993), and many others talk about.

An Intersubjective Example of Metasubjective Analysis: Moral Judgment in Kohlberg's Moral Dilemmas

The role in cognition of metasubjective factors, such as mental capacity, working memory, or structural learning, is already well recognized. It may be more useful to take an example from the intersubjective domain, an area little understood in cognitive science. The example I choose has been extensively studied in developmental psychology: moral judgment using Kohlberg's dilemmas.

The Heinz dilemma poses the following situation:

> Heinz's wife is gravely ill and will die unless she receives a specific drug that only a certain druggist possesses. The druggist asks a price for the drug that Heinz is unable to pay, and refuses any compromise. Having exhausted other means, Heinz considers stealing the drug to give it to his wife. The basic moral question is: "Should Heinz steal the drug? Why or why not?"

Stories such as this are read to children or adults who are then asked to address the basic moral question. From the subjects' replies (usually developmentally ordered in complexity), one may assume what semantic inference and which schemes (subjective units of knowledge) subjects used to generate their response. But to convey clearly this example of metasubjective analysis, I must give more details about schemes and silent hardware operators. Much of what follows is extracted from Stewart and Pascual-Leone (1992), with permission. A *scheme* is a neuropsychological unit: a unitized package of neuronal connections (a modular subsystem in the neuronal network) that functions as an intentional unit in a subject's praxis (Pascual-Leone & Johnson, 1991). Schemes have releasing conditions; and their activation weight is initially proportional to the degree of match between features of the situation and releasing conditions of the scheme (Pascual-Leone, 1984; Pascual-Leone & Johnson, 1991). When releasing conditions match features in the input, they become cues, which activate the scheme in question. This set of conditions is the releasing component of the scheme. The scheme's effecting component is the set of effects on performance when it applies. A scheme's activation by other schemes (that serve as cues) is explained by spreading of activation assumptions in the neuronal network (this explicates the "cue function"). Other dynamic principles of schemes, such as *Piaget's principle of assimilation* (i.e., the disposition of schemes to determine performance under minimal conditions of activation, unless prevented by other schemes) can also be explained in terms of well-known dynamic principles of the brain—for instance, the principle of neuronal summation of activation, and Sherrington's principle of the final common path (Pascual-Leone & Johnson, 1991). According to the *summation principle*, the activation of a neuron (or collection of neurons) grows with the sum of activation weights reaching it, irrespective of their source. According to the principle of *final common path,* performance is the combined product of all compatible "unitized" neuronal processes (i.e., all co-activated schemes) as they converge to reach appropriate motor or mental terminal centers. From this perspective the Piagetian principle of scheme assimilation becomes a *principle of overdetermination* (SOP principle), not unlike the one first formulated by Freud. In the version we use (Pascual-Leone, 1980, 1987, 1989, 1990a, 1995; Pascual-Leone & Goodman, 1979; Pascual-Leone & Johnson, 1991; Pascual-Leone & Morra, 1991), this principle says that final performance is co-determined by all clusters of compatible schemes (of any complexity) strongly activated at the time the performance is produced. This overdetermination principle, as I mentioned before, appears in connectionist networks as a "relaxation" mechanism (e.g., the Harmony function of Smolensky, 1988), which is used to calculate final activation weights in the nodes of a network.

Every phenomenologically distinguishable modality of experience and action (e.g., visual, auditory, affective, motor, mental, etc.) has its own prewired neurological origin and its own experiential determination. In addition to the largely content-based modalities, there are modes of processing, that is, generic functional categories of processing that run across most modalities and serve generic processing functions. Of particular importance for representing mental change, and thus for our task analyses, are the modes of processing that classify schemes into figurative (or declarative) versus operative (or procedural) versus executive. As mentioned above, figurative schemes are any processes that stand for states (e.g., objects, patterns, persons, situations). *Operative schemes* are processes that stand for transformations or procedures. *Executive schemes* carry the temporally structured plans and regulate mobilization and allocation of hardware operators (resources). Strategies, for example, are the result of executive processing. Suitable executive strategies, and the repeated application of suitable operative schemes, lead subjects to create complex figurative schemes—often called (core) *knowledge structures*—which embody relational or substantive characteristics of distal objects, persons, and situations in their dynamic interchange. Guided by the functional totalities that are knowledge structures, subjects pursue affective goals via the interplay of figurative, operative, and executive schemes in interaction with "resistances" of reality.

In interpersonal relations, these core knowledge schemes take a special form, which Mead (1934) anticipated with his notion of "generalized other," and Sartre (1960) and others may intend with the notion of "collective." These are the precursors of what I call *collective knowledge schemes* (henceforth abbreviated as collectives). A *collective* is a complex intersubjective representation, that is, an organization of interconnected knowledge schemes. Collectives mediate and govern interactions among persons within a given cognitive context, according to conditions stipulated by various circumstances (Pascual-Leone, 1983, 1990a, 1990b). In a collective, all persons, and the relations (roles, exchange formulas, scripts, operative expectancies, etc.) they play or have vis-à-vis each other within a given context, are indicated by more or less complex knowledge schemes. We can conceive a collective as being a symbolic complex; that is, an organization of symbolic structures that has been well learned but not automatized. In this symbolic complex of schemes the superordinate structure is the system (called *REL* below) of learned operative relations, which define person-characters in their mutuality vis-à-vis each other (within the context). The person-characters, the context, the objects, and the circumstances are parameters of *REL* in a collective.

We represent this manifold of symbolic schemes by abstracting the essential structure. Using an informal operator-logic formalism this structure appears as follows:

$$REL\ (CXT,\ SELF,\ OTHER1,\ OTHER2,\ldots,OBJ1,\qquad \text{(F1)}$$
$$OBJ2,\ldots CIRC1,CIRC2\ldots)$$

In this formula *REL* stands for experientially acquired (or conceptually known) operatives of all sorts that regulate exchanges among a person (*SELF*) and others (*OTHER1, OTHER2,* . . .), perhaps with regard to objects (*OBJ1, OBJ2,* . . .), according to the context (*CXT*) and its modifying circumstances (*CIRC1, CIRC2,* . . .).

REL is a functional category. It stands for relations or actions or roles held by the different characters at different moments of the collective's social and cognitive transactions. Notice that, in a collective, the parameter *CXT* stands for the general background of knowledge that serves to interpret the situation and relate the referents of person and object parameters to *REL*. Thus, *CXT* is the figurative correlate of *REL*, which in turn is the operative counterpoint of *CXT*. *CXT*, along with the concrete referents of the person and object parameters, functions as the releasing component of the collective. *REL* is the effecting component of this symbolic scheme. Since the collective is usually not fully automatized, all parameters involved must be boosted with mental attention during "cool" acts of moral judgment (see below). In intersubjective situations requiring acts of judgment, several collectives often are involved (as inferred from the response). Subjects must coordinate relevant collectives, by means of a dynamic mental synthesis, to reach a good judgment.

Let us now see how collectives can be used to analyze the Heinz dilemma given above. The collectives relevant here relate Heinz (= *SELF* . . . here abbreviated *H*), the wife (= *OTHER1* . . . abbreviated *W*), the druggist (= *OTHER2* . . . abbreviated *D*), the drug (= *OBJ1* . . . abbreviated *d*), the *consequences* of the act (= *CIRC1* . . . abbreviated *CONS*), and (last but not least) values the situation evokes in Heinz affecting his decision (*CIRC2* . . . abbreviated *VAL*).

Table 6.1 offers a semantic structuralist analysis of the Heinz dilemma in terms of collectives. The first column of this table gives the list of generic categories that constitute any collective. The second column gives a summary description of the dilemma using the categories. The other three columns decompose the dilemma into three elementary collectives. Mental capacity demand (*M*-demand) of a dilemma depends on the number of collectives, and the number of parameters within collectives, that it requires. The mental power of a child might then be assessed from the type of moral responses given, and the number of collectives and parameters required by them (see Stewart and Pascual-Leone, 1992, for a test of this prediction).

In the next few pages I use operator-logic notation informally to illustrate my analyses. These symbolic logical models help convey the complexity of emotion-bearing knowledge structures that mediate intersubjective human interactions. Intersubjective processes of this sort, rather than purely cognitive processes, might be a new frontier for cognitive science to investigate.

Table 6.1 General Formulation and Specific Collectives in the Heinz Dilemma

General Formulation	Heinz's Dilemma	Marital Relation	Wife's Illness[a]	Druggist's Management of Pharmacy
Relation	Evaluate: should *H* steal *d*?	*H & W*: vow of mutual love/help	*W* will die unless she takes *d*	*D* is "owner of" *d*
Context[b]	*CTX1*. *CTX2*. *CTX3*.	*CTX1*: . . . *H* loves *W*	*CTX2*: . . . No money to buy *d*	*CTX3*: . . . *H* asks *D* for drug
Self	*H* = Heinz	*H* = Heinz	*H* = Heinz	*H* = Heinz
Other 1	*W* = Wife	*W* = Wife	*W* = Wife	
Other 2	*D* = Druggist refuses *d* to *H*			*D* = Druggist refuses *d* to *H*
Object 1	*d* = drug[c] too expensive		*d* = drug[c] needed to heal	*d* = drug too expensive
Object 2	No money to spend		No money to spend	Money required to pay for *d*
Circum. 1	Consequences If steal, prison; if no steal, death of *W*		Consequence: *W* will die if she does not take *d*	
Circum. 2	Values[d]	Values[e]: Loyalty Love	*Value:* Life of *W*	*Value:* Property of *d* by *D*

[a]That is, Heinz's knowledge of his wife's illness.
[b]Notice that the three *CTX* of Heinz's dilemma (second column) are illustrated in the collectives of the last three columns.
[c]Need for the drug.
[d]Values relevant to stealing.
[e]Values relevant to marriage.

Source: This table is used with permission from Stewart and Pascual-Leone, 1992.

A Dimensional Analysis of Moral Reasoning in terms of Collectives: We count three elementary constituent collectives in the Heinz dilemma. The formula for the first one is:

$$REL\ (CXT,\ H,\ W,\ VAL,\ \ldots\ ) \tag{F2}$$

This represents the personal, marital relation between Heinz (*H*) and his wife (*W*). This collective might contain other persons (e.g., children), objects (e.g., a home, etc.), and life experiences that I am not representing. But they could be made available to the dynamic mental synthesis (i.e., "kept in mind") if the collective in question were strongly activated by affect (*A*), automatization (*L*), mental attention (*M*), etc., and thus were "unfolded" in working memory (these elements would otherwise be part of *CTX*). Any collective might be activated with a semantic focus on a particular character, such as Heinz. Linguistically, this semantic focus on the collective might appear as a relative clause referring, for example, to Heinz. We represent the semantics of these special circumstances by rewriting the same collective (F2) as follows:

$$H \text{ "}(REL(CXT, H, W, VAL, \ldots)) \text{"}$$ (F2a)

This formula could be read as meaning: "Heinz, that is, the person who occupies the place of *H* in the knowledge scheme enclosed in quoted parentheses. . . . "

The other two basic collectives of the Heinz dilemma are:

$$REL(CXT, H, W, d, CONS, \ldots VAL, \ldots)$$ (F3)

This represents Heinz's knowledge of the wife's illness and her need for the drug (*d*); and related consequences (*CONS*) such as the doctors' claim that she will die, that is, will lose her valuable (*VAL*) life, unless the drug is provided.

$$REL(CXT, H, D, d, CONS, \ldots)$$ (F4)

This formula stands for a largely generic collective relating Heinz, as a consumer, to stores of various kinds, such as pharmacies, their owners or attendants, and the rules governing these interactions. Consequences of actions such as stealing the drug—going to jail, etc.— would be represented there.

These formulas can be interpreted, if the reader prefers, as simplified representations of chosen regions in a modified (affect-loaded and top-down addressable) semantic network. Our formulas merely highlight a net of relations from the network—those assumed to stem from a particular collective scheme whose origin in experience we can study. When a subject first reads the Heinz dilemma, the probe questions activate various core collectives available in his or her repertoire. A dynamic mental synthesis enabling the subject to reach a decision might involve one or more of the constituent collectives. A decision that includes all collectives could be one leading to the reply given below. I assume the subject's reply is constrained (overdetermined) by all highly activated compatible collectives in his or her working memory when the reply is constructed. Thus, the subject's choice of words indicates these collectives.

Consider this reply:

"He should steal because a man should help his wife and she urgently needs the drug; he is in a moral conflict, but life is more valuable than property."   (R1)

A *truly novel* (Pascual-Leone, 1980, 1983, 1990a) synthesis of collectives might underlie this reply, which we represent thus:

$$EVAL: SHOULD\text{-}STEAL?(CXT, H\text{"}(F2a)\text{"}, W\text{"}(F3a)\text{"},$$ (F5)
$$VAL=(MORE\text{-}VALUABLE\text{-}THAN(CXT, VAL=LIFE\text{"}(F3a)\text{"},$$
$$VAL=PROPERTY\text{"}(F4a)\text{"}))).$$

where (F2a), (F3a), and (F4a) correspond to the formulas given above. For this response, they could be concretized as follows:

$$H\text{"}(SHOULD\text{-}HELP(CXT, H, W, VAL=marital\ duty, \ldots)) \text{"}$$ (F2a)

Here the personal relationship between Heinz and his wife is invoked.

$$WILL\text{-}DIE\text{-}UNLESS\text{-}SHE\text{-}TAKES(CXT, W, d, VAL=life).$$ (F3a)

This collective represents what Heinz has been told: the wife will die without the drug whereas she will live if she takes it. An alternative representation (unfolding) of the same collective is the following:

$d"(IF\text{-}TAKEN\text{-}CAN\text{-}HEAL(CTX, W, d, \dots)).$"

$IS\text{-}OWNER\text{-}OF(CXT, D, d, H, VAL=property)$ (F4a)

In (F4a) we represent the idea that the druggist (D) is the owner of d and Heinz has a social-contract obligation to pay the druggist for the drug taken from the store.

Subjects' replies do not always semantically imply all relevant collectives. When a collective is not implied, we assume that it has low activation in the subject's repertoire (is not "kept in mind"—in working memory) and so does not intervene in generating the reply.

These formulas should be seen as hypothetical semantic descriptions of complex patterns of co-activation (clusters of activated schemes) in the cortex. Since performance is determined by degree of activation (total activation weight) of one cluster of schemes relative to the activation of other conflicting clusters, an analysis of organismic sources of scheme activation (analysis of hidden resources—a metasubjective analysis) is needed to explain moral judgment tasks.

Table 6.2 summarizes the organismic operators that we regard as more important in moral development. I already discussed these constructs, and a more detailed discussion can be found elsewhere (Pascual-Leone, 1987, 1989, 1995; Pascual-Leone & Baillargeon, 1994; Pascual-Leone & Goodman, 1979; Pascual-Leone & Johnson, 1991; Pascual-Leone & Morra, 1991; Stewart & Pascual-Leone, 1992).

In this theory one might think of the metasubject, that is, the "psychological" organism of the subject, as representable by a modified connectionist network—a symbolic connectionist network—that differs from current networks in at least two major architectural features. Let me describe these architectural features (AF) in succession. ($AF1$) There are many functionally different repertoires of schemes (executive, action, logological, infralogical, affective, intersubjective, etc.) capable of dynamically interacting with each other. ($AF2$) There is a mental-attentional mechanism such as that described before: $<E, M, I, F>$ that is, a system constituted by the sub-repertoire E of executive schemes (the Executive) and three "hardware" operators M (mental energy), I (active central inhibition), and F (neoGestaltist field factor), controlled by E. This attentional system can be used to change "top-down" and, as demanded by the task, the activation of schemes where it applies. It can activate or inhibit schemes currently activated or previously associated with the Executive. In this theoretical model, different active regions of the network may be boosted in their activation by different (top-down or bottom-up) resources. But the final performance of the network always results from overdetermination by all activated node regions (all activated schemes—see Pascual-Leone & Johnson, 1991), irrespective of the silent operators boosting them. Thus, M capacity is only needed in tasks where relevant schemes are not sufficiently activated by other operators. In other words: the executive use of M is in a trade-off relation with the current dynamic strength and application of other situationally-driven operators (e.g., C, L, A, F). We have tested this important **principle of overdetermination trade-off** in many studies (e.g., Baillargeon, 1993; Johnson, Fabian, & Pascual-Leone, 1989; Pascual-Leone, 1970, 1978b, 1987, 1989; Pascual-Leone & Baillargeon, 1994).

To illustrate this dynamic overdetermination trade-off, and to show the sort of ontological (i.e., functional architectural) assumptions about the metasubject one has to make in the analysis of complex vital reason tasks, I now give an example related to the previous one but simpler.

A subject's reply to the Heinz dilemma might state:

"Anyone should steal because life is more important than property."   (R2)

The corresponding synthesis of collectives might then be as follows:

EVAL: SHOULD-STEAL?(CXT, H=anyone, VAL=(MORE-IMPORTANT-
THAN   (F6)

(CXT, VAL=LIFE"(F3a)", VAL=PROPERTY"(F4a)")))

Where (F3a) and (F4a) are the collective knowledge schemes illustrated above in the corresponding formulas. A metasubjective dimensional[7] analysis of this reply follows from the formula, after we add modifications that incorporate the symbols of hardware operators boosting different schemes. This is done by following a notational convention: we place in brackets { . . . } the name of any scheme that is assumed to be hyperactivated by a hardware operator other than *M*, and *subscript* to the second bracket the letter symbols of the hardware operators hypothesized to be boosting the scheme in question. And, when the source of hyperactivation of a scheme in the formula is another automatized scheme from the formula, the latter scheme is marked by *superscripting* to it a name such as *L2*, or *L3*, etc., which indicates that it is an *L*-structure with an automatized connection to the former scheme. Thus the *L*-structure can directly boost the first scheme via spreading of activation—provided that the *L*-structure is hyperactivated (perhaps by *M*-capacity allocation). In this case the principle of overdetermination trade-off will make it unnecessary for *M*-capacity to be allocated to the first scheme.

The metasubjective dimensional analysis of this reply can, using these conventions, be written as follows:

EVAL:SHOULD-STEAL?($\{CXT\}_{C,L,F}$ H=anyone,   (F6a)
VAL=(MORE-IMPORTANT-THAN^{L2}($\{CXT\}_{C,L,F}$
VAL=life^{L3}"($\{WILL-DIE-UNLESS-SHE-TAKES(\{CXT\}_{C,L,F}$ W, d,VAL=life)$\}_{L2,L3}$"),
VAL=property^{L4}"($\{IS-OWNER-OF(\{CXT\}_{C,L,F}$ D, d, H, VAL=property)$\}_{L2,L4}$")))

Notice that in formula (F6a) the constituent collective *VAL=life"(. . .)"* is the collective of formula (F3a) with a difference—the scheme *VAL=life* is now being emphasized. Similarly, the constituent collective of *VAL=property"(. . .)"* is that of formula (F4a) with property now being emphasized. The tacit assumption, congenial to connectionist networks, is that human processing can make reference to a collective knowledge scheme (e.g., (F3)) under different forms (such as (F3a), *VAL=life"(. . .)"*, etc.). This assumption is based on the principle of spreading of activation in well-learned knowledge schemes (i.e., in *L*-structures—see Pascual-Leone & Goodman, 1979). According to this principle, any *L*-structure will, as a functional totality, become activated whenever any of its constituent schemes are strongly activated. And, in turn, the activation of the structure tends to spread activation to its constituents (Pascual-Leone & Goodman, 1979). Similar structural assumptions about spreading of activation are often found in cognitive science theories (e.g., Smolensky, 1988). It is because of this principle that *L*-structures can sup-

Table 6.2   Silent Operators (Unconscious Capacities)

Name	Operator	Definition
A	Affective activation	Activated affective schemes that apply on task schemes to boost their activation weight
C	Content learning and its activation	Learning mechanism and content schemes that can boost activation weight of task schemes
F	Field factors	Gestaltist mental field biases such as the "Minimum Principle" and "S-R Compatibility" that can facilitate or hinder the application of task schemes
L	Logical learning and its activation	Structural learning mechanism and structural-logical schemes that can boost the activation of task schemes
M	Mental attention	Endogenous mental energy that can be used by the currently dominant executive to boost the activation weight of task schemes
I	Mental interrupt	Endogenous central mental inhibition or interrupt that can be used by excutives to inhibit irrelevant task schemes

plement the boosting power of the subject's *M*-operator, by adding their own *L*-operator boosting capacity.

In this formula (F6a) the situation of Heinz has been generalized to become a principle or directive to "anyone." *H* cannot be *L*-boosted (activated by *L*-structure) via operative scheme *EVAL:SHOULD-STEAL?*, because the former has been generalized to "anyone" and this generalization is not connected *(L*-structured) via the Heinz story with the latter. In contrast, we can assume that the subject who gave reply (R2) had a knowledge scheme (*MORE-IMPORTANT-THAN*) that compares the values of life with the values of property. And for him life is strongly connected with collective (F3a) via the gist of the Heinz story, and via automatized general knowledge about life and dying. And, for the same reason, property is strongly connected with collective (F4a). Consequently, we can expect that the *L*-structure that embodies value comparisons, that is, the scheme *MORE-IMPORTANT-THAN* will boost, with activation *L2*, the collectives (F3a) and (F4a); and also that the knowledge scheme of life will boost the collective (F3a) with activation *L3*. Similarly, property will boost the collective (F4a) with activation *L4*.

If we now simplify, by retaining only the constituent schemes that in (F6a) must be boosted by *M* capacity (the excluded schemes are only boosted by other hardware operators due to the overdetermination trade-off), the result is:

$$EVAL{:}SHOULD\text{-}STEAL?(\ldots H{=}anyone,$$
$$VAL{=}(MORE\text{-}IMPORTANT\text{-}THAN(\ldots VAL{=}life"(\ldots)",$$
$$AL{=}property"(\ldots)"))).$$

(F6b)

The number of schemes left in this formula gives the minimal *M*-demand of a dynamic synthesis generating reply (R2). At least five knowledge schemes belonging to different collectives must be simultaneously (top-down) boosted by *M*. This measure of mental capacity can only be found in 11- or 12-year-olds, or older subjects (Johnson, Fabian, & Pascual-Leone, 1989; Pascual-Leone, 1970, 1978b, 1987). Consequently, only children of this age can give replies to moral dilemmas exhibiting this sophistication. And visuospatial measures of *M*-capacity can predict the moral-judgment sophistication of these children (Stewart & Pascual-Leone, 1992).

Conclusions

The first part of the chapter has provided an epistemological critique of hard cognitive science, in particular its computer-reductionist view of the mind. I have proposed a new epistemological "test," the Mental Touring Test, to establish whether a unitized programing system (e.g., Newell, 1991), or a program or theory, can be regarded as a general model of the human mind. This epistemological test, together with theorems on the limits of formalisms, yield an unambiguous answer: unitized programming systems or programs and theories cannot be general theories of the mind—although they may illuminate or clarify aspects of it. This conclusion suggests two methodological prescriptions for cognitive science: (1) one must distinguish sharply between formal computation and mental processing, because mental processing is broader than and includes formal computation. (2) To investigate the broader mental processes, and explain them, we must have non-computational theories and methods designed to adopt the strict perspective of the subject's "psychological" organism: what I called metasubjective theories and methods. Much of the chapter has been devoted to explicating this idea.

The epistemologically formal, but computationally informal, methods of process or task analysis, which I referred to as metasubjective, were classified into four types: Objective, Subjective, Ultrasubjective (or mental-process analysis), and properly Metasubjective (or organismic-process) analysis. These methods are nested. The Ultrasubjective analysis adds to the Objective and Subjective analyses the real-time temporal perspective; and properly Metasubjective analysis incorporates causal-process modeling of the organismic resources (mental attentional capacity, central attentional interruption/inhibition, structural and content learning, field factors, etc.). These resources regulate the functioning of schemes. Schemes are the basic units of analysis, and have a clear neuropsychological interpretation. The "psychological" organism of the subject (i.e., the metasubject), might be modeled on first approximation as a modified connectionist network—a symbolic connectionist network—with two new architectural features: (1) There are many functionally different repertoires of schemes (executive, action, logical and conceptual, purely-experiential, affective, intersubjective, etc.), capable of dynamically interacting with each other; and (2) There is a mental-attentional mechanism constituted by a modular system $<E, M, I, F>$—where E is the set of currently dominant executive schemes, and M, I, F are three different "hardware operators" or prewired resources of the brain. These include a mental attentional capacity, an active inhibition or interruption, and an attentional closure mechanism—the neoGestaltist field factors. A strong hypothetical-deductive methodology along convergent lines of research, which uses different scientific perspectives concurrently, may be called for in this metasubjective investigation. This is particularly true in lines of research that incorporate aspects of humans (e.g., consciousness, affect, development, intersubjective processes) that computers have not been able to simulate with success.

Also, we should emphasize the study of human processes that are not computational, such as those of vital reason (reason applied to the issues of real life), of truly creative problem solving ("truly novel" performances of any sort), and of inter-subjective processes. I have examined an intersubjective example of human judgment: moral judgment using Kohlberg's dilemmas. This illustration conveys the extent to which ontological—functional-architectural—assumptions about organismic processes (e.g., existence of "collectives" and of principles such as the overdetermination trade-off) is implicated in any process or task analysis from the domain of vital reason. But the true ontology (functional architecture) of a human metasubject is not yet well established. This is another reason for why logico-mathematical formalization is not sufficient in human science. But it is necessary: the force of the real is made palpable by the foundering of *formal* thought.

Notes

1. Translated by the author from the original Spanish: *"El hombre seria, segun esto—y en varios sentidos del vocablo—un animal fantastico."*

2. I define praxis as goal-directed activity ultimately addressed to the environment in order to satisfy one's own affective or instinctive goals. (I accept praxis that is not fully conscious.) Goals are always ultimately affective. The cognitive representatives of these goals are plans embodied by *executive schemes* (see below).

3. My claim is that humans, often unwittingly, construct their own ontology to help in their praxis (coping with reality resistances). This is recognized by philosophers. For instance, Spain's major philosophers, Ortega y Gasset (1957/1980a) and Zubiri (1980, 1989), both insist, against Heidegger and many others, that ontology and Being are not expressions of reality as such but are rather human constructions—situated knowledge (Ortega's "circumstances"). Reality as such only has resistances, and interrelations among resistances.

4. Piaget called these units "infralogical" (inspired by Bertrand Russell's Theory of Types) because they stand for and constitute the particular objects of experience, and object parts themselves, that Logic may symbolize and represent using generic categories and relations. Others (Carbonell, 1978, 1982; Houde, 1992, 1994; Pascual-Leone, 1976, 1984) have called these units "mereological" (a reference to Lesniewski's part/whole logic for the construction of concrete objects) to signify that part-whole relations within concrete objects of experience are the main feature of these schemes, which make up the raw content (presentation without representation) of any particular experience.

5. Modalities are kinds of processing that differ qualitatively in their content (visual, auditory, somatosensory, affective, etc.). Modes are kinds of processing that are found across modalities, and differ qualitatively in their structural and functional characteristics (e.g., operative versus figurative; infralogical/experiential versus logological/conceptual, etc.).

6. Cognitive science of the computer-simulation variety does not often make these distinctions. In object-oriented programming one seems to find structures analogous to core knowledge structures, but they are largely missing in procedural programming. The distinction between executive processes and action processes also is not found in procedural programming. Connectionist network programming, of great interest in other regards, does not distinguish different *modalities* and *modes* of processing. Affective processing and intersubjective processing are to this day beyond the reach of computer simulation.

7. A metasubjective analysis is dimensional when it reviews the essential constituents of the process or task in question, including the hardware organismic operators involved. But it does so without modeling the "real time" unfolding. According to our definitions, a metasub-

jective dimensional analysis is not ultrasubjective but, because it is metasubjective, it could be unfolded to be so. Ordinary dimensional analyses do not offer this possibility (Pascual-Leone, 1993).

References

Baillargeon, R. H. 1993. "A Quantitative Study of Mental Development." Unpublished doctoral dissertation. Toronto: York University.

Baldwin, J. M. 1908. *Thoughts and Things: A Study of the Development and Meaning of Thought on Genetic Logic,* Vol. 2. New York: MacMillan.

Baldwin, J. M. 1894/1968. *Mental Development in the Child and the Race: Methods and Processes.* New York: Kelley.

Bates, E., Benigni, L., Brethelton, I., Camaioni, I., and Volterra, V. 1979. *The Emergence of Symbols: Cognition and Communication in Infancy.* New York: Academic Press.

Benson, N. J., and Pascual-Leone, J. *Mental Capacity Constraints on Early Symbolic Processing.* (Submitted.)

Bloom, L. 1993. *The Transition from Infancy to Language: Acquiring the Power of Expression.* Cambridge: Cambridge University Press.

Bloom, L. 1994. "Meaning and Expression," in W. F. Overton, and D. S. Palermo (eds.), *The Nature and Ontogenesis of Meaning,* pp. 215–35. Hillsdale, NJ: Erlbaum.

Bruner, J. S. 1975. "The Ontogenesis of Speech Acts." *Journal of Child Language, 2,* pp. 1–19.

Bruner, J. S. 1977. "Early Social Interaction and Language Acquisition," in H. Schaffer (ed.), *Studies in Mother-Infant Interaction,* pp. 271–89. London: Academic Press.

Bruner, J. S. 1990. *Acts of Meaning.* Cambridge: Harvard University Press.

Carbonell, S. 1978. "Classes Collectives et Classes Logiques dans la Pensée Naturelle." *Archives de Psychologie, 46,* pp. 1–19.

Carbonell, S. 1982. "Influence de la Signification des Objects dans les Activités de Classification," *Enfance, 35,* pp. 193–210.

Case, R. 1985. *Intellectual Development: Birth to Adulthood.* Orlando, FL: Academic Press.

Case, R. 1991. *The Mind's Staircase.* Hillsdale, NJ: Erlbaum.

Case, R. 1992. "The Role of the Frontal Lobes in the Regulation of Cognitive Development." *Brain and Cognition, 20,* pp. 51–73.

Crick, F. 1984. "Function of the Thalamic Reticular Complex: The Searchlight Hypothesis." *Proceedings of the National Academy of Sciences, 1,* pp. 4586–90.

de Ribaupierre, A., and Hitch, J. G. (guest eds.) 1994. *International Journal of Behavioral Development, 17*(1), (whole issue).

Edelman, G. M. 1989. *The Remembered Present: A Biological Theory of Consciousness.* New York: Basic Books.

Fuster, J. M. 1989. *The Prefrontal Cortex, 2nd ed.* New York: Raven.

Gibson, J. 1979. *The Ecological Approach to Visual Perception.* Boston: Houghton Mifflin.

Gonseth, F. 1936/1974. *Les Mathematiques et la Réalité.* Paris: Albert Blanchard.

Greenfield, P., and Smith, J. 1976. *The Structure of Communication in Early Language Development.* New York: Academic Press.

Halford, G. S. 1993. *Children's Understanding: The Development of Mental Models.* Hillsdale, NJ: Erlbaum.

Holland, J., Holyoak, K., Nisbett, R., and Thagard, P. 1986. *Induction: Process of Inference, Learning, and Discovery.* Cambridge: MIT Press.

Houde, O. 1992. *Categorisation et Developpement Cognitif.* Paris: Presses de Universitaires de France.

Houde, O. 1994. "La Reference Logico-mathemetique en Psychologie: Entre Methode Universelle et Rationalité Arrogante," in O. Houde, and D. Mieville (eds.), *Pensée Logica-mathematique: Nouveaux Objects Interdisciplinaires*, pp.47–119. Paris: Presses Universitaires de France.

Husserl, E. 1970. *The Crisis of European Sciences and Transcendental Phenomenology.* Evanston: Northwestern University Press.

James, W. 1892/1961. *Psychology: The Briefer Course.* (G. Allport, ed.). New York: Harper & Row.

Jaspers, K. 1971. *Philosophy of Existence.* Philadelphia: University of Pennsylvania Press.

Johnson, J. 1991. "Constructive Processes in Bilingualism and their Cognitive Growth Effects," in E. Bialystok (ed.), *Language Processing in Bilingual Children,* pp.193–221. New York: Cambridge University Press.

Johnson, J., Fabian, V., and Pascual-Leone, J. 1989. "Quantitative Hardware-stages that Constrain Language Development." *Human Development, 32,* pp. 245–271.

Johnson, J., and Pascual-Leone, J. 1989. "Developmental Levels of Processing in Metaphor Interpretation." *Journal of Experimental Child Psychology, 48*(1), pp. 1–31.

Kleene, S. C. 1967. *Mathematical Logic.* New York: Wiley.

Mandler, J. 1992. "How to Build a Baby: II. Conceptual Primitives." *Psychological Review, 99,* pp. 587–604.

Mead, G. H. 1934. *Mind, Self, and Society.* Chicago: University of Chicago Press.

Newell, A. 1991. *Unified Theories of Cognition.* Cambridge: Cambridge University Press.

Newell, A. 1992. "Precis of Unified Theories of Cognition." *Behavioral and Brain Sciences, 15,* pp. 425–92.

Ortega y Gasset, J. 1957/1980a. *El Hombre y la Gente.* Madrid: Alianza Editorial. p. 251.

Ortega y Gasset, J. 1944/1980b. *Sobre la Razon Historica.* Madrid: Alianza Editorial.

Pascual-Leone, J. 1969. "Cognitive Development and Cognitive Style: A General Psychological Integration." Unpublished Doctoral Dissertation, University of Geneva.

Pascual-Leone, J. 1970. "A Mathematical Model for the Transition Rule in Piaget's Developmental Stages." *Acta Psychologica, 32,* pp. 301–45.

Pascual-Leone, J. 1976. "Metasubjective Problems of Constructive Cognition: Forms of Knowing and their Psychological Mechanism." *Canadian Psychological Review, 17*(2), pp. 110–25.

Pascual-Leone, J. 1978a. "Computational Models for Metasubjective Processes: Commentary on Z. W. Pylyshyn's " 'Computational Models and Empirical Constraints.' " *The Behavioral and Brain Sciences, 1,* pp. 112–13.

Pascual-Leone, J. 1978b. "Compounds, Confounds and Models in Developmental Information Processing: A Reply to Trabasso and Foellinger." *Journal of Experimental Child Psychology, 26,* pp. 18–40.

Pascual-Leone, J. 1980. "Constructive Problems for Constructive Theories: The Current Relevance of Piaget's Work and a Critique of Information-processing Simulation Psychology," in R. Kluwe and H. Spada (eds.), *Developmental Models of Thinking,* pp. 263–96. New York: Academic Press.

Pascual-Leone, J. 1983. "Growing Into Human Maturity: Toward a Metasubjective Theory of Adulthood Stages," in P.B. Baltes and O.G. Brim (eds.), *Life-Span Development and Behavior,* Vol. 5, pp. 117–56. New York: Academic Press.

Pascual-Leone, J. 1984. "Attentional, Dialectic and Mental Effort: Towards an Organismic Theory of Life Stages," in M.L. Commons, F.A. Richards, and G. Armon (eds.), *Beyond Formal Operations: Late Adolescence and Adult Cognitive Development,* pp. 182–215. New York: Praeger

Pascual-Leone, J. 1987. "Organismic Processes for Neo-Piagetian Theories: A Dialectical

Causal Account of Cognitive Development," in A. Demetriou (ed.), *The Neo-Piagetian Theories of Cognitive Development: Towards an Integration,* pp. 531–69. Amsterdam: North-Holland.

Pascual-Leone, J. 1989. "An Organismic Process Model of Witkin's Field-Dependence-Independence," in T. Globerson and T. Zelniker (eds.), *Cognitive Style and Cognitive Development,* pp. 36–70. Norwood, NJ: Ablex.

Pascual-Leone, J. 1990a. "Emotions, Development and Psychotherapy: A Dialectical Constructivist Perspective," in J. Safran and L. Greenberg (eds.), *Emotion, Psychotherapy and Change,* pp. 302–35. New York: Guilford.

Pascual-Leone, J. 1990b. "An Essay on Wisdom: Toward Organismic Processes that Make It Possible," in R.J. Sternberg (ed.), *Wisdom: Its Nature, Origins, and Development,* pp. 244–78. New York: Cambridge University Press.

Pascual-Leone, J. 1991. "A Commentary on Boom and Juckes on the Learning Paradox." *Human Development, 34,* pp. 288–93.

Pascual-Leone, J. 1993. "An Experimentalist's Understanding of Children," *Human Development, 37,* pp. 370–384.

Pascual-Leone, J. 1995. "Learning and Development as Dialectical Factors in Cognitive Growth." *Human Development, 38,* pp. 338–48.

Pascual-Leone, J., and Baillargeon, R. H. 1994. "Developmental Measurement of Mental Attention." *International Journal of Behavioral Development, 17,* pp. 161–200.

Pascual-Leone, J., and Goodman, D. 1979. "Intelligence and Experience: A Neo-Piagetian Approach." *Instructional Science, 8,* pp. 301–67.

Pascual-Leone, J., Goodman, D. R., Ammon, P., and Subelman, I. 1978. "Piagetian Theory and NeoPiagetian Analysis as Psychological Guides in Education," in J.M. Gallagher and J. Easley (eds.), *Knowledge and Development,* Vol. 2: *Piaget and Education,* pp. 243–89. New York: Plenum.

Pascual-Leone, J. and Irwin, R. 1994. "Noncognitive Factors in High-Road/Low-Road Learning: I. Modes of Abstraction in Adulthood." *Journal of Adult Development, 1*(2), pp. 73–89.

Pascual-Leone, J., and Johnson, J. 1991. "The Psychological Unit and Its Role in Task Analysis. A Reinterpretation of Object Permanence," in M. Chandler and M. Chapman (eds.), *Criteria for Competence: Controversies in the Conceptualization and Assessment of Children's Abilities,* pp. 153–87. Hillsdale, NJ: Erlbaum.

Pascual-Leone, J., Johnson, J., Goodman, D., Hameluck, D., and Theodor, L. 1981. "I-interruption Effects in Backward Pattern Masking: The Neglected Role of Fixation Stimuli," in *Proceedings of the Third Annual Conference of the Cognitive Science Society,* Berkeley, California.

Pascual-Leone, J., and Morra, S. 1991. "Horizontality of Water Level: A Neo-Piagetian Developmental Review." *Advances in Child Development and Behaviour, 23,* pp. 231–76.

Piaget, J. 1985. *The Equilibration of Cognitive Structures: The Central Problem of Intellectual Development.* (Trans., T. Brown, and K. J. Thampy) Chicago: University of Chicago Press.

Polya, G. 1954. *Mathematics and Plausible Reasoning,* Vol. 2. *Patterns of Plausible Inference.* Princeton, NJ: Princeton University Press.

Sartre, J. P. 1960. *Critique de la Raison Dialectique,* Vol. 1. *Theorie des Ensembles Practiques.* Paris: Librairie Gallimard.

Sartre, J. P. 1966. *Being and Nothingness.* New York: Washington Square Press.

Scheler, M. 1973. *Selected Philosophical Essays.* Evanston: Northwestern University Press.

Searle, J. R. 1992. *The Rediscovery of the Mind.* Cambridge: MIT Press.

Smolensky, D. 1988. "On the Proper Treatment of Connectionism." *Behavioral and Brain Sciences, 11,* pp. 1–74.

Sperry, R. W. 1977/1992. "Forebrain Commissurotomy and Conscious Awareness," in C. Trevarthen (ed.), *Brain Circuits and Functions of the Mind: Essays in Honor of Roger W. Sperry*, pp. 371–88. Cambridge: Cambridge University Press.

Stern, D. 1977. *The First Relationship*. Cambridge: Harvard University Press.

Stern, D. 1985. *The Interpersonal World of the Infant*. New York: Basic Books.

Stewart, L., and Pascual-Leone, J. 1992. "Mental Capacity Constraints and the Development of Moral Reasoning." *Journal of Experimental Child Psychology, 54*, pp. 251–87.

Stuss, D. T. 1992. "Biological and Psychological Development of Executive Functions." *Brain and Cognition, 20*, pp. 8–23.

Thagard, P. 1989. "Explanatory Coherence." *Behavior and Brain Sciences, 12*, pp. 435–502.

Trevarthen, C. 1990. "Growth and Education of the Hemispheres," in C. Trevarthen (ed.), *Brain Circuits and Functions of the Mind*. New York: Cambridge University Press.

Vuyk, R. 1981. *Overview and Critique of Piaget's Genetic Epistemology 1965–1980*, Vol.1. New York: Academic Press.

Zubiri, X. 1980. *Inteligencia Sentiente*. Madrid: Alianza Editorial.

Zubiri, X. 1989. *Estructura Dinamica de la Realidad*. Madrid: Alianza Editorial.

7

Don Ross

Is Cognitive Science a Discipline?

I

An explicit theme in "Reassessing the Cognitive Revolution" is the question of the autonomy of cognitive science from other disciplines. This is most often treated as a question about *methodology*: is there a distinctive approach to the study of mind that sets cognitive science apart from its ancestor and component disciplines; and, if so, what is that method? In the following discussion, I will provide reasons for answering "No" to the first question, and will offer some diagnosis as to why the second one has seemed to many to have an answer. I conceive of this as part of a *defense* of cognitive science, against both its enemies (such as Searle) who construe attacks on what they take to be canonical cognitive science methods as attacks on the possibility and aims of cognitive science, and also against certain ill-advised friends of cognitive science, who, in seeking unnecessarily firm foundations for the discipline, make its status hostage to every passing conceptual shift.

I take it that the question as to whether cognitive science is a discipline arises, in the first place, because a number of philosophers, psychologists, computer scientists, and biologists find that they have more in common with one another than with their fellow department members, at least where their working concepts and research problems are concerned. If someone, a philosopher, say, ceases to care whether what he or she is doing is properly thought of as philosophy or psychology or biology, so long as it is recognized as cognitive science, then that individual is liable to start to see her or himself as a citizen of a new discipline. And if people in other disciplines feel the same way, then there are cognitive scientists; hence there must be cognitive science. But these cognitive scientists—at least, the philosophers among them—do not rest content with seeing their newfound discipline as simply the product of biographical convergences. Some seek to *justify* the existence of a distinct discipline of cognitive science. To judge from the most heroic of these attempts, by Pylyshyn (1984) and von Eckhardt (1993), the natural way to try to do this is by focusing on the tools—computational models of hypothesized internal mental processing—whose use distinguishes the typical cognitive scientist's working day from those of his or her predecessors in each of the ancestor disciplines. Both Pylyshyn and von Eckhardt insist that cognitive science is legitimated by its computationalist foundations,[1] where "computationalism" denotes a full-blown methodology, underwritten by deep and highly specific epistemological commitments. In Pylyshyn's case, this has already generated embarrassment (to the field, if not necessarily to Pylyshyn); since typical connectionist methods and assumptions violate the strictures of Pylyshyn's computationalism, connectionists are by his lights not doing cognitive science,[2] and this has led to a war of philosophical intuitions that is preposterously far ahead of the relevant empirical developments. This situation is symptomatic of a mistake, namely, seeking to build a philosophical/methodological exoskeleton for a discipline in the absence of a rich history of results and paradigmatic explanations around which to mould such a structure. Let me now expand on this charge.

II

I will begin with some very general reflections on the importance of methodology to questions about the structure of scientific inquiry. Much of the progress that the philosophy of science has made during the last century has consisted in the careful—I stress "careful"—abandonment of the view that science in general can be demarcated from non-science, and its special disciplines from each other, wholly or mainly by their methodologies. Rigorous substantiation of this view was the great hope of both logical empiricists and Popperians, but proved incapable of fulfillment. The project failed mainly because *good* methodologies—the strictures and procedural principles that successful scientists use to guide their research—consist in large measure of formless and inarticulate practical intuitions that resist the sort of regimentation that would be required if methodological classification were the basis for a philosophically interesting taxonomy of inquiry. The basis on which distinctions among scientific disciplines are justified, I would argue, is primarily ontological, not methodological. Microphysics studies the basic dynamics of matter and energy; astrophysics studies the large-scale structure of space-time; molecular genetics studies the means by which biological systems transmit their structures through time; microeconomics studies the nature of instrumental rationality in circumstances of scarcity. These disciplines arise because their objects of study constitute robust real patterns in the world (see Dennett, 1991), and they are *different* disciplines because the real patterns that they study are distinct. For any carefully specified methodologies, it would be *false* to say that any of the disciplines I have mentioned restrict themselves to a single methodology; and while one could obviously describe in very loose and general terms differences in the characteristic methods of, say, physicists and economists, this would have no bearing whatsoever on the *justification* for having, on the one hand, physics, and, on the other, economics.

I stress these points because they are the basis for a *respect* in which I don't attach much urgency to the question of whether cognitive science is an autonomous discipline. As the perspective on behalf of which I just pounded the table would have it, there is compelling motivation for a discipline of cognitive science if and only if there are minds, and if the kind of entity that minds are is not explicitly recognized *as* a kind by the ontological assumptions of another, better established discipline.[3] Even Paul Churchland, who has sometimes talked as if he thought that there were no minds, clearly believes that there is something ontologically distinctive about behavior and cognition; and this belief shows up, in his case, in his call for a discipline of neuropsychology that is *not* identical to neurophysiology and that includes cognitive robotics and connectionist AI (Churchland, 1989). Searle (1992, and elsewhere) denies that there is a justified place for cognitive science, because he denies my second requirement for distinctiveness. Because he rejects the central ontological article of functionalism, namely, that minds are in principle realizable in different sorts of physical substance, he concludes that questions about what the mind is and how it works are ultimately the business of neurophysiology. This conclusion rests on some very hasty reasoning.[4] However, Searle—and, by implication, the one or two other living reductive materialists that I know of—at least has a rationale, even if a defective one, for his denial that there is an autonomous discipline of cognitive science. But one of the many aspects of Searle's position that is unsatisfying is that he never tells us what his favored non-cognitive-scientists who study the mind should actually *do* that is different from what cognitive scientists do.[5] I frankly cannot even imagine what he might recommend. Until he or one of his ilk enriches the resources of my imagination on this matter, the question of whether cognitive science is or is not an autonomous discipline has no immediate bearing that I can see on our inquisitive practice, and thus is not very interesting.

I should add that the question of whether cognitive science is a discipline in its own right may, of course, be relevant to the organization of universities, granting agencies, and other parts of the research infrastructure, and in *this* way matters a great deal to what gets done in what order by whom. But I do not think that anything deep or philosophical—or, at least, involving reference to methodologies—lurks in this direction. I am confident that had cognitive science research reached its present level of activity during the 1960s, when academic resources were bountiful and regular institutional reorganization was financially feasible, all major universities would have separate departments of cognitive science. Administrators in that happy time would have been persuaded to establish such departments not by philosophical arguments, but by appeal to the fact that cognitive science has all of the other institutional trappings of an established discipline: its own journals, with a well-recognized prestige hierarchy among them, its own conferences, its own societies and sub-societies, and so on. In sum, it seems to me to be perfectly obvious that cognitive science is a discipline, whether or not this fact is reflected in university calenders, and that the evidence for it has nothing to do with issues of methodology.

III

In talking about Searle a couple of paragraphs ago, I suggested that denial of the thesis that the mind is multiply realizable could potentially figure as a premise in an argument against the autonomy of cognitive science, even if Searle himself has failed to find that argument. It might be supposed that, in that case, the concern with methodology must be more important than I have made out. The doctrine of multiple realizability is often associated, as a result of main themes in the philosophy of mind over two decades, with Turing-machine functionalism, classical computationalism and GOFAI; the viability of these approaches, it will be urged, is surely a crucial aspect, at least historically, of the enthusiasm for cognitive science. Furthermore, this viability is now under serious assault, from radical connectionists and others. The underlying idea here is that even if disciplines cannot be strictly demarcated on the basis of methodology, a discipline whose actual, paradigmatic methods have been rejected might thereby cease to exist. (Consider, for example, the reasons for the disappearance of the discipline of philology, whose ontology is perfectly respectable. This contrasts with the case of, e.g., alchemy, whose main problem was commitment to an ultimately worthless ontology.) We must therefore consider the possibility that while Pylyshyn is wrong to suppose that the only possible cognitive science is one cooked according to his computationalist recipe, he (or, more plausibly, von Eckhardt) is right about the assumptions that the *actual* discipline called "cognitive science" contingently depends on. And I take it that if the actual discipline of cognitive science is in the process of collapse, then there is an important sense in which the answer to the question of my title is "No."

Let us begin with the following premise: *if* there is any set of methods to which cognitive science has, in fact, tied its fate as a discipline, then that set comprises some form or other of computationalism (though the form in question is almost surely less narrow than Pylyshyn supposes). Let us also acknowledge as a very general fact about almost all existing models and explanations in cognitive science that they make use of at least *some* computational concepts in characterizing the dynamics of their objects of study. Having conceded this much, we will consider Pylyshyn's central question, which he answers in the affirmative: Is cognitive science committed to the claim that cognition *is* computation?

Almost all discussions of this question begin by recognizing that its answer depends crucially on how precisely "computation" is understood. An affirmative answer imposes no substantial constraints on cognitive science if "computation" is understood so loosely as to be *conceptually* equivalent to cognition. Taking "computation" more narrowly leads down the familiar paths of a vast field of philosophical literature. To what extent must we conceptualize cognition as operations on language-like elements? How and to what extent can semantic properties be bought with purely syntactic currency? And so on. These investigations have been serious and instructive. Their results have satisfied me, at least, that if one asserts that cognition is computation, *and* insists that "computation" be understood in some absolutely precise and immutably fixed way, then one thereby asserts something false.[6] Here, however, I want to raise a meta-question about this debate and its significance. Is there not something very odd about supposing that the viability of the *discipline* of cognitive science hangs in the balance while philosophers try to get clear about the concept of representational content? How could the hubris even of *philosophers* lead anyone to believe something so implausible?

Here's one (bad) way: One could suppose that unless cognitive science can be shown to have its foundations, in a strict sense, in computability theory, then there are no laws of cognitive science. This, however, would be a problem only for someone who believes that a scientific discipline *must* have laws; this belief is utterly at odds with the history of scientific practice. Another, more plausible, way of tying the role of the "computation" concept to questions about the viability of cognitive science has, I suspect, actually motivated (though usually only implicitly) most of the breast-beating that has gone on. Call this "Searle's association." It goes like this: If cognition is computation, then "strong AI" is possible. Much, or most, of the enthusiasm that has produced the discipline of cognitive science is fueled by the allure of strong AI. So if it can be shown that cognition is *not* computation, then the wind will be gone from the sails of cognitive science.

A standard response to this reasoning is to point out that it depends on two claims for which Searle has not provided sufficient argument. The first of these is that the institutional impetus of cognitive science depends on a widely shared belief that its pursuit will enable us to develop non-brain-based minds. Though I see no compelling grounds for accepting this claim, I also have little to say against it, since it is a sociological hypothesis with respect to which the relevant data, as far as I know, has not been collected. Let us therefore grant it for the sake of argument. We then move on to the second necessary premise, which is that the belief that there can be non-brain-based minds is, in fact, unwarranted. Note that although Searle has often tried (unsuccessfully) to establish the claim that there *cannot be* non-brain-based minds, his case against cognitive science does not depend on this very strong claim. *If* he is granted the first premise, about the socioeconomic motivations behind cognitive science, then he needs only the weaker claim that actual cognitive science research, as it now stands, need not be interpreted as leading toward what he calls "strong AI." But this claim depends in turn on the importance and viability of the distinction between strong and weak AI, where the former concept refers to the possibility of *reproducing* intelligence, and the second to the possibility of using computational models and simulations in developing a *theory of* intelligence. The reason for this is that Searle concedes the viability of (and, indeed, the *actuality* of) weak AI.

Recall that our original concern was with the relation between the concepts of cognition and computation. And with respect to the significance of that relation, Searle's distinction between weak and strong AI is too simplistic to be helpful. Everyone, including Searle, grants that weak AI may involve *simulation* of minds; one avoids commitment to strong AI so long as one does not imagine that simulation implies reproduction. However, there is no such thing as a simulation in which there are no actual properties in common between the simulated and the simulator, so the idea of "purely" weak AI makes no sense. Is there something *interesting* we might mean by weak AI? Consider the confrontational connectionist who, despite favoring the strongest anti-computationalist rhetoric, still agrees that computational theories are often useful in cognitive science, at least as theoretical descriptions in procedural form that describe input-output functions in extension. The anti-computationalism of some reasonable cognitive scientists consists just in their denying that a computational theory can describe all psychological processing phenomena by means of intensional functions. They deny this because their experience with analytically opaque models, such as very complex distributed connectionist models, leads them to doubt that all psychological processing phenomena are manifestations of capacities for which isomorphic, programmable, computationally tractable functions exist. Someone who thinks like this is likely to conceive of computational models as abstract procedural theories, which capture the phenomena only at a high level of idealization (Smolensky, 1988). Now, is a person who holds these views thereby an advocate of weak AI? The answer to this question is as indeterminate as the answer to a second question on which it implicitly rests: How detailed an account must a procedural theory give in order to pass beyond simulation to reproduction? Bit maps, I guess we'll agree, are clearly reproductions. But if pursuit of strong AI demanded that we try to implement bit maps of all the information, as coded in processing dispositions, embodied in some particular brain, then strong AI would obviously have nothing to do with real cognitive science, and Searle's association would thus be unjustified. So how detailed *should* our computational theories be? The answer, of course, is that they should be as detailed as we can make them, in light of the constraints that arise from both formal computation theory and from facts about the sorts of information processing of which brains turn out to be capable. At what level of detail will we have produced a reproduction, as opposed to a simulation, of a brain as an engine of thought? This is *not* the question addressed by the Turing test, which is just a proposed operationalization of the concept of intelligence. It is, instead, a question that is without scientific substance.

I end these reflections by voicing a concern that has often occurred to me while listening to philosophers and psychologists arguing in one another's company at cognitive science gatherings. It is a common piety in celebrations of the interdisciplinary character of cognitive science that the role of philosophers in the enterprise is crucial, because the conceptual problems confronted by cognitive scientists are so difficult. Of the conceptual difficulty there can be no doubt. But the proposition that a young science is better off for the constant attentions of professional, full-time conceptual investigators seems to me much more dubious. Early physicists and biologists wrestled with conceptual dilemmas whose difficulty can still be readily appreciated, and their resulting labors are a huge part of the shared history of science and of philosophy. But my impression (which may, I realize, be a distortion caused by hindsight) is that most developing sciences have focused only on conceptual questions that arose out of actual experimental and observational practice, and that were genuine obstacles to empirical progress. However, the incentives in place for contemporary philosophers encourage them to dig out and, sometimes, to invent, problems that make no present difference to the course of scientific investigation, but which are marketed in such a way as to risk cluttering the research agenda. I believe that the question of whether cognitive science is a justifiably autonomous discipline is a problem of this kind.

Notes

I would like to thank Paul Forster for his extensive comments on an earlier draft of this chapter, and William Rapaport, whose reflections on the character of cognitive science as a discipline originally inspired the ones here. I am also grateful for the financial support of the Social Sciences and Humanities Research Council of Canada.

1. It should be noted, however, that Pylyshyn's conception of computationalism is *much* narrower than von Eckhardt's.

2. According to Fodor and Pylyshyn (1988), *if* connectionists are doing anything worthwhile, then it is abstract neurobiology, or "implementation theory," as they sometimes say.

3. Does the existence of psychology then entail the nonexistence of cognitive science? This depends on interpretations of the history and philosophy of psychology. If one understands psychology narrowly as the study of human behavior, then one could argue that most psychology did not explicitly recognize the generic phenomenon of mind that cognitive science studies. On the other hand, little strain would attach to an argument that William James did something that is clearly cognitive science without needing to dress it up in a phrase that is redolent of physics envy, like "cognitive science."

4. The hastiness consists in the following. Searle never, as far as I know, provides an argument for his version of mind/brain identity theory, beyond his oft-used but question-begging analogy between thinking and photosynthesis (Searle, 1990). He seems to arrive at identity theory on the basis of two premises: that either functionalism is true or identity theory is true; and that functionalism is false. I believe that both premises are false. I am confident that most readers will agree with me as regards the first one.

5. Since Searle is not an eliminativist, they're presumably not supposed just to poke at brains and set aside their traditional conceptualization of behavior. Since he's also not a behaviorist, we may also suppose that they will take internal mental processes of *some* (noncomputational) kind seriously. So how are they not doing cognitive science?

6. I will gesture toward my reasons by expressing my sympathies with the conclusions reached by Millikan (1993).

References

Churchland, P.M. 1989. *A Neurocomputational Perspective*. Cambridge: MIT Press/Bradford.

Dennett, D. 1991. "Real Patterns." *Journal of Philosophy,* 88, pp. 27–51.

Fodor, J., and Pylyshyn, Z. 1988. "Connectionism and Cognitive Architecture: A Critical Analysis." *Cognition,* 28, pp. 3–71.

Millikan, R. 1993. "On Mentalese Orthography," in B. Dahlbom, (ed.), *Dennett and His Critics*. Oxford: Basil Blackwell.

Pylyshyn, Z. 1984. *Computation and Cognition*. Cambridge: MIT Press/Bradford.

Searle, J. 1990. "Is the Brain a Digital Computer?" *Proceedings and Addresses of the American Philosophical Association,* 64, pp. 21–37.

Searle, J. 1992. *The Rediscovery of the Mind*. Cambridge: MIT Press/Bradford.

Smolensky, P. 1988. "On the Proper Treatment of Connectionism." *Behavioral and Brain Sciences,* 11, pp. 1–74.

Von Eckhardt, B. 1993. *What Is Cognitive Science?* Cambridge: MIT Press/Bradford.

8

Ellen Bialystok

Anatomy of a Revolution

In 1938, Harvard historian Crane Brinton first published his influential analysis of four revolutions—the English, American, French, and Russian—extracting from them the common themes and mechanisms that united them as individual instances of a single process (Brinton, 1965). Revolution, according to Brinton, had structure, and its course was orderly and predictable.

In Brinton's analysis, every revolution consisted of four stages, each following inevitably from the previous. In the first stage, an old regime is in power, hanging on to authority despite internal stagnation and some measure of criticism by the intellectuals. A defining event occurs in which power, to the surprise of everyone, is transferred to those critics. This is the beginning of Stage 2. There is no dramatic seizure of power, only a ruling system that crumbles and yields to the new order. These are the halcyon days of the revolution, and they set off an irreversible process of modernization. No matter how the revolution plays itself out, the world will never be the same again. By the third stage, an inherent contradiction in the process is revealed. The new leaders are the same moderate intellectuals whose voices raised the spectre of revolution in the first stage and who were the unwitting recipients of power in the second. But they are now the establishment, conducting themselves with the same rationality that guided them when they were in opposition. The problem is that this very rationality that made the revolution possible is unsuitable for governing. A powerful force has been set in motion by the act of revolution, and power is quickly usurped from the rational moderates by those with more extreme visions. In the fourth stage, the prevailing authority is in the hands of the ideologues. The revolution has become the status quo.

The cognitive revolution now stands charged with abandoning its initial goals and becoming hostage to a specific agenda, namely, computational psychology. The evidence is the domination of research grants, publications in leading journals, prominent presence at conferences, and graduate student interest. The accusation brings with it the implicit assumption that there was a deliberate attempt by a group of scientists to redirect research attention in those ways. But all of this, it seems to me, is to credit scientists with both too much control and too much imagination. If Brinton was right, then these outcomes are part of the revolutionary process. The forces unleashed at Harvard University in the 1950s were subject to a greater power.

Brinton's analysis, loosely interpreted, provides a good model for describing the cognitive revolution. It presupposes the idea of paradigm shifts posited by Kuhn (1962), yet seems to capture more accurately the causes and stages for that shift than does Kuhn's more scientifically based interpretation. Amid the imperial rule of behaviorism, there was a growing discomfort that something was very wrong. If we were to understand anything that was essentially human, then mind and meaning could no longer be excluded from the table. The tentative formation of some alternative conceptions began to emerge, mostly along the banks of the Charles River. Stage one was set in motion, and the days of the old regime were numbered.

It is difficult to identify a seminal event that revealed the bankruptcy of the ruling behaviorists and precipitated the transfer of power to the new cognitivists, but a candidate must surely be the publication of Chomsky's (1959) review of Skinner's (1957) book, *Verbal Behavior*. This, in conjunction with exciting work being carried out by the group at Harvard, defined the beginning of the second stage. Power was now indisputably in their hands. The effect was overwhelming: the flood gates of intellectual excitement had been lifted and the torrent of new ideas seemed inexhaustible. We learned about thinkers whose work had been long suppressed, notably Piaget and Vygotsky, about language and its special place in mind, about memory and its delicate limitations, and about thinking as an intentional act. The questions were so rich and the answers so diverse that the most important and, it turns out, irreversible consequence of this second stage was that the seat of cognitive power no longer belonged exclusively to psychologists. From now on, there would be linguists, philosophers, mathematicians, computer scientists, anthropologists, and many others sharing the cognitive platform.

By the end of the third stage, the dust had settled and cognitive science seemed to be in the hands of a new ruling elite. The artificial intelligence establishment whose view of cognition was shaped by the computers they worked with had become the dominant force. Their metaphor was a compelling one, their computer programs dazzling, and the applied possibilities, apparently endless.

There was nothing new in the idea of using a metaphor to study cognition. A long list of such devices as steam engines, wells, telephone switchboards, and biological organs had served this noble function. But the choice of the metaphor is crucial in the development of the theory: Each metaphor suggests certain kinds of questions while obscuring others. What was new in this third stage of the revolution in which the computer metaphor took control was the tendency to confuse the metaphor with the thing studied.

We now sit in the fourth stage, a quiet time ruled by the radical usurpers of the previous stage. There is at present some excitement generated by the internecine squabbling between symbolic and subsymbolic paradigms, but, as Margaret Boden (1991) has pointed out, these are different colors but the same horse. In short, the computational approach has become the establishment and no longer seems to us to be radical. So was the revolution successful?

It is interesting that despite an appearance that a consensus has been established in cognitive science, there is immense diversity in the opposition to it. The new participants who were admitted as part of the revolution each seem concerned about the failure of cognitive science to achieve a rather different goal. Furthermore, each of these failures seems to be, in the minds of the critics, fatal.

Consider a few of the perspectives that challenge the achievements of the cognitive revolution. The concern from philosophy is that the philosophical problems that have been the energy for psychological inquiry, probably since the time of Aristotle, have not been resolved. We are probably no closer to a solution for the problem of other minds than Descartes was, in spite of our access to descriptions of certain aspects of mind so rich that Descartes could not dared have imagined them. Yet the ontological problems that have continuously haunted philosophers persist. How can we understand the relation between mind and brain without resorting either to a reductionism that collapses one into the other or an eliminativism that defines one out of existence? The cognitive revolution has not solved this problem.

Searle (1992) makes the strong claim that a computational psychology, no matter how sophisticated, ultimately will fail if it does not address the fundamental problems of consciousness, and of beliefs and desires. No machine, no matter how sophisticated, has ever dialed up another machine because of an inherent need to feel connected, spontaneously displayed data out of a need to express intentions, or plugged itself in to satisfy its desire to be charged. At the same time, Searle slips over the problems of how this conscious intentional mind gets on with the ordinary business of learning language, recognizing patterns, and attaining automaticity in skilled performance. The cognitive revolution has not found a way of including both the function of basic cognitive processes and a description of consciousness into explanations of human performance.

The concern from linguistics is the problem of innate knowledge. Language (like Quebec) has always demanded special status—a form of "sovereignty association with the mind." The learning principles invoked in other domains have never seemed adequate to overcome such barriers as "poverty of the stimulus" and "no negative evidence" to explain how children learn a language. The irony, however, is that attributing innate structures to language has not solved this problem of acquisition. The cognitive revolution has not settled the status of innate knowledge and its attendant issues.

The concern from anthropology is the problem of culture. Bruner (1990) elaborates this position in *Acts of Meaning* and poses the case for a transactional analysis rooted in the social and cultural community of human participants. However, this is a risky enterprise, for if behavior is only explained in this situated sense, how are we ever to discover what is universal and essential about humans? The cognitive revolution has made little progress in understanding, let alone modeling, these cultural influences and processes.

The concern from neuroscience is that our models are not sufficiently grounded in a physiology of the brain. Crick (1994) attempts to correct this by describing the relation between the structure and function of the brain, even to the level of individual cells and cognition, but our cognitive models are not typically rooted in these structures. Connectionist models give the impression of being neurologically valid. Rumelhart and McClelland (1986) claim that they are inspired by the architecture of biological brains and *are*, as opposed to implement, cognitive models. But as some critics point out, they are simply a different kind of mechanical simulation, like the digital artificial intelligence systems that preceded them. The cognitive revolution has given us little direction in how to incorporate the constraints of physiological systems into our conceptions of mental function.

To my mind, these various concerns stand as incontrovertible evidence that the cognitive revolution has been an unmitigated success. The questions jointly point to gaps in the current description of mind, but beneath the specific concerns is an unquestioned assumption of the centrality, complexity, and diversity of mind. These questions had no voice in the old regime. Indeed, they were inconceivable in the paradigm of that time. In Kuhn's revolution, recalcitrant data lead to new forms of interpretation. In Brinton's revolution, the irreversible process of modernization ignited by the revolution leads to the liberation of new ways of thinking and the boldness to ask new questions. This diversity of thought is a central legacy of the cognitive revolution.

Why, then, do we have the illusion that the revolution has failed? I believe there are two reasons. The first is the failure of a single paradigm to emerge, but it was probably unrealistic to expect that there could be one simple truth. Each of the competing perspectives is based on a construal of mind that is comprehensible within a particular metaphor. Each sets different constructs as invariant and each leads to the articulation of different questions. Why should we believe that human intelligence is ultimately reducible to a single metaphor?

The second reason is that we are seduced by the logical model in which we believe that events are the consequences of deliberate and reasoned choices that determine outcomes. In contrast to this scientific model of logic, Stephen Jay Gould (1989) presents what he calls the logic of historical contingency: Events occur because they follow other events, and a change of any incident in the chain would change everything. This, I believe, is similar to the narrative mode of thought described by Bruner. The cognitive revolution was set in motion and generated an evolutionary process based on this contingency. Ideas follow upon ideas, methods upon methods, each occurring because of what preceded it. On this logic, it makes no sense to ask if we chose wrong, or if we paid too much attention to computational modeling, because without the history happening precisely as it did, the questions we now face would not occur to us. The point is that our retrospective dissatisfaction with certain models is unjustified because what we learned from them is what now enables us to challenge them.

When we set out on the cognitive revolution we had no charts and little sense of what the goals were. We claimed that we were ready to explore mind in all its intricacies. What we found, I believe, is a mind that was far more enigmatic than anyone had imagined. It was part of the human condition, and the human condition was complex. In his book *Wonderful Life*, Gould (1989, p. 281) says:

We shall never be able to appreciate the full range and meaning of science until we shatter the stereotype of ordering by status and understand the different forms of historical explanation as activities equal in merit to anything done by physics or chemistry. When we achieve this new taxonomic arrangement of plurality among the sciences, then, and only then, . . . [shall we] . . . finally understand that the answer to such questions as "Why can humans reason?" lies as much and as deeply in the quirky pathways of contingent history as in the physiology of neurons.

References

Boden, M.A. 1991. Horses of a Different Color? In W. Ramsey, S.P. Stich, & D.E. Rumelhart (eds.), *Philosophy and Connectionist Theory*, pp. 3–19. Hillsdale, NJ: Erlbaum.

Brinton, C. 1965. *The Anatomy of Revolution.* Toronto: Random House.

Bruner, J.S. 1990. *Acts of Meaning.* Cambridge: Harvard University Press.

Chomsky, N. 1959. Review of "Verbal behavior" by B.F. Skinner. *Language*, 35, pp. 26–58.

Crick, F. 1994. *The Astonishing Hypothesis.* London: Touchstone Books.

Gould, S.J. 1989. *Wonderful Life: The Burgess Shale and the Nature of History.* New York: W.W. Norton.

Kuhn, T.S. 1962. *The Structure of Scientific Revolutions.* Chicago: University of Chicago Press.

Rumelhart, D.E., & McClelland, J.L. 1986. *Parallel Distributed Processing: Explorations in the Microstructure of Cognition.* Cambridge, MA: MIT Press.

Searle, J.R. 1992. *The Rediscovery of the Mind.* Cambridge, MA: MIT Press.

Skinner, B.F. 1957. *Verbal Behavior.* New York: Appleton Century Crofts.

PART TWO

Christina Erneling

COGNITIVE SCIENCE AND
THE STUDY OF LANGUAGE

This part on cognitive science and language focuses on the ideas of Noam Chomsky. This is not only, as Gardner points out, because "the history of modern linguistics is the history of Chomsky's ideas" (Gardner, 1985, p.185; see also Pinker, 1994), but also because of the impact Chomsky has had in promoting ideas that have been at the core of the cognitive revolution. Chomsky's review (1959), of B. F. Skinner's book *Verbal Behavior* (1957) paved the way for a non-behavioristic and non-inductivist approach to the mental. And, at least since the late 1950s and up to the chapters included in this volume, he consistently has argued for mentalism, computationalism, and the modularity of the mind—themes central to cognitive science as well as the study of language and its acquisition.

Chomsky's chapter in this part clearly illustrates his commitment to the fundamental tenets of the cognitive revolution. That is, he takes the view that the mind is modular, with language composing one of the clearest and most important of the separate modules. He insists that we need to understand language and other mental functions from an individual and internal point of view. By contrast, details of the actual public use of language and its socio-communicative aspects fall outside scientific investigations. This individualistic, and internal (one also could add intentional and innate) language Chomsky calls I-language to separate it from E-language—that is, the external and social use of language. Chomsky thinks, as is clear from his chapter, that questions having to do with E-language are questions we cannot approach scientifically, and which probably lack answers altogether. He compares them to questions of the type "How do things work?" Instead of discussing semantics and questions having to do with communication and normative aspects of language, he thinks that we should focus on the language system stored in the individual mind/brain. Chomsky thus clearly proposes to limit the study of language to a narrow range—that is, giving an account of the individual grammatical competence of an ideal speaker or listener.

Although Chomsky still is committed to several of the fundamental ideas that have driven his research program, many of his specific ideas have changed over the years. For example, he continues to be committed to the innateness of the internal language system. But he no longer claims that universal grammar in the form and sense of transformational rules is innate, and claims instead that this system consists of parameters that can be set at specific values depending on the experience encountered.

His views, although extremely influential, have not been uncontested. (See, e,g., Lyons, 1991, and Harris, 1993.) Furthermore, Agassi's chapter in this part is also critical of several aspects of Chomsky's approach. For instance, Agassi is skeptical of Chomsky's method of analyzing language, on the grounds that it does not take account of certain lessons taught by modern logic. He also criticizes another fundamental Chomskian tenet, when he throws doubt on the idea that linguistic (or, more specifically, grammatical) competence is innate. In contrast, this part's third chapter, by Green and Vervaeke, cites recent empirical evidence in support of Chomsky's general point of view.

According to Agassi, Chomsky's account of the innateness of language is very similar to the ethologist K. Lorenz's ideas of innate releasing mechanisms—that is, systems that only need one or a few limited experiences of a specific kind to begin functioning. The most well-known example is ducklings' inborn mechanisms that lead them to follow and be imprinted on the first moving object they encounter during a critical period in their development. Chomsky speaks (e.g., in 1988) of the innate systems being triggered but not changed by experience, thus inviting such comparisons. Agassi is apparently criticizing Chomsky's underlying idea of essentialism—namely, that there is a fixed and invariant language system unique to, and shared by, all humans. By contrast, Agassi sees learning as an open-ended quest where beliefs, etc. are replaced by other beliefs through the forming and testing of hypotheses in contexts that depend on the particular circumstances encountered. Learning a language is not unlike learning of any other sort, and utilizes the same general cognitive principles.

Thus, the issue here is not just whether language is innate, but also what exactly language is like. For instance, is language one of many different mental "systems" with its own mechanisms and ways in which it is acquired? This is the view of Chomsky and Green and Vervaeke, but not of Agassi. The answer to this question not only has consequences for how an individual learns language, but also for how language evolves, as shown in Green's and Vervaeke's chapter. Chomsky himself rejects speculation about the evolution of language as mere "story-telling" (see "Language and Cognition" in Part One). Yet, it seems to Green and Vervaeke that it is possible to tell a plausible story about this matter along Chomskian lines. (However, it also is clear that many different plausible stories can be told. See, e.g., Hattiangadi, 1987, Donald, 1991, and Wallman, 1992.) According to Green and Vervaeke, Bickerton's (1990) idea of proto-languages is a possible precursor of fully evolved language. In other words, the former are languages that lack grammar and only consist of content words like "tree" and "rock," without any function words like "the" and "by." This implies that language is not only a separate "system" or module, but can be broken down into many different sub-systems or separate parts.

Green and Vervaeke here are relying on the idea that it is possible to break down language into discrete modules—that is, proto-language and grammar. Although there are clear cases of persons who possess a proto-language (e.g., 18–30 month-old-children and speakers of pidgin languages), the question remains whether it is possible to acquire and use these proto-languages without the support and structuring of the actual communicative and social use of language. Clearly language is an extremely complex phenomenon involving many different abilities both innate and learned. The attempts of Chomsky, Bickerton and others (e.g., Lieberman, 1991) to isolate and study different aspects of exactly what is involved in language acquisition has contributed to our understanding. But the problem remains: How do all these different aspects come together in the skillful, creative, meaningful, and communicative use of language?

To paraphrase Meno's paradox of how a person can learn something he or she does not already know, let us ask how is it possible for an organism with many subsystems or mechanisms relevant to language, all of which lack meaning, to acquire a meaningful language. Can a study of the internal functioning of individual, "ideal" minds overcome this discontinuity? Is Chomsky's demarcation of language fruitful; or is it hindering progress by neglecting the task of trying to understand how concrete language skills develop? Can language and not only parts of it be understood in terms of individual computations; or is language essentially a social activity, which requires going beyond the biological and psychological aspects of an individual to include normative and social considerations as well? These questions are discussed in Part Five, which, like this part, focuses on language as a paradigm case, or chief example, for understanding mental life.

References

Bickerton, D. 1990. *Language and Species*. Chicago: University of Chicago Press.

Chomsky, N. 1959. "Review of B.F. Skinner's 'Verbal Behavior.'" *Language*, 35, pp. 25–58.

Chomsky, N. 1988. *Language and Problems of Knowledge*. Cambridge: MIT Press.

Donald, M. 1991. *Origins of the Modern Mind*. Cambridge: Harvard University Press.

Gardner, H. 1985. *The Mind's New Science: A History of the Cognitive Revolution*. New York: Oxford University Press.

Harris, R.A. 1993. *The Linguistic Wars*. New York: Oxford University Press.

Hattiangadi, J.N. 1987. *How Is Language Possible?* LaSalle, IL: Open Court.

Lieberman, P. 1991. *Uniquely Human: The Evolution of Speech, Thought and Selfless Behavior*. Cambridge: Harvard University Press.

Lyons, J. 1991. *Chomsky*. London: Fontana Press.

Pinker, S. 1994. *The Language Instinct*. New York: Morrow.

Skinner, B.F. 1957. *Verbal Behavior*. New York: Appleton-Century-Crofts.

Wallman, J. 1992. *Aping Language*. Cambridge: Cambridge University Press.

9

Noam Chomsky

Language from an Internalist Perspective

These remarks have a narrow and a broader concern. The specific focus is the character and scope of an approach to language that arguably underlies much current empirical work, and some questions about its status. The broader concern, which I will mention only briefly and inadequately, has to do with what lies beyond. Many questions remain untouched, but it is not clear that they are amenable to theoretical inquiry in any terms, at least for now. Perhaps they are of the kind that can be raised, but without serious intent, such as: "Why do things happen the way they do?," or "How do things work?"—pseudo-questions, to be understood only as indicating areas that one might want to explore, not as requests for a specific answer.

The approach I have in mind seeks to integrate aspects of the study of language within the natural sciences. We are investigating a certain biological system, seeking to discover its bounds, its nature, and its functions. Like any such inquiry, this one begins with certain phenomena and queries that seem to cohere, though we expect the boundaries of relevance to shift as inquiry proceeds. We find that in the respects that now concern us, Peter is rather like me and we both differ from Pierre and Wang, though in different ways. Furthermore, we all differ dramatically from chimpanzees and pigeons, and even more so from rocks. These characteristics relate to specific parts of the body, in particular, the brain; in fact, parts of the brain, though here we enter uncharted territory. We hope to discover the relevant properties of these parts of the brain, the structures and elements that enter into their functioning, the principles they observe, the species characteristics that shape the systems that develop and determine the range of their variation, the effects of environment, and so on.

Proceeding in this way, we discover an integrated system of elements and properties, one that seems unlike others found in the study of humans, or organisms generally, in interesting respects. We thus postulate the existence of a subcomponent of the brain, which we can call the human language faculty. It has a genetically determined initial state, common to the species with marginal variation that we may ignore. Adapting a traditional term to a somewhat new usage, we refer to the study of this initial state as "universal grammar" (UG). We find that this initial state changes in a systematic way through the early years of life, reaching a relatively steady state, which then undergoes changes of a very different kind, restricted largely to certain subtypes of lexical items. It may be that different processes and systems are operative here, yielding a product that is not homogeneous.

As in any rational inquiry, we put to the side a wide range of interfering factors. Thus, each person lives in a world of overlapping communities and will tend to adapt in one or another way to its structures of authority and deference, depending on personal factors and on shifting goals and interests, with variations so multifarious and chaotic that we have entered the domain of "How do things work?" Dismissing such issues, we focus attention on the initial and the steady states, the transition, and ways in which properties of these systems condition their use. Again adapting a traditional term, we call the theory of the steady state its "grammar."

Given the state of his language faculty, Peter expresses himself and interprets what he hears in certain ways, not others. He is rather like Mary in these respects, but quite unlike Pierre or Wang. What are these respects? We now proceed down a particular theoretical track, facing more empirical risk at each stage.

Proceeding to articulate the properties and structure of Peter's language faculty, we take its steady state to incorporate a certain generative procedure L that determines an infinite class of structural descriptions. Each structural description is a maximal system S of phonetic, structural, and semantic properties; maximal, in that L generates no supersystem of S (a superset, if S is as unstructured as a set). I will understand a "linguistic expression" to be such a structural description. Thus, a linguistic expression is the full array of phonetic, structural, and semantic properties determined by L. The expression "John is painting the house brown," for Peter, is one array of properties determined by L: properties relating to sound, meaning, and structure. I will use the term "signal" to refer to manifestations of Peter's linguistic expressions in particular circumstances (spoken, written, signed, whatever); and ambiguously, to the set of such manifestations of a given linguistic expression, repetitions relative to L; speech acts are manifestations of linguistic expressions in a broader sense.

We will say that Peter has the *I-language L*, where the terminology is chosen to indicate that this conception of language is internalist, individual, and intensional. It is internalist and individual in the obvious sense, and intensional in that it is concerned with the specific nature of the generative procedure. Mary might have a different I-language that generates the same set of linguistic expressions as Peter's. The correct grammars for Peter and Mary will differ accordingly.

Peter's I-language is a (narrowly described) property of his brain; we may think of it as a relatively stable element of transitory states of the brain. Each structural description determined by this I-language can be regarded as an array of instructions or conditions on actions such as articulating, interpreting, referring, and so on. Another part of the inquiry, then, is to discover the various performance systems among which the I-language is integrated. There is nothing intrinsic to the I-language that tells us that it should be taken to be a language. Some other organism might, in principle, have the same I-language as Peter, but embedded in a performance system that uses it for locomotion. We are studying a real object, a human being, whose I-language happens to be integrated into performance systems that yield such actions as articulation, expression of beliefs and desires, referring, describing, and so on. For such reasons, our topic is the study of human language.

In turning to performance systems, we have to resist a temptation to descend to vacuity. Thus, we might consider Peter as an interpreter in Donald Davidson's sense, that is, a person who seeks to figure out what Mary or Pierre or Wang have in mind when they speak. To achieve this end, Peter will use any information, artifice, background assumption, guesswork, or whatever, constructing what Davidson calls a "passing theory." Davidson concludes that there is no use for "the concept of a language" serving as a "portable interpreting machine set to grind out the meaning of an arbitrary utterance"; rather, we need something more mysterious and holistic, "the ability to converge on a passing theory from time to time." We thus are led to "abandon . . . not only the ordinary notion of a language, but we have erased the boundary between knowing a language and knowing our way around in the world generally." There is no linguistic competence because "there are no rules for arriving at passing theories." We "must give up the idea of a clearly defined shared structure which language-users acquire and then apply to cases." "There is no such thing as a language," so a recent study of Davidson's philosophy opens, with his approval.[1]

These observations about human interaction are accurate enough: interpreting what a person says is on a par with knowing one's way around in the world. But the startling conclusions do not follow. To ask "What is the nature of an interpreter?" is like asking "How do things work?" These are not genuine questions, but only guides to possible topics of inquiry. Exploring these topics, we hope to find real systems that enter in some way into the complexity and incoherence of mere phenomena. Most of these phenomena we will then discard as irrelevant, selecting some in theory-relative terms and often contriving them by experiment; in short, considering only data that can be taken as evidence for what we take to be real systems of the brain. These are our actual concern. We find good evidence that there is a clearly defined ("portable") system that language-users acquire and then apply to cases, though it is shared with others only in the sense of matching; that is the I-language. And we find elements of real systems of performance, including modes of expression and interpretation under idealized conditions. The study of human interaction generally is not a topic to be addressed in such an undifferentiated form, and from the fact that a passing theory may involve anything, we can conclude nothing.[2]

One standard hypothesis is that the performance systems include a parser, to be distinguished from a Davidsonian interpreter. An interpreter takes as input a signal and an array of circumstances that is open-ended and beyond inquiry; a parser takes as input a signal alone, and assigns to it a characterization (call it a "percept") that relates to the structural descriptions generated by the I-language in some manner to be determined. The parser incorporates the I-language along with other properties: perceptual strategies, memory structures, and the like. The evidence for the existence of I-language is quite strong; for parsers, less so. Whether this is the proper hypothesis is an open question. It need not be.

It is commonly asserted that a condition on UG is that it account for the fact that parsing is "easy and quick." But that is not true, so there is no such condition. Parsing is often slow, or in error, in the sense that the percept is markedly different from the structural description. There are many types of such "parsing failures"; one variety is what are called "garden path sentences." True, normal usage keeps to expressions that are readily produced and parsed, but that observation verges on triviality: you use what is usable. Some creature might exist with an I-language and performance systems that allow only restricted use of its linguistic capacities, or even none at all. We are, in fact, creatures of the former type, so it appears. A comparable error is common in the theory of language acquisition, with natural languages being defined as those that can be acquired easily and quickly under normal conditions. True, the existing languages will be of this type. But the initial state of the language faculty might, in principle, permit many I-languages that are not readily accessible, or not accessible at all. In fact, there could be a creature with human-type performance systems and a full-fledged language faculty determining a class of I-languages none of which is accessible to it. In that case, indirect means would be required to discover that the creature has a language faculty. Recent work in the so-called principles-and-parameters framework suggests that natural languages *are* readily accessible, but that is an empirical discovery, and a rather surprising one, not a matter of definition.

According to this approach, which I will henceforth assume, the initial state of the language faculty consists of a rich system of principles, with some parameters of variation. The fixed principles are highly restrictive. We may take an I-language to consist of a computational system generating the form of a structural description and a lexicon specifying the minimal word-like items that appear in it. Then it may be that computational systems are virtually identical for all languages. Variation appears to be restricted to the lexicon, in fact, to narrow parts of the lexicon. Jerry Fodor, for one, has argued forcefully that the substantive part of the lexicon tolerates little variation. There are, of course, differences in how phonetic, structural, and semantic features are associated in a linguistic expression, and it seems that there are certain semantic spaces that can be divided up differently, within limits; familiar examples are Spanish *ser* and *estar* versus the English *copula*; or French *savoir* and *connaître* versus English *know*. But more significant differences seem to be restricted to elements that belong to inflectional systems and the like. Languages also differ in how the computations "feed" the system of articulation.[3] Thus, there is good reason to believe that English has essentially the kind of case system one finds in Latin or Greek, but it is not overt: the computations involving cases do not feed the signaling apparatus, so the cases are only present to the mind, so to speak, though their consequences are manifest in many ways. Conversely, Japanese seems to share with English, perhaps every language, a rule that fronts question words to yield such expressions as "who do you think John met," but in Japanese the rule applies only in the mental computation, not feeding the phonological system. The overt form thus looks like the English "you think that John met someone," with the question word, which is like an indefinite, in place.

A rational Martian scientist, studying humans as we do other organisms, would conclude that there is really only one human language, with some marginal differences that may, however, make communication impossible, not a surprising consequence of slight adjustments in a complex system. But essentially, human languages seem cast to the same mold, as one would expect, considering the most elementary properties of the steady state and the manner in which it develops. The conclusion conflicts with what had long been believed, often with surprising dogmatism. A current variant is that while sound and structure may be subject to restrictive initial conditions, meanings are qualitatively different, largely shaped by environment and experience. There is no basis for such speculations, to my knowledge, wherever we have even limited understanding.

Summarizing so far, we have reason to believe that there is a narrowly circumscribed and richly structured initial state of the language faculty, leaving limited options for attainable I-languages. These are embedded in systems of use: for articulation, interpretation, expression of thought, and other human actions. About most of these, little is known.

The internally determined properties of linguistic expressions can be quite far-reaching. Take again the expression "John is painting the house brown," a certain collection of structural, phonetic, and semantic properties. We say it is the same expression for you and me only in the sense in which we might say that your circulatory or visual systems are the same as mine: they are similar enough for the purposes at hand. One structural property of the expression is that it consists of 6 words. Other structural properties differentiate "John is painting the house brown" from "John is painting the brown house," which has correspondingly different conditions of use. A phonetic property is that the last two words, "house" and "brown," share the same vowel; they are in the formal relation of assonance, while "house" and "mouse" are in the formal relation of rhyme, two relations on linguistic expressions definable in terms of their phonological features.[4] A semantic property is that one of the two final words can be used to refer to certain kinds of things, and the other expresses a property of these. Here too there are formal relations expressible in terms of features, for example, the relation between "house" and "building." Or, to take a more interesting property, if John is painting the house brown, then he is applying paint to its exterior surface, not its interior; a relation of entailment holds between the corresponding linguistic expressions.

Viewed formally, relations such as entailment have much the same status as rhyme; they are formal relations among linguistic expressions, which can be characterized in terms of their features. Certain relations happen to be interesting ones, as distinct from many that are not, because of the ways I-languages are embedded in performance systems that use these instructions for various human activities.

Some properties of an expression are universal, others language-particular. It is a universal phonetic property that the vowel of "house" is shorter than the vowel of "brown"; it is a particular property that the vowel in my I-language is front rather than mid, as it might be for you. The fact that a brown house has a brown exterior, not interior, appears to be a language universal; the fact that "house" is distinguished from "home" is a particular feature of the I-language. In English, I return to my home after work; in Hebrew, I return to the house.

Given the performance systems, the phonetic features impose narrow conditions on use (articulation, perception, etc.). The same is true of the semantic features. I've already mentioned one example: the expression "brown house" selects the exterior surface, not the interior one. The same is true of "container" words of a broad category, including ones we might invent: "box," "airplane," "igloo," "lean-to," etc. Other usages of such words are similar. If I see the house, I am seeing its exterior surface, not its interior one. If I am inside an airplane, I see it only if I can look out the window and see the surface of the wing, or if there is a mirror out the window that reflects its exterior surface. However, the house is not just its exterior surface, a geometrical entity. If Peter and Mary are equidistant from the surface, Peter being inside the house and Mary outside it, Peter is not near the house, but Mary might be, depending on the current conditions for nearness. So the house involves its exterior surface and its interior. But the interior is unspecified; it is the same house if I fill it with cheese—though if I clean the house I may interact only with things in the interior space. So the house is, somehow, the pair of an exterior surface and an interior space (with complex properties). Of course, the house itself is a concrete object; it can be made of bricks or wood, and a wooden house does not just have a wooden exterior. If my house used to be in Philadelphia, but is now in Boston, then a physical object was moved. In contrast, if my home used to be in Philadelphia, but is now in Boston, then no physical object need have moved, though my home too is concrete, not abstract, whether my home is understood as the house in which I live, or the town, or country, or universe (imagine a space traveler returning from a different universe); "house," in contrast, is concrete in another sense. The "house"-"home" difference has numerous consequences: I can go home, but not go house; I can live in a brown house, but not a brown home; in many languages, as partially in English too, the counterpart of "home" is adverbial.

Even in this trivial example, we see that the internal conditions on meaning are extremely rich, complex, and unsuspected—in fact, barely known. The most elaborate dictionaries do not dream of such subtleties. There seems at first glance to be something paradoxical in these descriptions. Thus, houses and homes are concrete, but from another point of view, are considered quite abstractly, though abstractly in very different ways. It is not that we have confused ideas, or inconsistent beliefs, about houses and homes, or boxes, airplanes, igloos, etc. Rather, lexical items provide us with a certain perspective for viewing what we take to be the things in the world; they are like filters or lenses, providing a certain way of looking at things.

The same is true wherever we inquire into lexical properties. London is not a fiction, but viewing it as London—that is, through the perspective of a city name, a particular type of linguistic expression—we allow that under some circumstances, it could be completely destroyed and rebuilt somewhere else, years or even millennia later, still being London, that same city. Two such perspectives can fit differently into Peter's system of beliefs, as in Kripke's puzzle. For some purposes, we construct a picture of the world that is dissociated as much as possible from these "common-sense" perspectives (never completely, of course; we cannot become something other than the creatures we are). In the natural sciences, one will not think of objects as homes, or houses, or cities, or persons; there will be no such thing as my home or London in this account. If we intermingle such different ways of thinking about the world, we may find ourselves attributing to people strange and even contradictory beliefs about objects divorced from the conceptions of the I-language in which the beliefs are expressed, a situation that will seem even more puzzling if we suppose that certain terms have a reference fixed in the "common language," of which people have only a partial grasp. Problems will seem to deepen further if we abstract from the background of individual or shared beliefs that underlie normal language use. All such moves go beyond internalist limits. I'll return to a few questions about them.

Pursuing the inquiry into lexical properties, we find a rich structure of purely internalist semantics, with some interesting general properties, and numerous formal semantic relations, including, in particular, analytic connections. Furthermore, a great part of this semantic structure appears to derive from our inner nature, determined by the initial state of our language faculty, hence unlearned and universal for I-languages. Much the same is true of phonetic properties and structural properties generally. In short, internalist semantics seems much like the study of other parts of the biological world. We might well term it a form of syntax, that is, the study of elements of symbolic systems. The same terminology remains appropriate if the theoretical apparatus is elaborated to include mental models, possible worlds, discourse representations, and other systems of postulated entities that are still to be related in some manner to things in the world.

When we move beyond lexical structure, the conclusions about rich universal principles are reinforced. Consider such expressions as "he thinks John is a genius" and "his mother thinks John is a genius." In the second, the pronoun may be referentially dependent on "John" in the sense clarified by James Higginbotham particularly; in the first, it cannot (though it might be used to refer to John, an irrelevant matter). The principles underlying these facts also appear to be universal, at least in large measure; they yield a rich array of conditions on semantic interpretation, on intrinsic relations of meaning among expressions, including analytic connections. Furthermore, in this domain we have theoretical results of some depth, with surprising consequences. Thus, the same principles appear to yield the interpretation of the expression "John is too clever to expect anyone to catch," which implies that we would not expect anyone to catch John, not that John would not expect anyone to catch us. That we have such I-languages is a matter of contingent fact; some other organism might have a system like ours but with principles that entail that the very same linguistic expression (with all the same intrinsic properties) has a different range of consequences.[5] It is interesting, if true, that these curious facts reduce in part to principles so general that they apply to the earlier examples as well.

In brief, the theory of I-language yields significant results concerning properties of expressions; and given the performance systems in which I-language is embedded, concerning the way language is articulated, perceived, used to communicate and refer, and so on. The internalist inquiry carries us far into the study of sound and meaning, still leaving other classical questions untouched.

We can clarify the notion of I-language by contrasting it with various others. One notion familiar from the study of formal systems is the set of well-formed signals. Let us call this E-language, where E is to suggest extensional and externalized, not a component of the brain, but on a par with the infinite set of structural descriptions generated by the I-language. The latter abstract entity has a physical interpretation; it is an articulation of a real property of the brain, in what appear to be the right terms, that is, those proper for an explanatory theory of the brain. It is widely assumed that E-languages also have some real status, but there is little reason to believe that. Despite some expository passages that, in retrospect, have been most misleading, it has always been assumed in the more technical work in generative grammar that there is no designated infinite set of well-formed signals, no E-language.[6] There have been no attempts to characterize "E-language," apart from some highly theory-internal early work, long abandoned. Furthermore, there is no known theoretical gap, no inadequacy of explanation, that would be remedied by constructing such a concept, again a reason for skepticism about its significance. Quine and others have suggested operational criteria to identify well-formed signals, but operational tests are a dime a dozen, and it must be shown that there is some interest in the concept they characterize. In the case of E-language, no such argument has been proposed. Until this deficiency is overcome, questions about generative capacity, about the alleged "richness" or "excessive richness" of various theories in this respect, and other topics that have been vigorously discussed, seem devoid of empirical content.

We may also contrast I-language to the notion "common language" or "shared language" that is assumed in virtually all work on externalist semantics. The status of such notions is highly questionable. A standard remark in an undergraduate linguistics course is Max Weinreich's quip that a language is a dialect with an army and navy; and the next lecture explains that dialects are also nonlinguistic notions, which can be set up one way or another, depending on particular interests and concerns. There are various subcultures with their particular practices, demands, and authority structures; there are colors on maps and ways of associating oneself with others for one or another purpose. Under some conditions, one can select privileged systems to which some might choose to conform. This can be done in virtually any way one likes, depending on interests and circumstances, often raising questions of great human importance. But there are no real entities to be discovered, a conclusion supported by both descriptive and theoretical considerations. The descriptive ones are familiar; the theoretical ones are, of course, theory-internal. If the principles-and-parameters approach is correct, then each language is a point in a space of possible parameter settings, and there is no metric that sets up privileged neighborhoods. Extrinsic conditions may vary as far as imagination carries us. In these respects, as Wittgenstein said, a language is a form of life; and questions about such an object are, again, on a par with: "Why do things happen?"

The problem is not one of vagueness; rather, of immense underspecification. To ask whether Peter and Mary speak the same language is like asking whether Boston is near New York or whether John is almost home, except far more obscure, because the dimensionality provided by interest and circumstance is far more diverse and complex. In ordinary human life, we find all sorts of shifting communities and expectations, varying widely with individuals and groups, and no "right answer" as to how they should be selected. It follows that no clear sense can be given to notions such as "partial knowledge of language," or "misuse of language" (apart from an I-language variant). We cannot evaluate, because we cannot understand, the idea expressed by Michael Dummett and many others that the "fundamental sense" of the concept of language is the sense in which Dutch and German are different languages, each of them a particular social practice "in which people engage," a practice that "is learned from others and is constituted by rules which it is part of social custom to follow," existing "independently of any particular speakers," who typically have a "partial, and partially erroneous, grasp of the language" (Dummett). These proposals will remain unclear until we are given some idea of what such a common language is supposed to be, and why the bounds are fixed one way or another. Such questions tend to be dismissed much too easily, in my opinion.

We may use such terms as "English" or "German" as we use "household pet" or "large animal." A term such as "giraffe" or "nematode" is far better defined. We may say, informally, that giraffes have evolved longer necks over time, and that the "wiring diagram" for the nematode neural system has been worked out, referring in the first case to a change in the distribution of traits in populations with a certain kind of historical connection, and in the second to some abstraction from individual nematodes. Similarly, we can say that English has changed over the centuries and that English differs from Chinese in the rule of question-formation; but we must be careful, far more than in the case of giraffes and nematodes, to avoid illegitimate reification.

It has sometimes been argued that the notion of common public language or "public object" is required to explain the possibility of communication. Thus, if Peter and Mary do not have a "shared language," with "shared meanings" and "shared reference," then how can Peter understand what Mary says? One forthcoming book argues that linguists can adopt an I-language perspective only "at the cost of denying that the basic function of natural languages is to mediate communication between its speakers," including the problem of "interstage communication" that enters into the transition from one to another I-language in so-called "language learning."

But these conclusions do not seem well-founded. Successful communication between Peter and Mary does not entail the existence of shared meanings in a public language, any more than physical resemblance between Peter and Mary implies the existence of a public form that they share. As for "the cost of denying that the basic function of natural languages is to mediate communication," it is unclear what sense can be given to an absolute notion of "basic function," for any biological system; and if this problem can be solved, we then ask why "communication" is the "basic function."

But there is no point pursuing this unappealing course, because there seems no problem in accounting for communication in internalist terms, insofar as an account is possible at all. Peter presumably listens to Mary on the assumption that she is identical to him, modulo M, some array of modifications. He then seeks to work out M (though much of this work is presumably automatic and unreflective). Sometimes it is easy, sometimes hard, sometimes hopeless. Any method can be used. Insofar as Peter succeeds, he understands what Mary says as being what he means by his comparable expression. Establishing shared entities has no more explanatory value than postulation of shared shapes.

As for the transition problem, that is no more mysterious than the problem of how I can be the person I am, given the stages through which I have passed.

Conformity to "community practices" is sometimes held to be essential for attribution of rule-following, but the arguments are not very persuasive. If Peter says "bring," "brang," "brung," we attribute to him the obvious rule. Similarly, if he understands "disinterested" to mean "uninterested," as most people do, or topicalizes in an unexpected way. Typically, we attribute rule-following in cases of notable *lack* of conformity to prescriptive practice. We can select a community of people like Peter in the relevant respects, or unlike him. In the former case, community practice makes no contribution to attribution of rule-following; in the latter, it gives the wrong conclusion. Since communities are free for the asking, we can always select communities that will be irrelevant or inadequate to the purpose of rule attribution, but these seem to be the only cases.

Alleged social factors in meaning generally have a natural internalist interpretation. If Bert complains of arthritis in his ankle and thigh, and is told that he is wrong about both, but in different ways, he may (or may not) choose to modify his I-language to that of the doctor's. Nothing substantive seems missing from this account.

Similarly, ordinary talk of whether a person has mastered a concept requires no notion of common language. To say that Bert has not mastered the concept "arthritis" or "flu" is simply to say that his I-language is not exactly that of people we rely on to cure us—a normal situation. If my neighbor Bert complains of his arthritis, my initial posit is that he is identical to me in this usage, and I will introduce modifications to interpret him as circumstances require; reference to a common language sheds no further light on what is happening between us. We gain no insight into Peter's improving his Italian by supposing that there is a fixed entity that he is approaching. Rather, he is becoming more like a wide range of differing people; both the modes of approximation and the selection of models vary as widely as our interests.

We also want to make sense of apparently valid arguments, e.g., the inference from "Peter understands English" and "Mary's language is English" to "Peter understands Mary's language." The inference translates as "Peter has I-language L and a functioning parser and interpreter; Mary has I-language L' and a functioning production system; and L and L' are sufficiently similar so that Peter, within some range, assigns percepts to Mary's signals that match (closely enough) Mary's structural descriptions. Nothing more can be said, or need be desired.[7]

For familiar reasons, nothing in this suggests that there is any problem in informal usage, any more than in the ordinary use of such expressions as "Boston is near New York" or "John is almost home." It is just that we do not expect such notions to enter into explanatory theoretical discourse.

It may be that a kind of public character to thought and meaning is guaranteed by the near uniformity of initial endowment, which permits only steady states that vary little in relevant respects. Beyond this, their character varies as interests and experience vary, with no clear way to establish further categories, even ideally. Appeals to common origin of language or speculations about natural selection, which are found throughout the literature, seem beside the point for reasons discussed elsewhere.

There are, of course, many differences among the various kinds of features of I-language and the ways they enter into performance systems. But there are analogies as well, and sometimes issues can be clarified by focusing on these. Thus, it is common to ask about the meaning of a linguistic expression, a question regarded as problematic. Comparable questions can often be raised about structure and sound, and in this context they seem less perplexing, a fact that may indicate, and may help us come to see, that some apparent mysteries of semantic inquiry are illusory.

Take, say, the question of meaning holism. A study critical of this thesis opens by explaining why it seems so plausible. Thus, the expressions "bark" or "flying planes" mean one thing in one context and another in a different context, and "Empedocles leaped" means that he jumped, in English, and that he loves, in German. So meaning is something that words have in sentences, and that sentences have in languages. The meaning of a word is determined by the whole language, or maybe even whole belief systems. We should, so it seems, abandon the quest for a theory of meaning as quixotic, at best on a par with determining why things happen.

The argument takes the linguistic expressions "bark," "flying planes," "Empedocles leaped," etc., to be objects lacking semantic properties (also of course finessing various questions about differences across I-languages). We then ask about the meaning of these objects, that is, we ask what semantic properties are associated with them. Quite analogously, we could construct objects from structural descriptions by eliminating their phonetic features, then asking about the sound of these objects. We now say that "bark" is sounded one way in sentence-medial position and a different way in sentence-final position, and that "Empedocles leaped" sounds one way in English and another way in German. We might also go on to construct an object with only some of the sounds dropped, just as some of the semantic features can be, thus considering "cat" with all its features apart from the final dental stop. We could then say that "cat" terminates /t/ in English. Proceeding in this manner, we arrive at a theory of sound holism: it is only the whole sentence, or the whole language, or maybe even the language plus performance systems, that really has sounds. So the study of phonology is hopeless, infected by lethal holism.

The latter argument impresses no one. By definition, it is a property of the I-language that features are associated in one way rather than another in expressions; it would be a different I-language if there were a different association. Of course the performance system treats expressions differently in different contexts. From such truisms we conclude nothing about sound holism; or about meaning holism, the argument being parallel. Maybe there is some sense to meaning holism, but arguments of this kind have no force.

Terminological artifacts may engender other unnecessary questions. Thus, the elements of structural descriptions are often called "mental representations": the expression "brown house" has the phonetic representation X and the semantic representation Y. There are reasons for the usage, but it can be misleading. Suppose we also adopt the common construal of a linguistic expression as a structural description without its semantic features (or as the corresponding signal). We are now tempted to raise a metaphysical problem about linguistics, which I borrow from a paper by Higginbotham.[8] The approach to language as a system of mental representations, hence as part of psychology, "is confused, because it conflates questions about what is apprehended, language, with questions about the means of apprehension." We do not confuse the mental representation of a tree with a tree, so we should not confuse the mental representation of a sentence with a sentence. The critique goes on to propose a "weak conceptualism," in which the structure apprehended is the "platonic shadow" of the representation; the properties of each are read off the properties of the other, and the difference between them is only "pedantically insisted upon," for the purposes of philosophical clarity.

The proposal at once arouses skepticism. Why should it be necessary to make the pedantic distinction at all? Why should we be led to set up a theoretical entity that has no role in explanation, that opens no new lines of inquiry and allows no new empirical questions, that in fact has no properties other than those determined by another entity that we take to be well established, with a real role in explanation and inquiry and no associated problems beyond the usual empirical ones of natural science? That seems odd, and I think it *is* odd; and unnecessary, the result of a line of thinking induced by misleading terminology.[9]

Suppose we divorce the notion "mental representation" from unintended connotations, taking these objects to be elements of the structural descriptions generated by the I-language, which interface with performance systems as a fact of human biology. Suppose we take a linguistic expression to be the collection of all the features determined by the I-language, not some subcollection, excluding semantic features. Now the problem just posed is unformulable. True, the mental representation of a tree is not a tree, but there is no question about confusing the mental representation of a sentence with a sentence. Rather, this would be to confuse the full system of features that constitutes a linguistic expression with one part of that system, excluding its semantic features. As for the question of what language Peter apprehends, the answer is that Peter has the I-language *L*, meaning that the relevant part of his brain is in that particular state. There is no more reason to seek platonic shadows than there would be in the study of the visual system or continental drift.

One can also be misled by the fact that we do, plainly, have beliefs about language. Thus, you and I believe that "house" rhymes with "mouse" and that "flying planes" can be interpreted in several different ways. These are beliefs about English, construed in some manner. One might then be tempted to conclude that there is something, English, about which we have these beliefs. And since we plainly do not have beliefs about I-language, English must be something else, perhaps a Platonic object that we apprehend.

But this line of reasoning is also faulty. You and I have beliefs about health, about style of dress or dining, and about national rights. Thus, we may believe that oat bran is good for you, that you should wear a jacket to dinner, that white wine goes with fish, or that the Welsh have the right of self-determination. But it does not follow that there are things—health, styles of dress or dining, national rights—that we have beliefs about. Such notions as "health," "style of dress or dining," "national rights" pick out certain topics of discourse and inquiry. Within these topics, we may ask particular questions, and sometimes even provide particular answers that may be true or false, typically in some way highly dependent on interests and circumstances. But, uncontroversially, we need not construct such Platonic entities as health to make sense of such discourse. The same is true of such notions as common shared language, or the meaning of an expression.

Consideration of the analogies of sound and meaning raises doubts about another standard approach to the theory of meaning, namely, that we should consider meaning as use, where the use of an expression is to be understood as some idealized array of dispositions. Suppose someone were to make a similar proposal about sound, proposing that the sound of an expression is its use, given by an array of dispositions to articulate the expression in particular ways under varying circumstances. We may therefore dispense with complicated inquiries that seek to determine universal phonological features, the laws governing them, the principles of syllable structure, etc. We need not ask why "house" and "mouse" rhyme, or why the vowel is longer in "brown" than in "house," or why the stress is on a different syllable in "residue" and "residual." All is exhausted when we have described the pattern of use.

That proposal would be dismissed at once. An argument has to be given to show that a similar proposal is more reasonable in connection with the strange semantic properties of "house," "home," or "John is too clever to expect anyone to catch." I doubt that it is. It seems to me, rather, that we see here the residue of a pernicious behaviorism that continually affects the domains traditionally assigned to the mental.

So far, I have kept to discussion of states of the brain. As we proceed to articulate their properties and determine the principles that hold about them, we construct theoretical entities that are justified, as usual, by the role they play in providing explanation and insight. Sometimes this level of discussion is called "mental." There is no reason to object to the usage, as long as we are not misled into supposing that some new domain is postulated apart from the physical, and that there is now a problem about finding the physical realization of the mental objects we construct. That problem could arise only if we are told what are the bounds of the physical or the material world, from which these objects are excluded. It is not clear what any of this might mean, either here, or in other areas of natural science.

Of course, we want to integrate various modes of description and understanding of the physical world. We have an account of phenomena of language in terms of I-language, structural descriptions, semantic and phonetic features, performance systems, and so on; we would like to integrate it with accounts in terms of the brain sciences. But this problem, call it the unification problem, is quite general, and arises everywhere, not just here. In ongoing inquiry in the natural sciences, the question is typically pointless. The problems of ontological commitment will be settled not by philosophical disputation, but by progress toward unification. This is by now pretty much taken for granted with respect to the hard sciences. It is a rare philosopher who would scoff at the weird and counterintuitive notions of modern physics as contrary to right thinking and therefore untenable. A more standard view is expressed by Michael Friedman, citing Kant's *Prolegomena*: "the philosophers of the modern tradition," from Descartes, "are not best understood as attempting to stand outside the new science so as to show—from some mysterious point outside of science itself— that our scientific knowledge somehow 'mirrors' an independently existing reality. Rather, (they) start from the *fact* of modern scientific knowledge as a fixed point, as it were; and their problem is not so much to justify this knowledge from some 'higher' standpoint as to articulate the new *philosophical* conceptions that are forced upon us by the new science."[10] It is not clear why this reasonable stance should be abandoned when we turn to the more obscure problems raised by emerging disciplines.

Analogies are never perfect, but they may be helpful, so let us consider a few. Consider an account of certain phenomena in terms of valence, the Periodic table, complex organic molecules, etc., and an account of other phenomena in terms of genes and alleles. Early in the century, these levels of description and explanation had not been connected with accounts of the world in terms of particles and fields, or connected with one another. More recently, the unification problem was solved in both cases, though in different ways. Much of the study of life was reduced to known biochemistry, while physics was radically changed and "expanded" to become unified with chemistry, which, in fact, provided guidelines for the modifications of physical theory.

Let's take the more mundane example I mentioned earlier: nematodes, little worms with a three-day maturation period, with a wiring diagram completely analyzed. One recent study treats them as abstract computational devices belonging to a special class of asynchronous interacting automata, implementing certain algorithms, with computational and control structures viewed abstractly, organized in terms of abstract constraints and underlying organizing principles, some of them general (universal biology), most of them unknown.[11] The approach is highly modular, with separate models for development, structure, and various functions, ranging in "level of resolution" from phenomenological to molecular. The investigator, Charles Rockland, dismisses connectionist models as inadequate for this 800-cell organism with 300 neurons, because they abstract much too far from physical reality. He proposes rather that we treat neurons as cells that "interact via a wide variety of chemical 'informational substances,' including neurotransmitters, neuromodulators, neuropeptides, etc., acting over multiple characteristic distance and time scales," in part not through conventional synaptic junctions. Multiple models select different aspects of what the cells are doing. These systems, he assumes, are "physically realized and metabolically supported," but how, no one knows, and we may not even know where to look. Putting it differently, this level of description of nematodes and explanation of their behavior is as yet unconnected with others, and the mode of connection may involve reduction, expansion, or modification of several levels: the unification problem may take any course.

In this case, no one is tempted to construct pale shadows or Platonistic biology; to insist that the inquiry must take account of the social life of nematodes; to claim that there is no truth of the matter because there will be infinitely many theories consistent with the results of some arbitrarily selected experiment or some stipulation about physical reality; to postulate a "common nematode system" to whose principles each worm only partially conforms; to dismiss all of this apparatus in favor of dispositions of nematodes to do this or that in particular circumstances; to hold that if a theory of nematode behavior is given, it adds nothing to insist that "some mechanisms . . . must correspond to the theory" (Davidson); etc. These are all paraphrases drawn from current critical study of language and meaning, having no more plausibility, in my view, than in the study of nematodes.

In each case, we pursue various levels of explanation and inquiry in the effort to gain understanding. We naturally seek to solve the unification problem, integrating these levels: chemistry with physics, biology with chemistry, asynchronous interacting automata with cellular biology, I-language with something about brains. In each case, prior to success, we do not know whether the proper move is reduction, expansion, or modification of all accounts. Ontological questions are generally beside the point, hardly more than a form of harassment. The problems will be solved as more is learned. And, as unification proceeds, it may well lead to empirical assumptions, including ontological posits, that now seem outlandish, as often in the past.

In the case of language and other cognitive functions, it is common to try to relieve the sense that something is amiss with such slogans as "the mental is the neurophysiological at a higher level," taken as a kind of definition. Others go further, arguing that we dispense with "the mental" entirely, turning to the study of neurophysiology, perhaps adopting the doctrine of "eliminative materialism." Others propose that the specific abstractions of connectionism will somehow avoid problems posed by accounts in terms of computational systems.

These are strange moves, as we see by transferring them to other domains. From a naturalistic perspective, there are just various ways of studying the world, with results of varying firmness and reach, and a long-term goal of unification. We assess the credibility of assumptions at various levels on the basis of explanatory success. In the case of language, the most credible assumptions about what the brain is and what it does are those of the computational theories. We assume, essentially on faith, that there is also an account in terms of atoms and molecules, though without expecting the operative principles and structures to be identifiable at these levels. And with a much larger leap of faith, we also tend to assume that there is an account in neurological rather than vascular terms, though a look at the brain reveals blood as well as neurons. It may well be that the relevant elements and principles of the brain have yet to be discovered. Perhaps the much more firmly based computational theories of the mind will provide the guidelines for the search for such mechanisms, much as nineteenth-century chemistry provided crucial empirical conditions for the radical revision of fundamental physics.

The familiar slogan about the mental and the neurophysiological should not be taken as a definition of the mental, but rather as a tentative hypothesis about neurophysiology: perhaps the neurophysiological is the mental at a "lower" level, perhaps not. As of now, we have more reason to feel secure about the mental than about the neurophysiological. The idea that cognitive psychologists should drop the inquiry into rule systems in favor of the study of neurophysiology seems about as reasonable today as a proposal that embryologists should drop their inquiries in favor of superstring theory; arguably less so, given the status of the theories. As for eliminative materialism, the very doctrine remains a mystery until some account is given of the bounds of the material, and given that account, some reason why one should take it seriously or care if successful theories lie beyond its bounds. Perhaps it will turn out that connectionist models are more adequate for a system of 10^{11} neurons than one with 300, but one awaits an argument. The discussion of what it would imply about rule systems if such theories were to become available is as interesting as a debate over what it would mean for embryology if it were shown that some unstructured system could achieve the explanatory power that biologists seek in terms of their complex notions. All of these discussions appear to reflect a lingering dualism, which should have been long ago abandoned.

As noted, an internalist approach to language leaves fundamental questions about sound and meaning unexamined. While a rich texture of phonetic and semantic conditions and instructions for performance systems comes to light in these inquiries, questions remain about how they enter into human action. Thus, how does Peter's structural description "brown house" enter into his speech production, or perception, or his acts of referring? As is familiar, the conditions are "open-textured"; it rarely makes sense to expect strict definition or complete specification of truth conditions. Conclusions about the use of language that derive from considerations of "common shared language" seem highly suspect, if even intelligible, when pressed beyond their home in ordinary usage, where we have nothing much beyond particular descriptions. Considerations deriving from facts about the world do not seem to me more persuasive; here too judgments vary with circumstances, so much so as again to raise the suspicion that we are finding a topic to explore, not real notions to clarify.

Suppose, for example, that twin-Oscar arrives on earth and, wanting to quench his thirst, asks for water as he points to a glass of H_2O and a glass of Sprite. He is misreferring in the latter case, but is there an absolute decision in the former? Suppose that XYZ on twin-earth is D_2O, so that, if we rely on experts as we are enjoined to do, we will say that he is now asking for "light water," the same substance as his "heavy water" (Jay Atlas's example). Is it really true that in ordinary usage water is a substance and milk is not, or that for the ancients, water was a substance and earth, air, and fire were not, because of the way science has turned out? If twin-Oscar is a Frankenstein's monster with no experience, is he referring at all when he asks for a glass of water? Suppose by some unknown law of nature, everyone is suddenly replaced by such a twin. Are none of us referring, even though the world goes on exactly as before—a kind of Cartesian nightmare? As interests and conditions shift, it seems that a good deal of room remains for choice and decision in such cases, and one may question whether there is a "right answer" to be found. Note that we cannot appeal to intuitions with regard to invented technical notions such as "extension" or "denotation" or "reference of words," which mean what one says they do (and may or may not be more appropriate for the study of the topics to be explored than an entity designated "health," in a study of our beliefs about health).

Much of the motivation for particular decisions in externalist discussion derives from an effort to make sense of the history of science. Hilary Putnam argues that we should take the early Niels Bohr to have been referring to the quantum-theoretic conception of electron, or we would have to attribute to him only false beliefs. The same is true of pre-Dalton chemists speaking of atoms. And perhaps, on the same grounds, we would say that chemists pre-Avogadro were referring to what we call atoms and molecules, though for them the terms were interchangeable, apparently.

The discussion assumes that such terms as "electron" belong to the same system as "house" and pronominal anaphora, which is hardly obvious; a proper theory of language should probably be more differentiated than that. Suppose we grant the point for the sake of argument. Agreeing further that an interest in intelligibility in scientific discourse across time is a fair enough concern, still it is hard to see why it is a basis for a general theory of meaning; it is, after all, only one concern among many, and not a central one for the study of human psychology. Furthermore, there seem once again to be internalist paraphrases. Thus, we could say that in Bohr's earlier usage, he expressed a set of beliefs that were literally false, because there was nothing of the sort he had in mind in referring to electrons; but his picture of the world and articulation of it was structurally similar enough to later usage and beliefs so that we can distinguish his beliefs about electrons from a belief in angels. What is more, that seems a reasonable way to proceed.

It is worth noting that the internalist paraphrase just outlined is consistent with the intuitions of respected figures. The discussion about "electron," "water," etc., projects backwards in time, but we can project forward as well. Consider a discussion in 1950 about whether machines can think (understand, plan, solve problems, etc.). By standard externalist arguments, the question should be settled by the truth about thought: What is the essence of what we refer to when we say that John is thinking about his children, or solving a quadratic equation, or playing chess, or interpreting a sentence, or deciding whether to wear a raincoat? But that is not the way it seemed to Wittgenstein and Turing, to take two notable examples. To Wittgenstein, the statement that machines can or cannot think is not an empirical statement: "We can only say of a human being and what is like one that it thinks," maybe dolls and spirits; that is the way the tool is used. Turing, in 1950, wrote that the question whether machines can think

> may be too meaningless to deserve discussion. Nevertheless I believe that at the end of the century the use of words and general educated opinion will have altered so much that one will be able to speak of machines thinking without expecting to be contradicted.

Both Wittgenstein and Turing reject the standard externalist account. For Wittgenstein, the questions are just silly: the tools are used as they are, and if the usage changes, the language has changed, the language being nothing more than the way we use the tools. Turing too speaks of the language of "general educated opinion" changing, as interests and concerns change. In our terms, there will be a shift from the I-languages that Wittgenstein describes to new ones, in which the old word "think" will be eliminated in favor of a new word that applies to machines as well as people. To ask in 1950 whether machines think is as meaningful as the question whether airplanes really fly. The considerations adduced to settle questions about what "atom" meant in the language of Dalton and Bohr should, it seems, also be used to settle similar questions about what "think" meant for Wittgenstein and Turing. The internalist perspective seems adequate throughout, not only to their intuitions, but to an account of what is transpiring; or other accounts, as circumstances and interests vary.

One might, in principle, argue that recent theories supersede earlier intuitions because of their explanatory success. But that does not seem a promising idea. Explanatory success is slight. In general, it seems to me that we have little reason now to believe that more than a Wittgensteinian assembly of particulars lies beyond the domain of internalist inquiry, which is, however, far richer than Wittgenstein, Austin, and others have imagined.

If this is on the right track, we have an internalist approach to language and its use that raises only problems of the kind familiar throughout the natural sciences. The approach reaches quite far toward the explanation of properties of language and language use. It leaves untouched many other questions, but it remains to be shown that these are real questions, not pseudo-questions that identify topics of inquiry that we might hope to explore, but little more than that, at least for now.

Notes

1. Davidson (1986); Ramberg (1989).

2. Note that there is no implication that a serious study of these issues *must* involve I-language. Rather, the considerations that Davidson adduces leave the question open.

3. Further explanation is required here. The assumption is that the generative procedure maps a *D*-structure object to an *LF* object; at some point in the process (*S*-structure), "branching" procedures map the object formed to the *PF* level (phonetic representations); languages differ in whether processes apply before or after this branching, and thus are or are not phonetically manifest.

4. Technically, we should speak of "I-rhyme," etc.

5. That is, the intrinsic semantic properties of the individual elements would be the same but the principles would impose different linkages among them (including the empty categories). When we speak of "the same" properties, we refer to structural similarity and matching, in the usual way.

6. Several issues are collapsed here. The weakest assumption is that each phonetic form permitted by UG—a well-defined set—is an element of some structural description for every language. One might propose the further empirical hypothesis that this is true only for some distinguished subset, determined by some choice of phonetic, semantic, or other features of the I-language. Without a substantive proposal (there are none), it is pointless to pursue the question. Note further that there is no paradox in the position that the I-language generates structural descriptions that "violate" principles embedded in the I-language, for example, that the linguistic expression "what do you wonder whether John read" is generated, identified as a (mild) subjacency violation, or that word salad is generated and identified as such.

7. Note that the original commonsense argument, under this interpretation, was only apparently valid; that is, it is strictly valid only under certain circumstances and specification of open texture.

8. Higginbotham (1991).

9. Note that there is no intention here to question the normal practices of the sciences with regard to abstract entities; merely to adhere to them, with all the familiar problems left to the side.

10. Friedman (1990).

11. Rockland (1989).

References

Davidson, D. 1986. "A Nice Derangement of Epitaphs," in E. Lepore (ed.), *Truth and Interpretation*. Oxford: Basil Blackwell.

Friedman, M. 1990. "Remarks on the History of Science and the History of Philosophy." Kuhn Conference. Published in P. Horwich, (ed.), *World Changes: Thomas Kuhn and the Nature of Science*. Cambridge: MIT Press, 1993.

Higginbotham, J. 1991. "Remarks on the Metaphysics of Linguistics." *Linguistics and Philosophy*, 14, 5.

Ramberg, B. 1989. *Donald Davidson's Philosophy of Language*. Oxford: Basil Blackwell.

Rockland, C. 1989. "The Nematode as a Model Complex System." Laboratory for Information and Decision Systems, MIT.

10

Joseph Agassi

The Novelty of Chomsky's Theories

Introduction

As a new field, cognitivism began with the total rejection of the old, traditional views of language acquisition and of learning—individual and collective alike. Chomsky was one of the pioneers in this respect; yet he clouds issues by excessive claims for his originality and by not allowing the beginner in the art of the acquisition of language the use of learning by making hypotheses and testing them, though he acknowledges that researchers, himself included, do use this method.

The most important novelty of Chomsky's work is his idealization of the field by postulating the existence of the ideal speaker-hearer, and his suggestion that the hidden structure of sentences is revealed by studying together all sentences that are equivalent to each other. Equivalence between two sentences is shown by the validity of two inferences, with one of them as premise and the other as a conclusion and vice versa. This is progress; but it is insufficient, as the study of the equivalence in question requires the comparison of the logical force or content of statements. For example, the structure of statements asserting relations is revealed by inferences whose validity rests on the transitivity of some relations (see below), and which do not conform to classical logic. The greatest shortcoming of Chomsky's view lies in his idea that every sentence has one subject, contrary to the claim of Frege and Russell that assertions involving relations (with two-place predicates) are structurally different from those involving properties (with one-place predicates). (See the Appendix to this chapter.)

Background Items to Ignore

The field to which this book is devoted is cognitive science, cognitive research, or cognitive psychology—cognitivism, for short. It is a new discipline, but as a field of activity it is at least as old as psychology. Modern psychology began by following the precepts of the traditional British empiricists, who found it advisable to pro-cure empirical information about the process of acquiring knowledge, so as to boost their theory that this process was empirical. Yet their theory of learning, by having evidence of the senses as input and theories as output—as well as by their theory of learning as constant conjunction (associationism)—make the modern computer an ideal learning machine. And so the program of Artificial Intelligence evolved, tra-ditional or connectionist, also known at times as computationism, the program of creating machines that can acquire language as well as new scientific knowledge. This program has blocked the way of progress in cognitive research. This needs stressing. Cognitivism impinges on different disciplines including, of course, the philosophy of science and linguistics, because the acquisition of language and also of science belongs there. But the relation cognitive science has to artificial intelli-gence differs from its relations with other fields, such as linguistics and the philoso-phy of science. Cognitivism shares both its subject matter and its theories with other disciplines. But it competes with artificial intelligence on basics, or the rules of research. The rise of cognitivism as a discipline is rooted in a rejection of traditional British empiricism. The antecedents of cognitivism in psychology are the two schools initiated by Oswald Külpe, the Würzburg and the Gestalt schools, as well as Jean Piaget and his associates. In the philosophy of science it is the Würzburgian Karl Popper whose lifework is the critique of traditional British empiricism and its modern offshoots—conventionalism and inductive logic. In linguistics no one has done more in this direction than Noam Chomsky, who criticized inductivism, asso-ciationism, and the view that computers of the kind already available, though more powerful, can acquire a language.

I find it hard to write on Chomsky, since I have very mixed feelings about his output. I find his sales techniques unattractive and distracting; but I appreciate his contributions even after I manage to overlook his oft exaggerated claims for them. In addition, I should say at once, my appreciation of his ideas is no sign of agreement. I take it to be a general truth that assent is a sign of appreciation only in a narcissis-tic mood, and that, likewise, dissent is a sign of downgrading only in a defensive mood. So let me say a few words on his sales techniques, and proceed to his contri-butions.

There is a great initial difficulty with almost all studies of linguistics that invites a radical reconsideration before criticism of any theory may be deemed fruitful. The field is vast. The students of language who follow the methodological prescriptions of traditional British empiricism—inductivism—simply get lost in details. They hope that salvation will come from the increase, not the reduction, of the number of data to consider before fruitful examination can begin. Moreover, they rightly see nothing but prejudice in all attempts to manage by the *initial* concentration on any part of the field. I have great admiration for the tenacity of these researchers; but I cannot for the life of me see how any one of them can theorize in an orderly, criti-cal manner. Those inductivists who do arrive at a theory simply do not follow their own prescriptions; and so their works tend to be methodologically murky. In some researches it does not matter; but in contemporary linguistics it does, no less than in cognitivism, if not more.

Arch anti-inductivist Sir Karl Popper says that the choice of an approach to problems does not matter at all, that anyone who ventures any hypothesis on any question, provided it is at all open to critical examination, will do well to suggest conjectures and attempt to criticize, and let the chips fall where they may. This attitude is not specific to Popper, of course. It is traditional enough to have a traditional name: It is called "hypothetico-deductivism." I do not know if Popper is right; and I have ventured to criticize him. Here my point is that the debates between linguists cannot even begin without settling the methodological debate; and then the inductivists are simply out of the debate. Here, as an exception, I do endorse Chomsky's critique of inductivism, especially of its classical, associationist variant, as amounting to imposing non-theorizing, which at best permits mere taxonomy or classification.

So much for the initial difficulty, which is greater than one may think, because whereas the inductivists share the merit of being consistent, some contributors to the field are both inductivists and hypothetico-deductivists. No smaller a figure than Quine has declared allegiance both to B.F. Skinner, whose operant conditioning theory is a version of inductivism, of course, and also to Popper's hypothetico-deductivism. I do not know how to approach this matter. When possible, I tend to ignore an author's sliding from hypothetico-deductivism to inductivism; but with the writings of such writers on linguistics as Jaakko Hintikka, I do not know how to do it.[1]

There is more to it than that. For my part I do not wish to contradict what Popper says on this score. I think he is in two minds, however, as to the following question, which is central to this discussion. I will present it while borrowing Einstein's terminology and his frankly metaphysical approach: What is a better strategy, to attack problems as the spirit moves one, or to offer some strategy, or a program proper? I suggest that this is decided largely, though not completely, by the facts of the matter. The diverse hypothetico-deductivists do have such research programs; indeed, their leaders differ in that they offer different programs. But the picture is not clear; and the different programs are more often expressed in different practices than in different explicit theories.

To give an example, before Chomsky, linguists seldom used inferences to elucidate the structure of a sentence. Chomsky repeatedly claims that the surface structure of a sentence is misleading, that it has a deep structure. And he proves it by showing that in two different senses, the sentence is logically equivalent to two different sentences possessing quite different structures. And as logical equivalence of two sentences merely involves inferences with one premise, this is a kind of inference Chomsky regularly employs. At times even expressions are so treated, like the famous "the shooting of the hunter," which may be described as an event in which the hunter is active, but also an event in which the same person is passive. I presume that readers can choose any sentence they like for the purpose of the Chomskyan exercise. This is a nice example, because it collapses many examples into one—all those which employ the expression in question. Very seldom does a linguist, especially Chomsky, employ inferences with more than one premise. This is essential for the structure of relational sentences. Consider an inference with three sentences: as premises take *a,* "Tom and Dick are brothers" and *b,* "Dick and Harry are brothers"; as a conclusion take *c,* "Tom and Harry are brothers." This inference can be analyzed either as an inference between categorical sentences, where being a brother is under-

stood in the sense of being monks, and where it is valid as it stands, or as an inference between relational sentences, where the inference is allowed in the light of the
sentence "a brother's brother is a brother," which holds for the relation of sibling-
hood but not for all relations—for example, not for the relation of friendship. Since
such inferences are not usually employed in the linguistic literature, it is not clear
how one should handle the situation. In particular, I do not know how Chomsky's
theory of the subject part and the predicate part of a sentence applies differently to
cases where two subject sentences can be reduced to two one-subject sentences and
to cases where this reduction is unavailable. There is, of course, a whole logic of
relations; and its significance for mathematics was never contested. But I do not
know what Chomsky—or others—think of the matter. Perhaps their strategy is simply to leave relations alone for a while (which may be wise). But if so, then not saying so out loud is not wise; they should say so explicitly—(though not doing so may
be good public relations).[2]

There is an enormous amount of material left aside, and this is quite legitimate.
But in deference to public relations it is done silently and therefore uncritically; and
this is a pity. For example, we might expect the translation of two equivalent sentences from two languages, each to its deep form, to show the identity of the deep
grammar in both, since Chomsky's chief thesis is that deep grammar is the same
in all languages. But there is no such study, not even one comparing two Indo-
European languages, let alone English and Chinese. Chomsky regularly advertises
his ideas while glossing over difficulties like these. He totally overlooks problematic aspects of the situation, instead of noting them and declaring his policy to leave
them unattended for a while. This is a perfectly legitimate policy, but it seems less
brilliant to the unindoctrinated. For example, he entirely ignores the question of how
a pre-verbal infant selects input from the verbal environment, even though he notes,
as he should, that according to his general schema, such a selection is essential for
language acquisition. He also argues too slickly, overlooking obvious objections to
his arguments. Thus, he proposes to support a major aspect of his theory, the claim
that language acquisition is not inductive, with the observation that infants acquire
perfect knowledge from imperfect and partial input. But this is objectionable: infants
do not possess perfect control over any language; no one does. The perfect speaker
of Chomsky's theory is an ideal speaker-hearer. Idealization is commendable; and it
is well within the scientific tradition ever since Galileo instituted it. Yet it always has
a price to pay; and Galileo stressed this. For example, Galileo's own idealization, the
one that ignores friction when describing gravity, is refuted by the fall of a feather or
a parachute, and by the flight of birds and gliders. Therefore it is not surprising that,
however admired it was, it was both corrected and supplemented with a theory of air
friction. I will not dwell on this point, but merely note that the correction, due to
Newton, is also idealized; and both the original and the corrected version can be
improved by the addition of friction factors. Not only the idealization has its costs;
the correction does also. The addition of the correction due to friction, even if correct, adds to complication and makes the theory at times too cumbersome to be of
any use. I stress all this in order to say that I do not object to Chomsky's idealization,
merely to his glossing over its loss when arguing against opponents, so as to give the
impression that much more has been accomplished than the facts warrant.

The pity of it is that Chomsky has contributed something to the idea of idealization that invites appreciation. His particular version of idealization is due to Sausure, of course, and Chomsky does full justice to him. He also has converted the distinction between language and speech into that between the ideal and the real language user (or speaker-hearer). Now the ideal language user differs from language: languages are ideal to begin with, in the sense that they are abstract entities. On the other hand, the ideal language user is a concrete entity minus some characteristics, but still is not so abstract as to be totally unobservable. For example, within the bounds of a given language, no matter how they are drawn, an ideal language user may be identified as one who speaks faultlessly—and we can idealize the speaker less and come up with one who speaks faultlessly a given dialect of that language. In other words, we can decide how much of reality we wish to iron out in the process of idealization. I stress this, as Chomsky does not: the observations made by the ideal language user are problematic, and so are the observations made by the student of language concerning the verbal conduct of the ideal language user—more problematic than the observations of frictionless free fall. The advantage of the ideal language user over language, therefore, is not obvious, since one has to take account of the problematic character of this presumed advantage. Hence, sales talk too has both its advantages and its disadvantages.

Chomsky advertises his ideas not only by presenting them in a superior light, but also by presenting his opponents' ideas in an inferior light. Consider his critique of the view that sentence formation does not proceed linearly, since there are, for example, nested sentences. Surely, no one ever said that all sentences were constructed linearly; and everyone knows that some sentences are constructed in this way, since we often say half a sentence without knowing as yet how we will construct the second half. What was presented was information theory; and information theory, as it handles signs, not symbols, cannot consider such things as nested sentences. It then was judged natural for it to take strings of signs linearly, though later on, with the rise of the need to select described portions from tremendously long strings, random approaches were introduced as well. Information theorists showed, in the early days of the subject, how surprisingly easy it was to generate linearly, strings of signs that resembled to us written texts full of misprints, if certain probability distributions and dependences were built into the algorithm for the generation of these strings. This is uncontested; and it was never presented as a description of any aspect of any natural language.

Chomsky did a similar thing when he criticized the linguistic theories of the Vienna Circle. It is very hard for me to criticize him on that account, since I appreciate his ideas and do not appreciate theirs. Yet the fact is, he was unjust to them. They developed a program of constructing artificial languages by developing syntax first and semantics after; that is, by first making rules of sentence formation, by which to generate well-formed formulas, and also rules of sign transformation, and only afterwards giving meanings and truth to signs and sentences. Surely this is cockeyed, and was meant as an abstract exercise, rather than anything natural. More than that, it is even distasteful when considered abstractly, since the rules for sign transformation include rules of inference, and without prior attention to semantics these rules never will become semantically acceptable. The uselessness of the exercise of starting with syntax and only then adding semantics is easily revealed by the absence of any syntactical ground for the endorsement of *modus ponens*, or the rule that all well-formed formulas are admitted into any inconsistent set (that is, any set which includes or admits any formula of the form "*a*-and-not-*a*," where "*a*" is any well-formed formula). Yet Chomsky has criticized this program, and from observations of natural languages which, *ab initio*, are not obviously relevant to the exercise.

The criticism itself is very amusing; and it has done much to popularize Chomsky. It is the observation of syntactic ambiguity: we do not first listen to the syntax of a sentence and then to its semantics, but at times decide its semantics and then decide on its syntax. This sounds odd, since it is like the case of the chicken and the egg: we need both. But Chomsky offers examples. The funniest and best known example (I do not know who invented it) is the famous sentence "time flies like an arrow." The first word "time" may be a noun, as most people at once will suppose. But also, instead, it may be an adjective qualifying the noun "flies," thereby offering us a compound noun "time flies" akin to the compound noun "fruit flies," while transforming the preposition "like" into a verb. And, finally, the word "time" may be a verb in the imperative mood, which makes the sentence into a request to time flies the way one times arrows.

Why is this exercise so special? After all, ambiguity is a very common phenomenon. The answer is that this is a syntactic ambiguity, cleared up by appeal to semantics, contrary to the Vienna Circle's claim that the road to semantics is *via* a complete syntax. However, members of the Circle never said this. Indeed, it is plausible to think they must have been familiar with many cases of syntactic ambiguity, since quite a few of them had a classical education and so were familiar with Latin, which has them in abundance.

The Place of Idealization in Science

I do not mean to dismiss Chomsky's criticism. Every abstraction, every idealization, has its price; and one undertakes it because one hopes the cost is outweighed by the benefit. So the question is: What is the benefit of the abstractions attempted by the Vienna Circle? The answer is that this benefit was entirely philosophical, not impinging in any way on reality. And because in that sense it was barren, it would have been better for them not to have undertaken it in the first place. I think Rudolf Carnap, the philosopher whose name was most closely associated with the program attacked by Chomsky, died upholding this program and hoping to see what he deemed to be its benefits. In the first volume of the once famous *Minnesota Studies in the Philosophy of Science*, he said he still hoped that an artificial language one day would be constructed, in which no metaphysical utterance could be well formed. That program is now defunct on many grounds (most relevant to us today is the Sapir-Whorf hypothesis); and Chomsky has said the last word on it.[3] I think the program finally was rejected as useless by Carl G. Hempel, perhaps the last of the Vienna Circle (though he claims he was too young for that epithet). I cannot say this definitely, since I am never clear about what Hempel says, as opposed to what he only suggests as a passing thought. (But at least he did say, for example, after Herbert Simon and his associates published their program BACON, that there now was a need to reconsider the standard objection to inductivism based on the claim that ideas do not emerge from factual information.) In any case, even if Hempel's remark was made only in passing, it shows that Chomsky's critique is not devoid of value. But to see that, one has to go into more detail than Chomsky has; and therefore his critique looks more brilliant than it really is. This brilliance is merely a matter of salesmanship.

Even apart from its being a critique, Chomsky's observations on syntactic ambiguities count as brilliant. The reason is that whereas the syntactic ambiguities found in the literature are there, staring us in the face, demanding to be removed, Chomsky-style ones are not like this at all: they are not noticed until shown. Why then should we show them? Because they indicate the way the human mind works in its attempts to make sense of what is said. It is a well-known fact that the whole context in which a sentence is said may enter into the way one comprehends that sentence. It also is well known that quoting out of context is always possibly a distortion. This is a standard mode of complaint by which people attempt to evade the charge that they said something they should not have said. In other words, the complaint is always possibly true. If Quine's hypothesis is correct that there never are any fully adequate translations and paraphrasing, then it might be the case that all such complaints are true *a priori.* Nevertheless, somehow, grammarians seldom have paid attention to all this. Perhaps they simply have not known what to do with it; and they may feel vindicated in their oversight, since their task is to discuss forms rather than contents. What role does syntactic ambiguity have in Chomsky's system other than as a criticism of the Vienna Circle?

Strangely, Chomsky himself did endorse the view that syntax precedes semantics. He did so because he thought he was obliged to assert that grammar is inborn; but he did not wish to assert the existence of inborn knowledge, only of inborn competence. This, translated into linguistic jargon, means that syntax is inborn but not semantics. What, then, did he do about syntactic ambiguity? He postulated that syntactic ambiguity applies to common or garden-variety syntax, but not to the inborn sort. The inborn one, I understand him to say, is unlike the common one in that it is crystal clear. He had to distinguish between the two anyhow since, naturally, the syntactic characteristics of observed languages differ. But if there is only one human nature, then there also can be only one inborn syntax. Hence, common language is sanctified by its very existence, whereas the hidden, inborn language is sanctified by its being indispensable for the evolution of common language. Unfortunately, this will not do—not even for deep grammar. It is clear that when the syntax is turned into a semantics, the verbal transformations syntactically sanctified also must be inferences semantically sanctified: they must be truth preserving. Chomsky has not started grappling with this difficulty. An alternative to Chomsky's approach does exist, which avoids the difficulties he wishes to avoid, but without presupposing his deep grammar.

The first to have noted this is William Fowler, the author of the celebrated *Modern English Usage,* where the dictum is reaffirmed that usage sanctifies use. This dictum already had been stated in the eighteenth century—by Dr. Samuel Johnson in the Preface of his classical *Dictionary,* and by Dr. Joseph Priestley. But they did not know how to handle the central problem involved with this dictum which is, indeed, quite problematic. There are many usages; and we observe incorrect as well as correct ones. So the question is, how do we distinguish the correct use observed from the incorrect ones also observed? Dr. Johnson spoke then of good taste, which is question-begging. But Fowler appealed to comprehensibility: if it is comprehended, then it is correct. It thereby even appeals to good taste, he added.

The test case for Fowler's dictum is a new rule; and I think it was he who discovered it: the rule of deletion. I do not know if Chomsky notices Fowler; but I will not enter matters of priority here. Suffice it to say that here, too, Chomsky uses the same essential idea in a remarkably far-reaching manner, claiming it is necessary to reconstruct and reinsert the omitted parts of speech in a sentence before it can be properly analyzed.

There is one more preliminary item to notice before I can come to my point. It concerns Chomsky's review of B.F. Skinner's *Verbal Behavior*, which many think is his—Chomsky's—most brilliant piece. (As a public relations ploy, it certainly did much to enhance his reputation.) One reason for the impressiveness of Chomsky's critique was that it showed that the competition here was between two whole research programs.[4] Thus, the only question was: Which of the two programs was better? Today this question no longer is in doubt. After Karl Popper demolished the idea of induction in general, and Konrad Lorenz demolished the idea of operant conditioning in particular, and Quine's attempt to apply these ideas to language acquisition proved barren, it is now reasonable to agree that Chomsky's program was the more promising.[5] But was it so understood at the time? At the time, the common sentiment was expressed by Bar-Hillel, who understood the review as a disproof of Skinner's theory rather than the rejection of his program. (Incidentally, Bar-Hillel thought—erroneously—that the claim was correct and the disproof valid.)

The chief characteristic of Skinner's program was to avoid speculations about the mental processes that go on inside one's skull. Since Skinner was usually considered a materialist, and rightly so, readers regularly overlooked the fact that his research project was indifferent to all questions of brain-mind interaction. Here, let me stress, Chomsky is in exactly the same boat. He too is a materialist, and he too presents a program indifferent to the matter of the brain-mind interaction. The chief difference between Skinner and Chomsky is this: Skinner is an inductivist and he thinks learning by induction is operant conditioning, and learning a language is inductive. Chomsky rejects all operant conditioning, and has a different theory of learning altogether. What is it? Does he agree with Skinner that humans and other animals learn the same way? Is his theory of learning one's first language the same as his theory of learning? These questions deserve investigation.

Significant Background Items

Chomsky stresses that classical empiricist learning theory, from David Hume to B.F. Skinner, does not distinguish between the learning process of human and of other animals. He rejects that view, and takes his cue from the all too obvious fact that only humans possess abstract language. This is a bit ambiguous. Of the various species that are capable of learning, every one of them exhibits different learning capacities. There are tasks that a dog can learn and a horse cannot, and vice versa.

This may, but need not answer the question, are their learning processes alike? Hume and Skinner say Yes; and so they are not disturbed by the fact that only humans can acquire abstract language. How does Chomsky stand on this issue?

Chomsky is the thinker who has stressed the difference between *a priori* knowledge and innate dispositions or capacities, including the capacity to learn. It is not that Chomsky identifies knowledge with dispositions to act, as some followers of Wittgenstein did. Rather, he does not want to assume that inborn knowledge is knowledge of facts, since this will land him in the same problems and impasses that befell Kant's system. The difference between nativism and *a priorism*, Cartesian or Kantian, can best be described by means of a metaphor: We want the human mind to be as programmable as possible, with as much capacity for memory as possible, but also with as few innate programs as possible. Of course, some inborn programs are essential for learning. So the question is, what is the initial program needed for humans to be able to pick up a language? Initially Chomsky said the whole of grammar and the whole capacity to acquire language must be inborn. Later he said that if a human infant was able to learn one item and use this item to increase its learning capacity, then that would do. Evidently, however, humans are able to react to some speech by means of speech. Dogs, too, are born with the disposition to react to animal speech, and they can learn to react to human speech; but their reaction to speech is not speech. Why?

It is the greatest compliment to Chomsky that he effected a radical change on this matter. The same question was asked long ago, for example, by La Mettrie. But traditionally, this question was not taken seriously—at least in the sense that it did not lead anyone to undertake empirical research. Under Chomsky's influence, just such a study was conducted. Indeed, one of the animals on which the resulting experiment was conducted was a chimpanzee called Chimpsky. The experiment was conducted to test the hypothesis that the limitations on the apes was not in respect to their learning capacity, but rather their performance capacity. When they were given artificial means by which to use abstract language, the claim is that they were able to do this. The claim is controversial, of course; and it also does not help solve the question but forces us to restate it in gradual terms rather than in black and white. It is not: Do other animals possess the capacity to acquire an abstract language? Instead, it is: What makes it possible for them to acquire the abstract language that they observably do learn?

I do not wish to make my appreciation seem an assent. In particular, I think it is an error to suppose that ascribing inborn knowledge to humans will land us in traditional philosophical problems. Today we speak freely of the inborn knowledge of many different animals, without imagining that our calling this knowledge instinctive solves any problem. The main thing to realize is that inborn knowledge is not necessarily true; but it is sufficiently true for its possessor to survive—or else it dies. There are two important differences in knowledge: First, some animals are capable of modifying their knowledge in the light of new experience. Second, human knowledge is conceptual. The idea that there is inborn knowledge capable of correction is more in line with what Chomsky wants his theory to do than what he himself says about these matters. His first suggestion was unbelievably strong: that newborn humans have inborn syntax that totally captures the deep syntax of all (natural?) human languages, as well as the full language acquisition apparatus. Later he reduced the amount required for the newly born, not in an attempt to respond to empirical criticism, but out of recognition that the assumption is stronger than logic requires.

The idea that humans are born with a complete verbal apparatus, like a new computer with no program in it, seems very much in accord with the imprinting theory of learning of Heinroth and Lorenz. The learning mechanism, according to this hypothesis, is complete at birth (or perhaps even before), and all that is needed for the acquisition of knowledge is the trigger. Moreover, the items learned, the data, are the same as the trigger. For example, a newly hatched bird of a certain species learns to identify its mother by sound; and the first sound triggers both the learning mechanism and the ability to identify the mother by the sound she emits. Another example: in a different bird species the trigger for the same learning process is the sight of movement; and the moving object is seen and identified as the mother. The second example is a bit more complex, since it is not motion but the image of the moving object, possibly her color, that identifies her. One way or another, the important point about the knowledge thus gained is that although it evidently is not inborn, the learning process it involves is not empirical in the classical empiricist sense. Its importance lies in its ability to account for the fact that learning sometimes is the result of one observation, and at times is beyond reach no matter how many times the facts present themselves. The process of learning by repetition—the classical idea of learning by induction, and of learning in school—is a rare phenomenon.[6] The weakness of the imprinting theory is that it leaves no room for learning afresh. Lorenz himself asked the question, can an animal relearn? He showed that it can, though with great effort and only when no choice was left but that it either should learn or die. Yet he did not know how to incorporate this into his theory of learning by imprinting.

It was Chomsky's debt to Lorenz that made him declare that one learns one's mother tongue best. (However, bilingualism is a counter-example to this point. Even adults can become genuinely bilingual; and at times they do.[7])

Perhaps Chomsky views language as a whole as inborn in the wish to agree with those philosophers who consider languages as wholes. The generally accepted view of language altered after 1900, when the "new logic" was forged; and this was the basic idea behind the change. It is clear that Leibniz had entertained the idea of a universal language. But he had not thought of natural languages as separate and distinguishable from the universal one, and so did not say, what Russell later suggested instead, that logic offers the structure of all human languages, that is, the essence of language. This idea, that there is a single essence of language, has stayed with Chomsky to the end; and he now uses the imprint theory with a vengeance. Humans are born, he says, with a template that, once triggered, creates in the child the whole of language in one go, namely, the child's mother tongue.

To repeat, what I have tried to do in this chapter is present reasons for doubting that Chomsky's theory is either entirely new or entirely correct. Since, so far, Chomsky has not responded to my arguments to show this, I can only leave things as they stand until he or someone else from his entourage decides to do so.[8]

Appendix

Chomsky's technique of analyzing sentences in order to elicit their "deep" structures is too complex to present here in anything like a systematic manner. But it will be useful to make several points about it. First, he takes a great deal of liberty just by choosing the (kinds of) sentences he wants to analyze and omitting those he prefers not to deal with. (Although he often presents his choices in a critical mood, they soon become paradigmatic —without any explanation.)[9] Second, both his disciples and opponents usually follow his choice of examples uncritically. Third, there is no discussion, justification, or systematization of the (kinds of) sentences he has chosen to analyze, much less of how these sentences reflect the vagaries of his speculations. (Often he conveys the impression that there is so much to do that any item is welcome. But this is too facile.)[10] Yet this much has to be admitted: he usually contrasts the structure of one sentence with that of another, nearly identical with it, in a way that makes it clear that—so far as those two particular sentences are concerned—his analysis is not arbitrary.

A famous example is his contrast of the structures of two nearly identical sentences, where the only difference between them is the verbs "promise" and "persuade." That is, compare "x promises y to do z" on the one hand, with "x persuades y to do z" on the other. Assuming the identity of structure of all simple sentences with one word for a subject and one for a predicate, a person might suppose that the same things hold for sentences of the two forms cited. Yet one can make their difference in structure manifest by translating them into more explicit sentences that (according to the hypothesis) ought to be *semantically equivalent*—for example, "x promises y that x will do z" and "x persuades y that y will do z." The questions of why these two expanded sentence forms are different, and of why the extended ones but not the original ones represent the deep structure, are relatively easy to answer in relatively simple cases. Therefore these questions are unjustly neglected. But here is not the place to elaborate on this matter. Rather, the point of this Appendix is to offer a more obvious analogue to this exercise—one that Chomsky systematically overlooks, possibly because (correctly or not) it apparently runs contrary to his fundamental assumption that every sentence in a natural language has a subject part and a predicate part (and nothing else!).

Compare "x and y are suitors" with "x and y are lovers." The first sentence is equivalent to the conjunction of two sentences of the form, "x is a suitor." Not so the second: the conjunction "x is a lover" and "y is a lover" will not suffice, as it does not yet inform us that the lover x is identical with the lover of y (or even the weaker sentences "x has a lover" and "y has a lover" that follow from that identity). To mention an example used before, consider "x and y are brothers," where the classical analysis holds, if "brother" means a member of a religious order, but not if it means sibling—or at least not according to Frege and Russell, whose starting point was that the structure of simple two-subject sentences differs radically from that of either simple one-subject sentences or compound two-subject sentences. Both these philosophers postulated the existence of n-place predicates that are irreducible to one-place predicates, and also claimed that sentences employing these predicates have n subjects each. This was meant to reflect the essence of language, and so to represent any natural language (in idealized form). In this respect, then, Chomsky's conflict with Frege and Russell is clear-cut. Nevertheless, he avoids any mention of it.

(Semantically, for a symmetrical relation R, "xRy" is synonymous with "yRx," whereas for an anti-symmetric relation R', "$xR'y$" is synonymous with "ySx," where S is the inverse of R. Chomsky cannot avoid making regular use of this trivial part of the logic of relations. What he overlooks is that this renders both x and y equally legitimate subjects of the two sentences—"xRy" and "yRx" in the symmetric case and "$xR'y$" and ySx" in the anti-symmetric case—in the "deep" sense of the word, as he regularly employs the word "deep." Hence, "xRy" has two subjects.)

This, however, is not to deny that deep structure is different from the structures that Frege and Russell discussed. This is clear from the study of the reduction of two-place predicates to one-place ones: compare "x runs," "x eats," and "x eats y," all of which are quite uncontroversially well formed. Frege and Russell would not say that intrinsically the one-place predicate "x eats" resembles more the two-place predicate "x eats y" than the one-place predicate "x runs," even though the one-place predicate "x eats" is definable by the use of the two-place "x eats y." By definition, "x eats" if and only if there exists a y such that "x eats y." As far as Frege and Russell are concerned, "x promises y to do z," as well as "x persuades y to do z," are simply three-place predicates; and they left it at that. So Chomsky has a point in showing that the two are different. This, however, raises the question of the uniqueness of the deep structure. Chomsky never discussed this uniqueness, though he implicitly ascribes it to any sentence (and is so understood). Nor does he discuss the transition from one deep structure to another, as from "x eats y" to "x eats."

Notes

1. See note 6.

2. It was Barbara Hall Partee who contrasted Yehoshua Bar-Hillel's suggestion to use the insight of formal logic in linguistics with Chomsky's rejection of this suggestion. See Partee (1973, p. 309).

3. See Chomsky's introduction to Adam Schaff (1973), where he says the metaphysics implicit in language à la Sapir and Whorf is not obligatory. He offers a stunning example by applying an analysis akin to that of Whorf to the tenses in English, and concludes that English has no future tense, and this is why it takes recourse to an auxiliary verb.

4. I am not eager to defend Skinner's ideas, and will not dwell on them. But I invite anyone interested in the case, to compare Skinner's book (1957) with Chomsky's review of it (1959), and see that Skinner is much more tentative and cognizant of the incompleteness of his ideas in that book than one would surmise from Chomsky's review.

5. As to Lorenz, clearly he was an inductivist. But never mind that: his imprinting theory is certainly non-inductivist.

As to Quine, what is said here of him accords with the common reading of his works. I do not know if that reading is true. I have too little information about Quine's view of psychology, and the little I know puzzles me. He endorses Popper's view of science as refutable, however reluctantly. (See, for example, Schillp, 1986.) Yet he says we need to endorse behaviorism, since all we have as empirical evidence is overt behavior. (See his 1987.) Now all evidence, being public, must indeed be of overt behavior. This alone, without the theory of operant conditioning, does not amount to behaviorism, or else Popper is a behaviorist too.

6. The empiricist or inductivist characterization of induction as the repetition that, they say, is used in scientific research is so traditional that it is very hard to document it. Nor need one trouble oneself to do so. The strange fact is that the theory that learning is by repetition was denied by Sir Francis Bacon, the father of modern inductivism, who called it "induction by enumeration" and denounced it emphatically as unsure and childish. (See 1620/1960, "Distributio Operis," paragraph 9, and Book I, Aphorisms 69 and 105.) Nevertheless, it was accepted—on the authority of Newton. One can, of course, rescue inductivism by calling another process induction. This way one can alter one's views without admission; and it has, to use Russell's immortal expression, all the advantage of theft over honest toil. To clinch matters, Jaakko Hintikka ascribes Popper's methodology to Newton, and dismisses Popper's critique of Newton's inductivism as "without any force whatsoever" (1992, pp. 24–43, 40). Perhaps this marks the end of an era; but it is also truly thought provoking.

7. See Agassi (1976), pp. 33–46.

8. For more details of my critique of Chomsky, see Agassi (1977), chap. 2.

9. Chomsky apparently meets the problem of how to choose paradigms by demanding a universal grammar, which would render grammar independent of all paradigms. See Harris (1993). In particular, see the Index Article, "Restrictiveness and Relational Grammar."

10. See the previous note.

References

Agassi, J. 1976. "Can Adults Become Genuinely Bilingual?" in A. Kasher (ed.), *Language in Focus: Bar-Hillel Memorial Volume*, Boston Studies in the Philosophy of Science, 43.

Agassi, J. 1977. *Towards a Rational Philosophical Anthropology*. Dordrecht: Kluwer.

Bacon, F. 1620/1960. *Novum Organon*. New York: Liberal Arts Press.

Chomsky, N. 1959. "A Review of B.F. Skinner's *Verbal Behavior*." *Language*, 35, pp.26–58.

Fowler, H.W. 1968. *A Dictionary of Modern English Usage*. Oxford: Oxford University Press.

Harris, R.A. 1993. *The Linguistic Wars*. New York: Oxford University Press.

Hintikka, J. 1992. "The Concept of Induction in the Light of the Interrogative Approach to Inquiry," in J. Earman (ed.), *Inference, Explanation and Other Frustrations*. Berkeley: University of California Press.

Partee, B.H. 1973. "The Semantics of Belief-Sentences," in K.J. J. Hintikka et al. (eds.), *Approaches to Natural Language*. Dordrecht and Boston: Reidel.

Quine, W.V. 1987. "Indeterminacy of Translation Again." *Journal of Philosophy*, January, 1987.

Schaff, A. 1973. *Language and Cognition*. New York: McGraw-Hill.

Schillp, P.A. (ed.), 1986. *The Philosophy of W.V. Quine*. La Salle, IL: Open Court.

Skinner, B.F. 1957. *Verbal Behavior*. New York: Appleton-Century-Crofts.

11

Christopher D. Green & John Vervaeke

But What Have You Done for Us Lately?

Some Recent Perspectives on Linguistic Nativism

The problem with many contemporary criticisms of Chomsky and linguistic nativism is that they are based on features of the theory that are no longer germane; aspects that either have been superseded by more adequate proposals, or dropped altogether under the weight of contravening evidence. It is notable that, contrary to the misguided opinion of many of his critics, Chomsky has been more willing than the vast majority of psychological theorists to revise and extend his theory in the face of new evidence. His resistance to the proposals of those of his early students and colleagues who banded together under the name "generative semantics" was not, as is widely believed, a matter of his unwillingness to entertain evidence contrary to his own position. Rather, it was a matter of his unwillingness to entertain vague, ambiguous, and inconsistent theoretical claims. His ultimate victory over generative semantics was grounded squarely in his willingness to alter his theory to bring it more in line with new evidence; and the theory now bears only a modest resemblance to that which he developed in the 1950s and 1960s.

Thus, matters of "deep structure" and "transformations" of the sort described in *Aspects of the Theory of Syntax* (1965) are of only historical interest now; and criticism of them that claims to damage the current theory is simply off the mark. As Chomsky himself has written,

> For about 25 years, I've been arguing that there are no phrase structure rules, just a general X-bar theoretic format. . . . As for transformational rules, for about 30 years I've been trying to show that they too reduce to an extremely simple form, something close to "do anything to anything" (interpreted, to be sure, in a very specific way). (Personal communication, 1994)

It seems that many of those who were opposed to a formal theory of grammar in the 1960s (and their students) are still opposed to it today, but continue to trot out the same old critiques without bothering to keep up with new developments. Some have argued that Chomsky's continuous revision of his theory makes him impossible to "pin down," and imply that he is engaged in some sort of ruse to keep his critics "off balance." A close reading of the revisions, however, shows them to be well motivated by and large, and indicative of a progressive research program, in Lakatos' (1970) sense. We take his modifications, far from being some sort of ruse, to be simply good scientific practice; a practice that more psychological theoreticians should consider adopting.

In this chapter, rather than rehashing old debates voluminously documented elsewhere, we intend to focus on more recent developments. To this end, we have put a premium on references from the 1990s and the latter half of the 1980s. It is our hope that, in so doing, we can shift the debate somewhat from the tired old bickering about whether John is eager or easy to please (Chomsky, 1965), to a more fruitful and relevant line of discussion. It is not that we believe Chomsky to be infallible, or his theory to be definitively or ultimately True. Rather, it is that arguing about issues that were either resolved or dissolved in the 1970s can serve little purpose other than to cloud the important issues that face us today.

We cannot hope to cover all the issues surrounding the innateness of language in the short space allowed us here. Thus, we have decided to focus on a small set of questions that have produced the most interesting work lately. First, we will describe exactly what is now thought to be innate about language and, second, why it is thought to be innate rather than learned. Third, we will examine the evidence that many people take to be the greatest challenge to the nativist claim: ape language. Fourth, we will briefly consider how an innate language organ might have evolved. Fifth, we will look at how an organism might communicate without benefit of the innate language structure proposed by Chomsky, and examine a number of cases in which this seems to be happening. Finally we will try to sum up our claims and characterize what we believe will be the most fruitful course of debate for the immediate future.

1. What Is Innate?

When faced with the bald claim that "language is innate" most people immediately balk, and with good reason. It is obvious to everyone that babies born into English speaking families acquire English because that is what they hear and, loosely speaking, that is what they are taught. *Mutatis mutandis* for babies born to Turkish families, Cantonese families, Inuit families, etc. Of course, the problem is that Chomsky's claim never has been that all aspects of language are innate. It was, rather, that grammar is innate. Grammar is, very roughly speaking, syntax and as much of semantics as can be accounted for by the structure of language. Phonology and morphology have, of course, been the subjects of Chomskyan-style analysis as well, and are related in interesting ways to syntax and semantics; but these are not the focus of this chapter. Chomsky does not claim—or, more important, his linguistic theory does not imply—as some of his critics have imputed to him, that all knowledge of meaning (semantics) is innate, or that knowledge of appropriate language use (pragmatics) is innate.

There is a systematic ambiguity in the term "grammar," an ambiguity that has caused much confusion among Chomsky's critics. It refers both to the knowledge that is hypothesized to be "in the head" of the human, and to the linguist's theory of that knowledge. In an effort to stem the confusion somewhat, Chomsky recently has adopted the term "I-language" to refer to the knowledge of language the human is thought innately to have, the "I" standing for individual, internal, and intensional. Once again, the point here is not to reject the possibility that there are also social, external, and extensional aspects to language. It is only to identify those features of language that Chomsky is interested in addressing.

So just what is thought to be innate? Not, as was once thought, a vast array of transformation rules for converting the deep structures of sentences into their surface structures. Chomsky long since has rejected that research program, seeing, as did many of his early critics, that such an approach leads to a wildly unconstrained proliferation of rules to cover increasingly rare grammatical cases. This was precisely the tack taken by advocates of abstract syntax in the 1960s and of the early generative semanticists of the 1970s (before they abandoned the possibility of a formal theory of language at all), and led to Chomsky actually working to reduce the generative power of his proposal for fear of falling into the same difficulties (Harris, 1993).

The theory now proposes that there is a relatively small set of linguistic principles for which given parameters are given values by the early linguistic environment of the child. For instance, all languages seem to be either what is known as "head-first" or "head-last." That is, nouns and verbs either come before their complements (head-first) or after them (head-last). To give an example, in English, a head-first language, we say "Malcolm caught the *train that was going to Winnipeg.*" Note that the object noun, "train," comes before its complement, "that was going to Winnipeg." In Japanese, a head-last language, the complement comes before. One can mock this up in English for the purposes of an example: it would be like saying, "the was-going-to-Winnipeg train." There is some controversy over whether languages are head-first or head-last *tout court* or whether the parameter has to be set separately for each major grammatical category (nouns, verbs, adjectives, etc.). Most languages seem to be consistent across categories. There are some apparent exceptions, however.

Notice just what is being claimed to be innate here: simply the fact that languages are either head-first or head-last. The ultimate value of the parameter is set by the environment. Put conversely, what is being claimed is that young children do *not* have to learn that languages are head-first or head-last. That part is, as it were, innate. Of course, as with many cognitive activities, most notably perception, innate here does not necessarily mean "present at birth." It may take a few months or even years after birth for the brain to develop to the point where the mechanism is fully in place. What the environment contributes is the information necessary to establish the parameter's value; to set the parameter's "switch" to the correct position for the language being acquired. Good introductions to this material include Chomsky and Lasnik (1993), Cook (1988), Cowper (1992), and Radford (1988).

Chomsky's associate at MIT, Robert Berwick (1985), has been working for over a decade on precisely what parameters are needed to account completely for the principles and parameters underlying the structures of all languages. He is currently working with computer simulations that employ 24 parameters, and believes there may be another 12 to be worked out before the basic structures of all languages are accounted for (Berwick, 1993). Other researchers such as David Lightfoot (1991) have been working on just how environmental stimulation is able to set parameters without the conscious attention of the child. Among Lightfoot's discoveries is the apparent fact that the "triggering" stimuli must be robust and structurally simple, and that embedded material will not set linguistic parameters.

There is also some evidence that the "switches" are set to certain "default" positions at birth, and that only certain kinds of linguistic stimuli can change these default settings. There is a fair bit of debate about this claim, however. Although it would explain some of the most interesting features of creole languages, as we explain below, the default hypothesis has some, probably *not* insurmountable, trouble accounting for cases of deprived children not learning language at all.

2. Why Must It be Innate?

The immediate question of doubters is, why must the principles be innate? Why can't the child figure out for him- or herself what the relevant structures of language are through general cognitive mechanisms, the way he or she might figure out that clouds mean rain or that blood means pain? The responses to such questions are quite straightforward.

2.1. Complexity of Language

First of all, the structures of language are very complex and, in large part, domain-specific. Constituent hierarchical structure, an almost definitional feature of language, is just not something, by and large, that we come up against in the everyday world; and even when we do, it is darned hard, even for the best and brightest among us, to figure it out. Witness, for instance, the struggles of linguists themselves to characterize language adequately. Moreover, in using general cognitive procedures, children (and adults, for that matter) are usually consciously aware of the hypotheses they are testing and rejecting. Bruner's and Wason's cognitive tasks of the 1950s and 1960s are the paradigmatic examples (see, e.g., Bruner, Goodnow, & Austin, 1956; Wason & Johnson-Laird, 1972); and it is clear to everyone that whatever children do to learn language, it bears little relation to what those subjects were doing.

Compare, for instance, children's learning of language to linguists' learning *about* language. In effect, the general cognition hypothesis says children are "little linguists" trying to discover the rules that govern the structures of the utterances that are allowable within their language community. The fact is, however, that linguists have been unable to discover exactly what the rules are, even after dozens (one might argue hundreds or even thousands) of years of research. By contrast, virtually every child does it within a few years (with far less in the way of specialized cognitive machinery, and control over the quality of the incoming data, it is worth pointing out, than a Ph.D. in linguistics).

Some argue that the hypothesis-testing model of cognition is inadequate, and have offered various alternatives, such as the "Idealized Cognitive Models" of George Lakoff (1987) and his associates. Although we recognize the difficulties of hypothesis-testing models of cognition, and are open to new, more adequate models of thinking, we find most of the current alternatives vague and/or ambiguous; they simply don't have the formal rigor to be adequate theories of language acquisition (see, e.g., Green & Vervaeke, in press; Vervaeke & Green, in press). Of course, the bottom line with people like Lakoff is that you just *can't* have a formal theory of language as you can, of say, the movements of the planets. We concede that there are difficulties, but we are inclined to believe that Chomsky has cut up language into just about the right parts: one part where a formal theory might be applied fruitfully (grammar), and another where such a theory probably will not be (pragmatics and some aspects of semantics). We are not willing just yet to give up on the formalist project altogether. Simple scepticism is rarely the last word on anything.

Connectionism is often suggested as an alternative model of learning. Undoubtedly it has enjoyed many explanatory successes over the last decade. However, there are some mistaken assumptions commonly attributed to connectionist models that bear on its relevance to linguistic nativism. Chiefly, it is often assumed that connectionism implicitly favors an empiricist account over a nativist one. This is simply untrue. Although connectionism was embraced early by psychological empiricists, all connectionist networks begin life with a given, pre-set computational architecture, and with given activation and learning rules. These features are essentially innate and they ultimately dictate what can and cannot be learned by the network.

It might be argued that this sort of nativism is very weak because the particular features of given computational architectures do not strongly constrain the learning capacities of the networks, and the initial connection weights are set to random values. Certain problems have come to light in connectionist research, however, for which plausible non-nativist solutions seem, at present, to be in profoundly short supply. Particularly in work on language acquisition, connectionist networks are often unable to generalize correctly to new cases of the linguistic structures they supposedly have learned unless exposed to a "phased training" regime (Elman, 1991), in which training cases are presented in a specified order engineered ahead of time by the researcher explicitly to prevent the network from getting "lost in space(s)" as Andy Clark (1993, p. 142) has put it so aptly.

Whether or not such extraordinary procedures ultimately get the machine to generalize properly, they cannot seriously be countenanced as plausible theories of language acquisition. It is precisely such carefully constrained training regimes that children are not given when learning their first language. A system that cannot robustly capture the right linguistic generalizations despite being trained on multiple arrays of widely diverse input sets simply fails to do whatever it is that kids do when learning language. Put more plainly, the single most important datum to capture when modeling language learning is that children virtually never fail to learn language correctly, regardless of what kind of linguistic data they are exposed to early in life. Models that don't capture this feature are simply wrong.

The importance of the native aspects of connectionist networks is just now beginning to be recognized. Some researchers have built innate structure directly into their networks (e.g., Feldman, 1994). Paul Smolensky (1995) has explicitly endorsed a strong linguistic nativism, and is collaborating with a traditional linguist (Prince & Smolensky, in press) who was once highly critical of connectionist approaches to language learning (Pinker & Prince, 1988) in an effort to integrate connectionist and Chomskyan approaches to language.

Among the most interesting efforts to combine nativism and connectionism are "artificial life" models in which a large set of networks are trained, all beginning with random weights. Only those few that learn fastest and best, however, are allowed to "reproduce." The new generation of networks starts out learning with the weights with which their "parents" began (plus a bit of randomness); but the competition is now a lot tougher than it was for the "parents." Again, only those that learn fastest and best are allowed to "reproduce." By iterating this procedure, a set of weights optimal for learning the task at hand can be developed in an evolutionarily realistic manner (see, e.g., Nolfi & Parisi, 1991). Clark (1993) has argued that these efforts constitute only a "minimal nativism," but there is no real reason to call it "minimal"; the result is a richly articulated native structure the absence of which would make reliable learning of particular and important kinds (such as of language) effectively impossible. Different as it may be from what Chomsky and his followers envision, it is quite strongly nativist in its thrust. Others have recently come to agree with this verdict. Quartz (1993), for instance, has argued that "PDP models are nativist in a robust sense" (p. 223).

2.2. Developmental Dissociation of Language and General Cognition

Another reason to believe that the principles of grammar are innate is that children all seem to learn the structures of language in pretty well the same order, and at very close to the same ages. Moreover, people who have, for one reason or another, been prevented from learning language at the usual time of life seem, later, to be virtually incapable of learning it despite having intact general cognitive abilities. Such cases include Genie, the girl from Los Angeles who was essentially locked in a closet until the age of 12 and, thereafter, was never able to learn to speak anywhere near adequately. Although the facts of her case are suggestive, her life circumstances were such that one is wary of drawing firm conclusions from her particular case. More recently, better evidence has come from congenitally deaf children who were not taught sign language until adulthood. Such people typically never learn to employ the structures of their sign languages with the fluidity of those who learn it from birth (appropriately controlling, of course, for the simple length of time the sign language has been known to the person). Interestingly, however, their own congenitally deaf children often do become syntactically fluid (see Pinker, 1994).

Even more important, Grimshaw, Adelstein, Bryden, and MacKinnon (1994) recently reported a case of a congenitally deaf boy from Mexico, E.M., who was not taught a sign language at home, but whose hearing was restored at the age of 15. Because there are no signs of abuse in his early life, unlike Genie, his linguistic progress is of crucial relevance to followers of the critical/sensitive period hypothesis. Grimshaw, et al. report that

> After 2 years of exposure to verbal Spanish, E.M.'s language production is limited to single words, combined with gesture. His articulation is very poor. He has a large receptive vocabulary, consisting largely of concrete nouns, adjectives, and a few verbs. Despite his language limitations, he appears to function well in everyday situations. . . .

E.M. has particular difficulty with pronouns, prepositions, word order, verb tense, and indefinite articles. (1994, p. 5)

Thus, he is having great difficulty learning anything but the most rudimentary grammar. His progess in coming years will be interesting to watch.

In addition to cases of children with relatively normal intelligence, but severe language deficits, there recently have been reported cases of severely cognitively handicapped children whose syntactic abilities nevertheless seem to remain substantially intact. The most fascinating recent case is a woman called Laura (Yamada, 1990), whose testable IQ is in the mid-40s. She cannot even learn to count, and surely is incapable of learning the structures of language by means of her general cognitive capacities, yet she freely uses a wide array of complex linguistic structures, including multiple embeddings (Yamada, 1990, p. 29). More interesting still, although her general cognitive handicap seems to have left her syntactic abilities relatively intact, her semantic capacities are seriously affected. This makes it highly unlikely that linguistic structures develop as a result of semantic learning, for Laura has mastered pretty well the structure in spite of a serious impoverishment of the semantics. Moreover, Laura's pragmatic linguistic ability—her ability to converse competently—is grossly impaired. Thus, over and above the problems Laura poses for someone who thinks language acquisition can be explained in terms of general cognitive abilities, she also generates significant difficulties for the language theorist who considers communicative function—that is, pragmatics—to be the basis for structural and semantic aspects of language (Yamada, 1990, p. 72). This includes, of course, the popular Anglo-American neo-Wittgensteinian approaches to language that are often extended in an attempt to account for linguistic structure, as well as many related Continental approaches (e.g., Habermas, 1976/1979). Whether they are useful in explaining aspects of language *other* than grammar is, of course, a matter that might be decided on independent grounds.

The last several paragraphs evince a double-dissociation between general cognition and language ability. On the one hand, we have cognitively able individuals who are unable to learn standard linguistic structures that are picked up by any normal three-year-old. On the other, we have a woman who has the general cognitive abilities of a toddler, but is able to employ the linguistic structures of an adult. Such anomalies must be explained by the advocate of general cognitivist accounts of language acquisition (e.g., Bates, 1976).

2.3. Comparison to Other Innate Aspects of Cognition

Third, the development of children's linguistic abilities bears a strong relation to the development of other aspects of cognition—most notably in vision—that are widely thought to be essentially innate, but delayed in their onset due to maturational factors. For instance, one does not learn to see binocularly but, given sensory stimulation generally available early in life, cells in the visual cortex develop to mediate binocular stereopsis. Notice that even though there is environmental involvement (and no one denies this in the case of language either), children are not explicitly taught to see binocularly, nor are they able to learn to do so later if they somehow are robbed of normal visual experiences early in life. The development of binocular stereopsis occurs in the natural course of development because of the innate presence of certain brain structures in the human. The current evidence is that they are built to mediate quite specific functions and, if not stimulated by the early environment, degenerate. In almost exactly the same respects as Chomsky claims certain aspects of language to be innate, binocular vision is innate as well.

3. Ape Language

For decades now there have been attempts to teach language to a wide variety of higher animals, primarily chimpanzees. One motivation for this project has been to show that there is no special organ unique to humans that mediates the learning of language. The reasoning is that if a chimpanzee, our closest evolutionary relative, can learn language then one is forced either to impute the same language-learning organ to chimps, or forced to say that language learning is mediated by general cognitive abilities that we share with chimps. Since chimps never learn language on their own, nor with the same degree of competence as even a human five-year-old, it seems doubtful that we have the same language-specific resources. Thus, it is argued, the reason humans learn language more naturally, and better, is that we have stronger general cognitive resources than the chimps, but nothing important that is specific to language (other than, perhaps, articulatory organs).

Everyone, by now, knows the story of how Herb Terrace (1979) tried to teach language to Nim Chimpsky, but later, after reviewing the videotapes of his research, reversed himself and argued that Nim had never really learned to do much more than imitate, and had certainly never learned syntax. For a while, that appeared to spell the end of ape language studies; but recently there has been a flurry of excitement over a bonobo named Kanzi (trained by Susan Savage-Rumbaugh) who seems to understand a wide array of English language commands, and can produce sentences by using a touch-sensitive symbol board. In one experiment (Savage-Rumbaugh, 1988) Kanzi was faced with 310 sentences of various types. There were (in order of frequency) action-object sentences (e.g., "Would you please carry the straw"), action-object-location sentences (e.g., "Put the tomato in the refrigerator") and action-object-recipient sentences (e.g., "Carry the cooler to Penny"). Of the 310 sentences tested, Kanzi got 298 correct. Savage-Rumbaugh concluded that Kanzi's sentence comprehension "appears to be syntactically based in that he responds differently to the same word depending upon its function in the sentence" (cited in Wallman, 1992, p. 103). Wallman (1992) argues, however, that this conclusion is ill founded. Almost all the sentences with which Kanzi was presented were pragmatically constrained so that the relationships between agent, action, and object were clear simply from a list of the nouns used in the sentence. (For instance, how likely is it that Kanzi would be asked to carry Penny to the cooler? Or put the refrigerator in the tomato?) With respect even to those sentences that were pragmatically ambiguous enough to require a syntactic analysis, one still must be cautious about rejecting the pragmatic account because the sentences were given in everyday (i.e., not experimentally well-controlled) situations over a three-month period. Without knowing the contexts in which the sentences were presented, it is difficult to know what can be validly concluded about Kanzi's behavior. As for the matter of Kanzi's sentence production, similar doubts arise. According to Wallman (1992, p. 95),

Kanzi's most frequent string consists of "only one lexigram in combination with one or more deictic gestures."

Consequently, Kanzi does not show evidence of the kinds of grammatical knowledge that would pose a serous threat to the Chomskyan view. In fact, as we see below, Kanzi's behavior is precisely what one would expect from an intelligent animal who is attempting to communicate, but does not have the grammatical resources normally available to human beings. What is truly fascinating about Kanzi's performance is its remarkable similarity to the behavior of human beings who, for one reason or another, have access to some basic linguistic vocabulary, but not to the structure of the language to which the vocabulary belongs. This is discussed extensively in the next section.

4. The Evolution of Language

The question of how humans happened to evolve a complex language organ, such as the one proposed by Chomsky and his associates, looms very large. Chomsky (1988) himself, as well as noted evolutionary biologist Stephen J. Gould (1987, cited in Pinker & Bloom, 1990), have suggested that language may be something of an evolutionary accident, a by-product of relatively unrelated evolutionary factors. This is grossly unsatisfying to most people. Bates, Thal, and Marchman (1989), for instance, have argued that if humans were equipped with a language organ such as that proposed by Chomsky, one would expect to find precursors of it in animals that are closely related to us, such as chimpanzees. If Chomsky is right that we do not see any important linguistic abilities in chimps, then, Bates argues, the Chomskyans are left with a sort of "Big Bang" theory of language development that we, of all species, somehow were accidentally blessed with a fully functioning prefabricated language organ to the exclusion of all other animals. Such a picture of the evolution of language, argues Bates, is hardly plausible; many others have concurred. (Notice, ironically, that the real Big Bang theory is, as far as we now know, true. Bates, however, tries to invoke it rhetorically *against* Chomsky.)

The main problem with Bates' argument is the picture of evolution implicit in her account. If, in fact, humans were the direct descendants of chimpanzees, then you might expect the kind of evolutionary continuity for which she plumps. Even then, however, the continuity would be structural, not necessarily functional. That is, you would expect to see physiological continuity in the structure of the brain, but evolution has shown over and over again that a little bit of structural change can bring on a great deal of behavioral alteration, particularly if a number of small changes begin to interact (see, e.g., Pinker & Bloom, 1990).

All this notwithstanding, the basic premise of Bates' argument is demonstrably false. Humans are not direct descendants of chimps. Chimps and humans share a common ancestor. We are cousins, to extend the familial metaphor, and fairly distant cousins at that. There has been fair bit of evolutionary water under the bridge since we and chimps went our separate evolutionary ways: three or four species of *Australopithecus*, and perhaps the same number of *Homo* species as well. What went on, linguistically speaking, during those thousands of millennia is difficult to pin down exactly, but it may well be the gradual development that Bates expects to find. More likely, assuming a more modern evolutionary view like punctuated equilibrium, there were a few jumps in the development of language: perhaps first words, then some structural rules for putting them together more effectively, and so on. In any case, the assumption of gradualism does not imply that the evidence of gradual change must be written on the faces of currently surviving species. As Steven Pinker (1994) has mused, if some gigantic cataclysm had wiped out all mammals between humans and mice, making *them* the animal most closely related to us, would we then be inclined to look for evidence of precursors of all human abilities in mice, and base negative conclusions on a failure to find such precursors? Of course not.

5. Protolanguage

Derek Bickerton (1990) has put forward a persuasive account of how language might have developed in the human species. What makes it especially fascinating is that it makes empirical predictions about what sort of verbal behavior one would expect not just from "cave men," but also from people today who do not have access to normal grammatical knowledge. Bickerton believes that before full-fledged grammatically complex language developed, people spoke something he calls "protolanguage." Protolanguage contains many of the concrete content words of modern language (e.g., rock, sky, father, go, etc.), but few of the function words (e.g., the, to, by, of, etc.). The function words are crucial to making a grammatical language operate. They are what distinguishes, for instance, "Buy the painting by Philip" from "Buy the painting for Philip." Thus, according to Bickerton, before the development and dissemination of function words, people resorted to combining concrete nouns and verbs together with gestures and signs to get their meanings across.

Bickerton has found evidence of protolanguage being used today in situations in which people have access to a rudimentary vocabulary, but little in the way of grammar or function words. For instance, when one is in a foreign country, one must often resort to shouting a couple of nouns, repeatedly, and pointing. More interestingly, this is pretty well the level of language attained by Genie and E.M., discussed in Section 1 above. E.M.'s language production, it will be recalled, "is limited to single words, combined with gesture" (Grimshaw, et al., 1994, p. 5). This just is protolanguage. Perhaps even more interestingly, this also seems to be precisely what the chimps of the ape language studies attain after their linguistic training. To repeat Wallman's conclusion about Kanzi's language production: Kanzi's most frequent string consists of "only one lexigram in combination with one or more deictic gestures." This is protolanguage to a tee.

Indeed, protolanguage seems to pop up almost everywhere that someone is communicating but does not have the grammatical resources to use full-blown language. Consider, for instance, Radford's (1990) analysis of a corpus of over 100,000 early child English utterances. Children who are beyond the single-word stage, into the multi-word stages (ages 18–30 months, roughly), seem to speak protolanguage. To draw material directly from Radford's table of contents, they have not yet developed, or have only incompletely developed, a determiner system, a complementizer system, an inflection system, and a case system. These are precisely the things that are missing from protolanguage.

In fact, Bickerton himself independently points out the similarity. The relation does not seem to be casual either. Bickerton shows that it stands up to a formal analysis and, moreover, that other kinds of language deficits, such as aphasia, seem to result in qualitatively different kinds of verbal behavior. That is, protolanguage is not just a grab-bag of language deficits; it seems to be a relatively well-defined *kind* of verbal behavior that occurs in a variety of situations that are tied together only by being instances in which language is robbed of grammar.

By far the most fascinating part of Bickerton's work, however, has been his examination of pidgins and creoles. Pidgins are languages, of sorts, developed by adults who are forced to operate under circumstances where no one around them speaks their native language, and they do not have access to resources that would allow them to learn properly the language of the place in which they find themselves. This was a quite common situation during the days of the colonial slave trade. All pidgins have features that are theoretically relevant. First, pidgins are syntactically very impoverished. They lack functional vocabulary; they evince no systematic use of determiners, no systematic use of auxiliary verbs, no systematic use of inflections, no subordinate clauses, no markers of tense or aspect, no prepositions (or very few with clear semantic content used unsystematically), and even the verbs are frequently omitted. Second, pidgin word order is extremely variable and unsystematic; this is not, of course, compensated for by the presence of an inflectional system, as is found in many natural languages. Finally, pidgin utterances are restricted to short word strings because, without syntax, longer strings quickly become too ambiguous for unique interpretation.

Thus, pidgins provide extremely impoverished environments in which to learn syntax. In a very real sense there is no syntax present in a linguistic environment dominated by a pidgin. This is important since it bypasses the old debate about how impoverished the linguistic environment of the child really is. Here we have a situation in which it is clear that the linguistic environment is extremely impoverished. It is, thus, truly fascinating what children brought up in such an environment do. In a single generation (though not necessarily the first) they develop a creole. A creole is a full-blown language with its own complete syntax, showing none of the deficits of its "parent" pidgin. Its vocabulary is drawn from a pidgin, but its grammatical principles are not drawn from a pidgin because, to repeat, pidgins do not possess such principles in the first place. Nor does the creole, as Bickerton (1984) takes great pains to demonstrate, derive its grammatical principles from the target language (that of the "masters") nor from any of the substratum languages (one spoken by some subgroup of the laborers). The children appear not to have been using either of these as sources of syntactic information. Although the creole grammar does not reflect either the target or substratum grammars, Bickerton found that the grammars of creoles all around the world are remarkably similar even though they arise independently. Attempts to explain this in terms of similarities between target languages or substratum languages do not work. To take one example, creole languages are predominantly SVO (subject-verb-object word order) languages. One might account for this by saying that all the colonial powers were European and therefore used Indo-European languages. Although this might explain English- and French-based creoles, it will not explain Spanish- or Portuguese-based creoles because both these languages frequently exhibit VSO patterns, nor will it account for Dutch-based creoles, since Dutch is SOV. If one tries to explain the creole word order system on the basis of a common African language family, one is faced with the fact that recent research suggests that "the underlying order of a number of West African languages may well be SOV" (Muysken, 1988, p. 290). Finally, there is the particular case of the creole Berbice Dutch, spoken in Surinam. The substratum language it is directly related to is Ijo, a completely SOV language, and its target language was the SOV language, Dutch. Yet Berbice Dutch is "still a straightforward SVO system" (Muysken, 1988, p. 290). Such evidence indicates that creole similarities are not based on features common to target languages or substratum languages.

What we have with creoles, then, is a situation in which children are acquiring syntax in an environment that is impoverished to such a degree that an account of acquisition based on inductive learning is implausible to the point of incomprehensibility. The burden of proof is on the person who believes that general cognitive abilities account for language acquisition. Such a person must explain the similarity of creole grammars separated by huge distances in space and time. The Chomskyan, however, has an easy and apparent answer: Children possess something like a universal grammar, a set of innate principles with parameters that are pre-set to default values. Barring the regular occurrence of the kinds of grammatical inputs that will reset the "switches" (as one would be unlikely to get in a pidgin environment, given the irregularity of the pidgin utterances) the defaults become fixed and, thus, all creoles have similar grammars.

Pinker (1994) points out that Bickerton's work has been "stunningly corroborated by two recent natural experiments in which creolization by children can be observed in real time" (Pinker, 1994, p. 36). Until recently, there was no sign system for the deaf in Nicaragua. After the revolution the children were brought into schools for instruction. However, the instruction there followed the oralist tradition in which the children were not taught a sign system but were forced to learn to read lips, etc. In the school yards the children communicated with a sign system they cobbled together from the makeshift sign system they used at home (see Jackendoff, 1994, for a discussion of what is called "home sign"). The system used is called *Lenguaje de Signos Nicarauense* (LSN). It is used with varying degrees of fluency by young adults who developed it when they were 10 or older. LSN lacks a consistent grammar, however; it depends on suggestion, circumlocutions, and context. "Basically," says Pinker (1994, p. 36), "it is a pidgin."

Children who were four or younger when they joined the school were exposed early to the already existing LSN. As they grew, however, their mature sign system turned out to be quite different from LSN. Most important, it has a consistent grammar. It uses syntactic devices common to other developed sign languages. It is highly expressive and allows for the expression of complex abstract thought. Due to these significant differences, this language is recognized as distinct from LSN and is referred to as *Idioma de Signos Nicaraguense* (ISN). Basically, ISN is a creole that the deaf Nicaraguan children spontaneously created from the pidgin LSN.

Pinker (1994) also describes an unpublished study, by psycholinguists Singleton and Newport, of a family made up of two deaf adults who acquired ASL late (and therefore use it in a fashion similar to pidgin speakers) and their deaf son, who is known by the pseudonym Simeon. What is astounding about Simeon is that, although he saw no ASL but his parents' defective ASL, his signing is far superior to theirs. He consistently uses grammatical principles that his parents use only inconsistently, and makes use of syntactic operations that appear not to be part of his parents' grammatical competence. As Pinker (1994, p. 39) puts it, with Simeon we see "an example of creolization by a single living child."

Simeon's case, and that of ISN, provide excellent corroborating evidence for Bickerton's thesis while also dealing with some potential criticisms of Bickerton's work. All this recent work on language acquisition in unusual circumstances provides important new evidence for the linguistic nativist that the general cognitivist is going to have a great deal of difficulty explaining. The overwhelming effect of all of these arguments is to shift the burden of proof onto the general cognitivists. They no longer can assume that their position has, *prima facie*, superior plausibility. The cognitivists' claim to hold the more reasonable position requires new argumentation and evidence.

6. Summary and Conclusions

The primary aim of this chapter was to bring the nativism debate about language up to date because far too much time is spent, in our opinion, pointlessly debating issues were resolved or dissolved some 20 years ago. A great deal of fascinating research is currently underway, however, that was not available when those debates began; evidence that bears directly on the current status of linguistic nativism. We have tried to review and integrate some of that evidence here.

First, we argued that debates about "deep structure," "transformations," and the like are no longer simply germane. Chomsky has long since given up on these concepts, largely as a result of penetrating criticism. Instead he, and associates such as Berwick, have advanced a small set of principles and parameters as the bases of universal grammar. It may be that the values of such parameters are set to defaults at birth, but that these can be changed across a small range of values by certain linguistic experiences.

Second, several lines of evidence suggest that these principles are innate and are relatively independent of general cognitive abilities. (1) Their complexity makes it unlikely that they could be learned by small children in the same way as facts about the natural world are learned. (2) The uniformity of both the order in which, and the time at which, children learn them across a widely diverse array of early linguistic and more broadly cognitive experiences suggest that innate factors are crucially involved. Moreover, the dissociations between linguistic and general cognitive abilities found in several individual cases belies the idea that they both spring from the same psychological ground. (3) The similarity of the developmental pattern of structural aspects of language to other uncontroversially innate aspects of psychological development, such as vision, suggests that these also might be innate.

Third, we argued that the evidence from Kanzi, which has provoked so much interest of late, is actually quite weak from the standpoint of demonstrating that an ape can learn complex linguistic structures on the basis of general cognitive capacities. Fourth, we showed that, contrary to widespread opinion, there is nothing particularly unusual, from an evolutionary standpoint, about claiming that humans possess a specific linguistic capacity that no other currently living species seem to have.

Finally, we described the main features of protolanguage, as advanced by Bickerton, and noted that it seems to characterize not only what early humans may have done to communicate, but also the behavior of children who have been raised without language, and animals in ape language experiments, as well as the utterances of small children before full-blown grammar has developed, and of speakers of pidgin languages. We also noted, with Bickerton, the rapid transformation of pidgins to creoles in children who are raised entirely within a pidgin environment, and the grammatical similarity of creoles the world over.

We believe that accounting for facts such as these is crucial to any adequate theory of language acquisition. As far as we can tell, only the Chomskyan program has come even close. This is not by any means to say that the job is done. Language is more complex than we could have imagined when we first began studying it in earnest. There is still a lot of work to do.

References

Bates, E. 1976. *Language and Context: Studies in the Acquisition of Pragmatics*. New York: Academic Press.

Bates, E., Thal, D., and Marchman, V. 1989. "Symbols, and Syntax: A Darwinian Approach to Language Development," in N. Krasnegor, D. Rumbaugh, M. Studdert-Kennedy, and Schiefelbusch, R. (eds.), *The Biological Foundations of Language Development*. Oxford: Oxford University Press.

Berwick, R. C. 1985. *The Acquisition of Syntactic Knowledge*. Cambridge: MIT Press.

Berwick, R. C. 1993. Interview with Jay Ingram, in *The Talk Show. Program 2: Born to Talk*. Canadian Broadcasting Corporation.

Bickerton, D. 1984. "The Language Bioprogram Hypothesis." *The Behavioral and Brain Sciences,* 7, pp. 173–221.

Bickerton, D. 1990. *Language and Species*. Chicago: University of Chicago Press.

Bruner, J. S., Goodnow, J. J., and Austin, G. A. 1956. *A Study of Thinking*. New York: Wiley.

Chomsky, N. 1965. *Aspects of the Theory of Syntax*. Cambridge: MIT Press.

Chomsky, N. 1988. *Language and the Problems of Knowledge: The Managua Lectures*. Cambridge: MIT Press.

Chomsky, N. and Lasnik, H. 1993. "The Theory of Principles and Parameters," in J. Jacobs, A. von Stechow, W. Sternfeld, and T. Vennemann (eds.), *Syntax: An International Handbook of Contemporary Research,* pp. 506–569. Berlin: Walter de Gruyter.

Clark, A. 1993. *Associative Engines: Connectionism, Concepts, and Representational Change*. Cambridge: MIT Press.

Cook, V. J. 1988. *Chomsky's Universal Grammar: An Introduction*. London: Blackwell.

Cowper, E. A. 1992. *A Concise Introduction to Syntactic Theory: The Government-Binding Approach*. Chicago: University of Chicago Press.

Elman, J. 1991. *Incremental Learning, or the Importance of Starting Small*. (Tech. report 9101). San Diego: University of California, Center for Research in Language.

Feldman, J. A. 1994, July. "Structured Connectionist Models." Paper presented at the First International Summer Institute for Cognitive Science, Buffalo, NY.

Green, C. D. & Vervaeke J. (in press). "The Experience of Objects and the Objects of Experience." *Metaphor & Symbol*.

Gould, S. J. 1987. "The Limits of Adaptation: Is Language a Spandrel of the Human Brain?" Paper Presented to the Cognitive Science Seminar, Center for Cognitive Science, MIT.

Grimshaw, G. M., Adelstein, A., Bryden, M. P., and MacKinnon, G. E. 1994, May. *First Language Acquisition in Adolescence: A Test of the Critical Period Hypothesis.* Poster presented at the 5th annual conference of Theoretical and Experimental Neuropsychology/Neuropsychologie Experimentale et Theoreticale (TENNET), Montreal, Quebec.

Habermas, J. 1979. *Communication and the Evolution of Society* (T. McCarthy, Trans.). Boston: Beacon Press. (Original work published 1976.)

Harris, R. A. 1993. *The Linguistics Wars.* Oxford: Oxford University Press.

Jackendoff, R. 1994. *Patterns in the Mind.* New York: Basic Books.

Lakatos, I. 1970. "Falsification and the Methodology of Scientific Research Programmes," in I. Lakatos and A. Musgrave (eds.), *Criticism and the Growth of Knowledge,* pp. 91–196. Cambridge: Cambridge University Press.

Lakoff, G. 1987. *Women, Fire, and Dangerous Things: What Categories Reveal about the Mind.* Chicago: University of Chicago Press.

Lightfoot, D. 1991. *How to Set Parameters: Arguments from Language Change.* Cambridge: MIT Press.

Muysken, P. 1988. "Are Creoles a Special Type of Language?" in F. Newmeyer (ed.), *Linguistics: The Cambridge Survey,* Vol. 2, pp. 285–301. Cambridge: Cambridge University Press.

Nolfi, S. and Parisi, D. 1991. *Auto-Teaching: Networks That Develop Their Own Teaching Input.* (Tech report PCIA91-03). Rome: Institute of Psychology, CNR.

Pinker, S. 1994. *The Language Instinct.* New York: Morrow.

Pinker, S. and Bloom, P. 1990. "Natural Selection and Natural Language." *Behavioral and Brain Sciences, 13,* 707–784.

Pinker, S. and Prince, A. 1988. "On Language and Connectionism: Analysis of a Parallel Distributed Processing Model of Language Acquisition," in S. Pinker and J. Mehler (eds.), *Connections and Symbols,* pp. 73–193. Cambridge: MIT Press.

Prince, A. and Smolensky, P. (In press). *Optimality Theory: Constraint Interaction in Generative Grammar.* Cambridge: MIT Press.

Quartz, S. R. 1993. "Neural Networks, Nativism, and the Plausibility of Constructivism." *Cognition, 48,* 223–242.

Radford, A. 1988. *Transformational Grammar: A First Course.* Cambridge: Cambridge University Press.

Radford, A. 1990. *Syntactic Theory and the Acquisition of Language.* London: Basil Blackwell.

Savage-Rumbaugh, E. S. 1988. "A new look at ape language: Comprehension of vocal speech and syntax." In D. W. Leger (Ed.), *Comparative Perspectives in Modern Psychology,* pp. 201–255. Lincoln, NB: University of Nebraska Press.

Smolensky, P. (1995). "Reply: Constituent structure and explanation in an integrated Connectionist/symbolic cognitive architecture." In C. MacDonald & G. MacDonald (eds.), *Connectionism: Debates on Psychological Explanation* (pp. 223–290). Oxford: Basil Blackwell.

Terrace, H. S. 1979. *Nim.* New York: Knopf.

Vervaeke, J. & Green, C. D. (in press). "Women, Fire, and Dangerous Theories: A Critique of Lakoff's Theory of Categorization." *Metaphor & Symbol.*

Wallman, Joel. 1992. *Aping Language.* Cambridge: Cambridge University Press.

Wason, P. C. and Johnson-Laird, P. N. 1972. *Psychology of Reasoning: Structure and Content.* Cambridge: Harvard University Press.

Yamada, J. E. 1990. *Laura: A Case For the Modularity of Language.* Cambridge: MIT Press.

PART THREE

David Martel Johnson

CONNECTIONISM

A Non-Rule-Following Rival, or Supplement to the Traditional Approach?

At one time—inspired by Alan Turing's conception of a universal computing machine—many philosophers and scientists assumed that a mind always had to take the form of a computational system—that is, a series of procedures exhaustively specified in terms of explicitly formulated rules. After all, these people reasoned, minds were essentially problem solvers; and Turing had shown that what it meant to solve any puzzle or answer any question was simply to produce a list of rule-directed steps that led from one abstractly specified state ("the problem") to another state of exactly the same sort ("the solution").

However, this attitude of "what else can it be?" no longer is universally accepted. Furthermore, in a style typical of the last half of the twentieth century, people's gradual abandonment of this view and their switching to a different conception of mentality has been more the result of concrete technological developments than of abstract arguments. Traditional cognitive scientists (also known as functionalists) only gained the confidence to claim to have discovered a viable alternative to behaviorism, once digital computers and their programs had become widely employed. Similarly, those theorists who now refer to themselves as connectionists only were able to mount a credible challenge to functionalism in turn, when practical experience made it clear that digital computers operating in standard ways could not offer as effective solutions to problems of certain types as informational resources organized along different lines. In particular, some of these latter resources ("neural nets") neither employed nor presupposed the previously sacrosanct notion of following rules.

In general, connectionists do not claim to be able to displace traditional cognitive science in the same wholesale, "revolutionary" way cognitive science itself once proposed to replace behaviorism. Rather, they modestly say that their view provides a means of dealing with special cases that traditional cognitive science never was equipped to handle in the first place.

What sort of cases am I talking about? The answer is that they are those where it is necessary to arrive at a solution by progressive elaboration and refinement of certain initially vague indications, traces, or patterns, rather than by following algorithmic steps to a supposedly inevitable conclusion. Examples of such cases are the problems of learning how to recognize a face or distinguish multiple star systems from single systems on the basis of fragmentary and fluctuating optical and microwave evidence, as contrasted with problems like dividing one number into another, or determining the first prime number that occurs after 10,000,000,000.

I want to make two related points about so-called connectionism. First, despite facile talk by materialists and other sorts of reductionists about "brain/mind," there is an important difference between these two, supposedly identical terms. For example, it is appropriate to think of a standard digital computer as a "mind machine," because its designers explicitly built it to perform tasks that a person (that is, a mind) might accomplish by following rules (for example, intentions and plans) that one might express in words and other symbols. More correctly—someone "might" do this, if only the person were much quicker, more careful, exhaustive, and thorough than in fact is the case. On the other hand, brains are organized along quite dissimilar lines, as shown by the fact that no human being—including the most sophisticated surgeon or neurophysiologist—has any *intuitive* idea of how his or her brain manages to accomplish various tasks—for example, forming constantly updated representations of external temperature, internal blood pressure, the layout of the local environment, and so on and on. At least in the latter sense, it seems appropriate to think of a connectionist problem-solving system as a "brain machine."

Second, the brain carries out its characteristic activities by employing resources that are massively parallel, rather than organized serially. This implies, for example, that unlike a digital computer, a brain cannot contain any central processing unit (or, to use Daniel Dennett's alternative metaphor, "Cartesian theatre") where awareness of everything it does is collected, and from which it is controlled. For instance, learning a second language is a typical "mind task," since a person typically does this by following rules, all of which, in principle, could be set down in a textbook of grammar. But—as Chomsky has reminded us—the same is not also true of learning a first language, as shown by the fact that this does not require similar types or levels of intelligence, sophistication, or ability to follow explicit procedures.

A connectionist system involves the three general levels of (1) input units or "nodes," (2) hidden or processing nodes, and (3) output nodes. Such a system is able to learn to solve problems by a "training-up process," in which it is subjected to a regime of trial and error. The result of each trial is immediately "corrected" on the basis of feedback considerations about whether and how much each of the system's parts—especially the "weight" assigned to each connection between an input or output node and a middle or processing node—has (or has not) contributed toward a correct solution. This allows the system gradually to modify or "shape" itself, until it finally reaches a state where it is capable of transforming new, untested inputs into desired outputs. For example, given a certain group of sounds, the system can report that the sounds are the voice of *X*; or given a particular set of colored lines and patches, it can correctly identify ("recognize") them as a visual representation of *Y*'s face.

The basic ideas here are not so remote from ordinary experience as some have supposed. Thus, theorists sometimes describe walking, talking, and standing as brain tasks, on the grounds that a person does all these things without having the slightest idea of how he or she accomplishes them. But this is not literally true, since we at least have some inkling of how to influence and change such activities consciously and deliberately—although indirectly. Thus, think of a biofeedback machine in a doctor's office. Suppose I desire to relax my stiff and painful neck muscles, but am ignorant of any steps to take in order to accomplish this. The doctor then uses electrical wires to connect the muscles in my neck to the machine in some appropriate way. Next, the doctor instructs me hypothetically to assume various (undescribable) positions and postures in more or less random order. After each trial, the doctor consults the dials or read-outs on the machine, and says simply "Yes" or "No." This procedure may allow me to train myself to relax my neck, without knowing any methods or rules I have followed in order to accomplish this learning.

Let us be more specific about how the trick is done. I remember one of my junior high-school teachers who constantly tinkered with the seating arrangements of her class, in order to bring about a special result. That is, she treated everything she said to the class as a kind of "input" to a connectionist-like informational system. The "output" she wanted was that the class always should give her words careful (and quiet) attention; and she wanted to avoid the opposite situation where the class was frivolous, noisy, and inattentive. Interactions among students were an important part of the processing by which input was transformed into output. Physical distance between students partly determined what sorts of interactions were bound to take place. Therefore, the teacher continually experimented to correct each "mistake" of the class in this respect—that is, the activities and situations she judged to have led to a wrong output in each test-case—by changing where people sat. For instance, if she formed the impression that two girls were constantly chatting in a way that disturbed them and others around them, she would separate them, and by this means weaken the strength of their connection. On the other hand, if she thought there was a preoccupied, bored, dreaming boy sitting in the back row, she would move him closer to the front, in order to create a stronger connection between him and herself. And so on. Similarly—if this analogy deserves to be taken seriously—the method the brain (as contrasted with the mind or person) employs to learn to recognize a particular face is experimentally to strengthen some connections (for example, synapses) among neurons and weaken others until "the right mix" appears.

In the following part, Andy Clark, William Bechtel, and Itiel Dror together with Marcelo Dascal explain and illustrate connectionist systems in more complete and abstract terms than I have done here. Sidney Segalowitz and Daniel Bernstein sound a cautionary note, warning that it is easy for theorists to deceive themselves, whenever they engage in the popular game of comparing connectionist computer systems with complexes of neurons—that is, the working parts of actual, living brains. Finally, Timothy van Gelder suggests that it is a mistake to suppose in the usual way that the most important distinction today among contemporary theories of mind is between computational (rule-following) and connectionist (non-rule-following) systems. Rather, he says the crucial division falls between computational systems on one side and dynamic systems on the other (commonly proposed connectionist systems are not especially clear or typical examples of the latter). Then van Gelder argues for a conception of mind as modeled on certain more or less ordinary physical objects (for example, Watt's governor for steam engines), which retains many advantages widely thought to be characteristic only of connectionist analyses. The crucial point, according to him, is that theorists should emphasize the developing, dynamic character of the physical objects they choose as a model for the mental, in a way that earlier philosophers like Aristotle, John Locke, and Gilbert Ryle did not.

References

Churchland, P. 1988. *Matter and Consciousness: A Contemporary Introduction to the Philosophy of Mind*, rev. ed., pp.156–65. Cambridge: MIT Press.

Churchland, P. 1989. *A Neurocomputational Perspective: The Nature of Mind and the Structure of Science*. Cambridge: MIT Press.

Clark, A. 1993. *Associative Engines: Connectionism, Concepts, and Representational Change*. Cambridge: MIT Press.

Port, R. and van Gelder, T. (eds.) 1995. *Mind as Motion: Explorations in the Dynamics of Cognition*. Cambridge: MIT Press.

12

Andy Clark

From Text to Process

Connectionism's Contribution to the Future of Cognitive Science

What's Special About Connectionism?

Connectionist approaches have, in the space of about a decade, transformed the scope and practice of cognitive science. Yet despite this intense activity it is sometimes hard to see what truly *fundamental* features distinguish the connectionist approach from its rivals. In what follows I try to isolate three such fundamental features. These features together characterize the class of connectionist models of (I claim) greatest *philosophical* interest. The three features are:

1. Superpositional Storage
2. Intrinsic Context-sensitivity
3. Strong Representational Change

Superpositional storage and context sensitivity are treated in the first two sections. The next section addresses the question of whether a network can employ *only* context-sensitive representations and concludes (somewhat surprisingly) that it can. The last section addresses the remaining property of strong representational change.

Superposition—A Key Feature

The key determinant of the class of interesting connectionist models lies, I believe, in their use of superpositional representations. For here we locate the source of both their genuine divergence from so-called "classical" approaches, and of the complex of features (representation/process intermingling, generalization, prototype-extraction, context-sensitivity, and the use of a "semantic metric"—more on all of these below) that make the models psychologically suggestive. The basic idea of superposition is straightforward. Two representations are fully superposed if the resources used to represent item 1 are coextensive with those used to represent item 2. Thus, if a network learns to represent item 1 by developing a particular pattern of weights, it will be said to have superposed its representations of items 1 and 2 if it then goes on to encode the information about item 2 by amending the set of original weightings in a way that preserves the functionality (some desired input-output pattern) required to represent item 1 while simultaneously exhibiting the functionality required to represent item 2. A simple case would be an auto-associative network that reproduces its input at the output layer after channeling it through some intervening bottleneck (e.g., a small hidden unit layer). Such a net might need to find a single set of weights that do multiple duty enabling the net to reproduce any one of a whole set of inputs at the output layer. If *all* the weights turned out to be playing a role in *each* such transition, the representation of the various items would be said to be *fully* superposed. (In most real connectionist networks, the totality of stored representations turns out to be only *partially* superposed.)

This general notion of superposition has been rendered precise in an extended treatment due to van Gelder (1991). Van Gelder terms a representation R of an item C *conservative* just in case the resources used to represent C (e.g., a set of units and weights in a connectionist network) are equal to R, that is, just in case *all of R* is involved in representing C. It is then possible to define fully superpositional representation as follows:

> A representation R of a series of items C_i is fully superposed just in case R is a conservative representation of each C_i. (van Gelder, 1991, p. 43)

As van Gelder notes, this definition, though quite general, can be seen to apply directly to familiar cases of connectionist superposition. These range from the partial superposition found in so-called coarse coding schemes (in which individual units process overlapping inputs) to the total superposition created by tensor product approaches (Smolensky, 1991) in which the representation of a structured item is achieved by the addition (or sometimes multiplication) of the vectors coding for the constituents. The most basic case of connectionist superposition, however, is the standard information storage technique of multilayer feedforward networks using distributed representations. The details of learning and representation in these systems are by now no doubt boringly familiar to many readers and I will not attempt to reproduce the necessary detail[1] here. The fundamental features responsible for the superposition are, however, worth spelling out. They are the *combination* of (1) the use of distributed representations with (2) the use of a learning rule that imposes a *semantic metric* on the acquired representations. A word, then, on each.

Consider a network of units and weights. A representation may be said to be *local* in such a network if it names the content associated with the activity of a single unit (e.g., a network in which a single unit represents grandmothers). It will be said to be *distributed* if it names a content associated with the joint activity of several units. This standard sketch of the distinction is, however, potentially misleading. For as van Gelder (1991) points out, distribution, conceived as the mere *extendedness* of a representational vehicle, is neither distinctive of connectionist models, nor the source of their attraction. Instead, what counts is the use of *internally structured* extended representations. A representation can be said to have internal structure if it is in some sense a *non-arbitrary construction*. An example will help.

One way to represent the letter "A" in a connectionist network would be to have a single unit stand for that letter. In such a system, a different unit could stand for "B," and for "C," etc. This is clearly a localist representational scheme. But now consider a second scheme in which the letters are represented as *patterns* of activity across 78 units. And the encoding scheme is as follows. The joint activity of units 1, 2, and 3 is the representation of "A," the joint activity of units 4, 5, and 6 is the representation of "B," and so on. Clearly, despite the intuitive extendedness of the representations of each letter, the scheme is *still effectively localist*. This is because the representations, although spread out, do not *exploit* that extendedness in any semantically significant way. Contrast, finally, the following scheme. In this scheme, individual units (or groups of units) stand for features of letterforms in a given font. Thus, some will code for the vertical bar common to F & P, others for the horizontal stroke common to F & E, etc. The system's representation of the letter "A" can then be just the joint activity of the various features that distinguish it. Likewise, for "B," "C," etc. Here, at last, we are dealing with distributed representation in an interesting sense. For notice that, in such a scheme, the fact that the letterform "E" shares more features with "F" than it does with "C" will be *reflected* in the system's use of resources to code for the letters. The "E" representation will involve the activation of many of the same units involved in the representation, whereas it may be almost (or even completely) orthogonal to the pattern associated with the letter "C." This is what is meant by speaking of such a system as imposing a *semantic metric*. The semantic (broadly understood) similarity between representational *contents* is echoed as a similarity between representational *vehicles*. Within such a scheme, the representation of individual items is *non-arbitrary*. A new letterform, say, "Z," would have to be represented by a vehicle (a pattern of activity across the set of units) that reflected its position in the relevant similarity space. The upshot is that

> the particular pattern used to represent an item is determined by the nature of that item, and so similarities and differences among the items to be represented will be directly reflected in similarities and differences among the representations themselves. (van Gelder, 1991, p. 41)

The great achievement of connectionism is to have discovered learning rules that cause networks to *impose* such a semantic metric as the natural effect of the process of learning (see footnote 1). This effect is visible in, for example, NETtalk's (see Sejnowski & Rosenberg, 1987) discovery and encoding of phonetic features, and is the prime source of the attraction of such approaches as a means of modeling knowledge of concepts (see Clark, 1993, Chap. 5). The tendency of such networks to represent semantically similar cases using overlapping inner resources (i.e., to use superpositional storage techniques) is also the root of the important properties of prototype-extraction and generalization. We may round off our treatment of superposition by commenting briefly on these.

Connectionist models are usefully seen as deploying *prototype-style* knowledge representations. The idea of a prototype is just the familiar idea of an especially typical example of an item that falls under a given category or concept. Thus, a robin may count as an especially typical bird whereas a penguin does not. A major advantage of organizing knowledge around prototypes is the easy explanation of *typicality judgments*. We judge that a robin is a more typical bird than, say, a penguin because the robin instance shares more features with the prototypical bird than does the penguin. The limiting case of such feature sharing is the case where the instance is identical to the prototype, though we need not suppose (see below) that the prototype always corresponds to any concrete exemplar of the category. A related advantage is that we can judge deviant cases nonetheless to fall under the category just so long as they exhibit enough of the prototypical features to raise them above some threshold.

How, then, do connectionist models come to embody prototype-style knowledge representations? The basic idea is that the distributed, superposed encoding of a set of exemplars results in the features common to the exemplars becoming most strongly associated (i.e., by powerful mutually excitatory links). The natural mechanisms of connectionist learning and superpositional storage immediately yield a system that will extract the *statistical central tendency* of the exemplars. That means that it will uncover which sets of features are most commonly present in the learning set. It will also learn *commonly* occurring groupings of features. To see how superpositional storage can figure in this, consider that a network that uses superpositional storage techniques must amend existing resources if it is to encode new information that overlaps (regarding, say, some subset of semantic features) with information already stored. Connectionist learning techniques do precisely this. The result is that semantic features that are statistically frequent in a body of input exemplars come to be both highly marked and mutually associated. By "highly marked" I mean that the connection weights that constitute the net's long-term stored knowledge about such *common* features tend to be quite strong, since the training regime has repeatedly pushed them to accommodate this pattern. By "mutually associated" I mean that where such features co-occur, they will tend to become encoded in such a way that activation of the resources encoding one such feature will promote the activation of the other. The joint effect of these two tendencies is a process of *automatic prototype extraction*: the network extracts the statistical central tendency of the various feature-complexes and thus comes to encode information not just about specific exemplars but also about the stereotypical feature-set displayed in the training data. The idea of knowledge being organized around such stereotypical feature-sets is an important theme in recent psychological literature. The description of connectionism as crucially involving the organization of knowledge around representations of prototypes has recently been pursued in detail by Paul Churchland (see, e.g., Churchland, 1989, especially Chaps. 6 and 10).

Prototype extraction, thus conceived, is of a piece with *generalization*. A net is said to generalize if it can treat novel cases sensibly, courtesy of its past training. Any net that responds to a barrage of training exemplars by extracting a prototype will be well placed to succeed in future, novel cases. For as long as such cases display some of the central themes (e.g., sub-complexes of features) extracted from the training corpus, the network's response will be sensible: for example, a novel instance of a dog (say, one with three legs) will still be expected to bark just so long as it shares enough of the doggy central tendencies to activate the knowledge about prototypical dogs. Such a "dog-recognition network" is detailed in Chapter 17 of McClelland, Rumelhart, and the PDP Research Group (1986) vol. II. We can use it to introduce one last property of connectionist-style knowledge of prototypes, namely, its *flexibility*.

The dog recognition network (see also Clark, 1989, Chap. 5) is trained on a variety of cases in which descriptions of correlated sets of dog-features (supposed to correspond to individual dogs) are fed to the network. These descriptions are obtained by selecting one set of features and stipulating that these describe the prototypical dog. The training cases are then derived by creating a series of deformed versions of this description (i.e., by changing individual features). The network is then trained *only* on these deformed instances: it never "sees" the "prototypical" dog. Nonetheless, it is able (courtesy of the learning rule) to act as a statistical "signal averager" and hence extract the general pattern exemplified in the overall set of deformed instances. The result is that not only can it recognize the deformed cases as cases of dogs, but it can also reproduce the pattern of activation for the prototypical dog (which, recall, it was never exposed to) if it is fed a portion of that pattern as an input cue. It can also generalize, that is, categorize new but deformed exemplars on the basis of similarity to the prototype. In short, the net has come to encode a multidimensional feature space organized around a central point (the sipal average, prototype, hot spot, harmony maximum, call it what you will). In addition, such a network can find assignments of weights that enable it to store knowledge about *several* categories in a single set of weights. As a result the network

> does not fall into the trap of needing to decide which category to put a pattern into before knowing which prototype to average it with. (McClelland, Rumelhart & Hinton, 1986, p. 185)

That is, distributed connectionist encoding provides for a high degree of flexibility in the system's use of the correlated-feature information it stores. It is not forced to "decide," in advance, on a particular set of prototypes. Instead, it can settle into "blended responses" that in effect mix up elements (or sets of elements) from two (or more) prototypes. Thus, the dog recognition network, if it also encoded information about cats, could (in response to ambiguous inputs) produce a response blending elements of both (op. cit., p. 188). Likewise, a network that encodes featural information that amounts to knowledge of prototypical room contents (e.g., the contents of a normal kitchen, bedroom, etc.) can, if the context-fixing input is sufficiently peculiar (including, say, a *bed* feature and a *sofa* feature), complete to a pattern of activation that departs from both the typical bedroom and the typical living room. Instead, it fills in features appropriate to a "large fancy bedroom" (Rumelhart, Smolensky, McClelland, & Hinton, 1986, p. 34). This ability depends on the network's encoding only *local* information: information about correlations and inhibitions between groups of features. It has not decided in advance just how this local information should be used to carve up the world. Instead, such decisions can be made according to the context of use (see below) and are thus maximally sensitive to input information.

To sum up, the use of superpositional storage techniques combined with the exploitation of distributed representations, yields

Non-arbitrary representations
A semantic metric among representations
Automatic prototype extraction
Generalization
Flexible deployment of prototypes.

We next consider a powerful consequence of this organization of *non-arbitrary representations* into a *semantic metric*, namely, the deep *context-sensitivity* of these kinds of connectionist representations.

Intrinsic Context-Sensitivity

Suppose we concentrate, for now, on *activation patterns* as the locus of representational activity in a network. (The idea of weights on connections as representations will be addressed later.) We can then observe that the representation of a specific item will involve a distributed pattern of activity that contains sub-patterns appropriate to the feature-set involved. Recall also that such a network will be able to represent several instances of such an item, which may differ in respect of one or more features. Recall finally that such "near neighbors" will be represented by *similar* internal representational structures, that is, the vehicles of the several representations (activation patterns) will be similar to each other in ways that echo the semantic similarity of the cases — this is the semantic metric (see above) in operation. One upshot is that such a network can learn to treat several inputs; this results in subtly different representational states (defined across the hidden units) as prompting outputs that have much in common — for example, they could all activate a common label such as "dog." The property of context-sensitivity, as I will understand it, relies on essentially this kind of process, but taken in *reverse*. Thus, a net, exposed to a label like "dog," will need to fix on one of the several inner states that are associated with such a label. To do so, the net relies on contextual information. Thus (to adapt the kind of example given in McClelland & Kawamoto, 1986), the representation of "dog" in the context of "wore a woolly protector" might be driven to a position, in the overall representational space, appropriate to a poodle feature-complex, whereas in the context "mauled the burglar" it might activate a Rottweiler feature complex. This variability in the inner vehicles and (going hand in hand) in the detailed content represented, leads Smolensky to claim that:

> In the symbolic paradigm the context of a symbol is manifest *around* it and consists of *other* symbols; in the subsymbolic paradigm the context of a symbol is manifest *inside* it, and consists of subsymbols. (Smolensky, 1988, p. 17; original emphasis)

The standard example of the kind of context-sensitivity just described is Smolensky's infamous coffee case. Briefly, Smolensky (1991) describes a connectionist representation of "coffee" that is distributed across units or groups of units that code for various features of coffee-involving scenarios. Consider now the distributed representation of "cup with coffee." And suppose (for simplicity's sake) that the set (or vector) of active hidden units is just that set comprising the units coding for the microfeatures "upright container," "burnt odor," and "brown liquid contacting porcelain." What, then, constitutes the network's representation of the conceptual constituent coffee? A suggestion (due originally to Zenon Pylyshyn) is that to isolate the representation of coffee you just take the one for "cup with coffee" and *subtract* the cup parts. But if we do this, an interesting effect is observed. Subtracting the cup microfeatures (e.g., "upright container") leaves us with a set of microfeatures that include a number of *contextually biased* items like "brown liquid contacting porcelain." Now imagine that instead of starting with "cup with coffee" we had started with "can with coffee." In that case (as Smolensky, 1991, p. 207–212 points out) the representation of coffee would include a contextual bias for the can scenario (e.g., "granules contacting tin"). The upshot is that there need be no context-independent, core representation of coffee. Instead, there could be a variety of states linked merely by a relation of family resemblance (i.e., State *A* overlaps with *B*, and *B* overlaps with *C*, but *A* and *C* need not have any members in common). The unit level activation profile that characterizes the system will thus not map neatly and accurately on to a conceptual level specification. A single, recurrent conceptual level item will have a panoply of so-called "sub-conceptual" realizations; and which realization is actually present will make a difference to future processing. This feature (of multiple, context-sensitive sub-conceptual realisations) makes for the much vaunted fluidity of connectionist systems and introduces one sense in which such systems merely *approximate* their more classical cousins. For classicists picture the mind as manipulating context-free symbolic structures in a straightforwardly compositional manner. Connectionists, not having context-free analogues to conceptual level items available to them, have to make do with a much more slippery and hard-to-control kind of "compositionality" in which the compositionality consists in the mixing together of context-dependent representations. Thus, Smolensky writes, of the coffee example, that:

> The compositional structure is there, but it's there in an *approximate* sense. It's *not* equivalent to taking a context-independent representation of coffee and a context-independent representation of cup—and certainly not equivalent to taking a context-independent representation of the relationship in or with—and sticking them all together in a symbolic structure concatenating them together to form syntactic compositional structures like "with (cup, coffee)." (Smolensky, 1991, p. 208)

The most radical description of this rampant context-sensitivity would be that (these) connectionist systems *do not involve computations defined over symbols.* Instead, any accurate (i.e., fully predictive) picture of the system's processing will need to be given at the numerical level of units and weights and activation-evolution equations, while more familiar symbol manipulating computational descriptions will at most provide a rough guide to the main trends in the global behavior of the system. The proposal, then, is just this: that there are no syntactically identifiable elements that both have a symbolic interpretation and can figure in a full explanation of the totality of the system's semantic good behavior, that is, "There is no account of the architecture in which the same elements carry both the syntax and the semantics" (Smolensky, 1991, p. 204). This is what is meant by Smolensky's description of connectionism as a "two-level architecture," namely, that:

> Mental representations and mental processes are *not* supported by the same formal entities—there are not "symbols" that can do both jobs. The new cognitive architecture is fundamentally two-level; formal, algorithmic specification of processing mechanisms on the one hand, and semantic interpretation on the other, must be done at two different levels of description. (Smolensky, 1991, p. 203)

Mental processes, according to Smolensky, are best understood by reference to the numerical level descriptions of units, weights, and activation-evolution equations: the elements at this level defy semantic interpretation. On the other hand, the larger scale activity of such systems *allows* interpretation, but the patterns thus fixed on are not capable of figuring in accurate descriptions of the actual course of processing. (See Smolensky, op. cit., p. 204.) This is because the *interpreted* patterns (e.g., the groups of vectors associated with a single conceptual level item like "coffee") paper over those more microscopic differences, which nonetheless *make a difference* to the future course of processing.

Since the coffee-example is now stale (as it were), it may help to introduce a further, slightly more complex case. We can then get down to the real question all this raises, namely, in what sense, if any, the idea of a subsymbol or a subconceptual constituent amounts to anything more interesting than the idea of *smaller symbols*. Here, then, is a second case.

Elman (1991a, 1991b) describes a net that aims to categorize words according to lexical category (e.g., verb, noun, etc.). The goal was to learn, by exposures to sequences of linguistic input, something about the classes and categories of the words presented. A simple recurrent architecture was used that consisted of (1) a standard three-layer feedforward network and (2) an additional set of context units connected to the hidden unit layer. The context units are set up to copy the activation at the hidden unit layer and (on the next time-step) to feed that information back to the hidden unit layer, which is thus simultaneously receiving the external input (from the first layer of the standard three-layer net) *and* the "temporal context" information, that is, the copy of its own previous state.

The network's task is prediction. That is, it must take a succession of input words and predict (by output layer activation) the next word in the sequence. A lexicon of 29 nouns and verbs was used, and these were composed into a training corpus of 10,000 two- and three-word sentences. The structure of these sentences reflected properties of subclasses of the lexical items, for example, "only animate nouns occurred as the subject of the verb eat, and this verb was only followed by edible substances" (Elman, 1991a, p. 348). Naturally, there is no unique answer to the question "What word should come next?" in this task. Nonetheless, there are certainly words that should not come next, and the network's performance was evaluated on that basis, that is, on whether or not it identified a valid succession *class* word.

The network proved fairly successful at the task, and a subsequent analysis of the internal representations at the hidden unit layer revealed that it was indeed partitioning the space into recognizable lexical categories, for example, it displayed very similar hidden unit activation for mouse, cat, and dog, thus warranting a cluster label of "animal," and at a coarser grain it displayed a similarity in its treatment of all nouns in the lexicon, and a similarity in its treatment of all verbs. In a sense, then, it has also "discovered" the categories "noun" and "verb." Mid-range sensitivities indicated groupings for animate and inanimate objects, food and breakable objects.

Two of Elman's observations are especially pertinent to our discussion. First, the categories are "soft"—there can be genuine borderline cases and membership is always more or less rather than all or none. Second,

> In this simulation the context makes up an important part of the internal representation of a word. Indeed, it is somewhat misleading to speak of the hidden unit representations

as word representations in the conventional sense, since these patterns also reflect the prior context. As a result it is literally the case that every occurrence of a lexical item has a separate internal representation. (Elman, 1991a, p. 353)

Thus, imagine that the network is dealing with information concerning an individual, John. Nonetheless, Elman insists:

We cannot point to a canonical representation for John; instead there are representations for John1, John2 . . . John$_n$. These are the tokens of John, and the fact that they are different is the way the system marks what may be subtle but important meaning differences associated with the specific token. (Elman, 1991a, p. 353)

The various tokens of John will be grouped together insofar as they involve very similar patterns of hidden unit activity. But the subtle differences between them will *make* a difference; they will build in information concerning the current context of occurrence of the word.

Elman's model is thus deeply *dynamic*. Against a classical picture in which a language master stores context-free representations of lexical items that are retrieved when the word is heard and that "exist in some canonical form which is constant across all occurrences" (Elman, 1991a, p. 377). Elman urges a fluid picture in which:

there is no separate stage of lexical retrieval. There are no representations of words in isolation. The representations of words (the internal states following input of a word) always reflect the input taken together with the prior state. . . . The representations are not propositional and their information content changes constantly over time in accord with the demands of the current task. Words serve as guideposts which help establish mental states that support (desired) behavior; representations are snapshots of those mental states. (Elman, 1991a, p. 378)

The hard question we have gently been leading up to can now be formulated. The basic issue is: "How far can such context-sensitivity go?"—and, in particular, mustn't it "bottom out" in some set of context-free representations that (*pace* Smolensky's claim quoted earlier) really do deserve the title of symbols? The question is best taken in two parts, namely:

1. Could all of the system's *word-level* knowledge be thus context-sensitive?
2. Even if that makes sense, mustn't there still be some context-free content-bearers that, although not identical in meaning to any public language *words*, still deserve the title of "symbols"? (e.g., the *real* microfeatures out of which so-called "hokey" ones, like "burnt odor" are built up.)

I will argue that the system's representations could be context-sensitive at all levels, and hence that the answers to the questions are "yes" and "no," respectively. That is to say, I think there is a clear sense in which the whole idea of context-free inner symbols ("building blocks" of thought) is inapplicable to connectionist systems. The apparently paradoxical air of supposing that a system's representations can be context-sensitive "all the way down" is caused by our persistent tendency to think in terms of static, "classical" symbols that persist as unaltered, stored syntactic items and are *retrieved* (rather than constructed) in the course of processing. In the next section we develop an alternative model.

Are Subsymbols Just Little Symbols?

In a posting to the e-mail newsgroup "Connectionists-request" I raised the problem concerning context-sensitivity like this:

> Current wisdom in philosophical circles depicts a major sub-class of connectionist representations as context-sensitive. Standard examples include Smolensky's (in)famous coffee story [see above] in which the occurrent state which represents coffee is composed of a set of microfeatures which vary according to the overall context. Thus in a "spilt" context, the distributed pattern for coffee might include a "contacting tablecloth" feature. Whereas in a "cup" context it includes, for example, a "contacting porcelain" feature.
>
> Now things, clearly, can't stop there. For we need to ask about, for example, the feature "contacting tablecloth" itself. Presumably it too is context-sensitive and is coded as a bag of microfeatures whose exact make-up varies according to context (e.g., it includes "paper" in a "transport cafe" context, cloth in a "Hilton hotel" context).
>
> My question, then, is this: Must this process bottom out somewhere in a set of microfeatures which are genuinely SEMANTIC (genuinely contentful) but which are NOT prone to contextual infection? If the process DOES bottom out, don't we have a kind of language of thought scenario all over again—at least insofar as we have systems which systematically BUILD new (context-dependent) representations out of a basic stock of context-free atoms? (Quoted from message to . . . "Connectionists-request," February 1992.)

But if it is supposed instead that the process does *not* bottom out, isn't there a puzzle about how the system builds appropriate representations *at all*? Context-sensitivity seems intelligible if it involves combining *basic* representational resources as some input probe suggests. But without any such basic resources it seems like the *ex nihilo* creation of complex representations. (Such creation seems to occur in a learning phase proceeding from random weights to encoded knowledge. But contextual shading of occurrent representations does not involve learning as such.)

I raise this question because it seems unclear which (if either) of the following two claims best characterizes the highly distributed "sub-symbolic" connectionist vision, namely:

1. Really Radical Connectionism (RRC)
   Representation is context-sensitive all the way down.
2. Radical Connectionism (RC)
   The representation of all daily *conceptual level* items ("folk" objects, states, and properties) is context-sensitive *but* this sensitivity consists in the systematic combination of unfamiliar (but genuinely contentful) micro-representations that are context-free content bearers.

Finally, it is of course true that representation itself must stop at some point. My question is not "Is there a *non*-representational level?" but rather "Is there a level of *context-free representation*?"

The response was enormous, and enormously divided. Many working connectionists opted for the semi-classical vision (RC) in which context-sensitivity is conceived as the combination of context-free representational primitives. But just as many (including, notably, Jeff Elman whose work we described above) opted for context-sensitivity all the way down (RRC). I now believe (see below) that the root of the trouble lay in my phrasing the (apparently exhaustive and exclusive) options in the vocabulary of *symbols* and *combination*. I was in fact offering only a Hobson's choice or no-win situation. Once the problem is conceived in such terms, the "context-sensitivity all the way" option is indeed pretty well unintelligible. But we do not have to think in those terms.

Thus, consider the very *idea* of a "context-free atom." What can it be? One reasonably clear definition is:

> Something is a "context-free atom" iff it is (a) a syntactic item (i.e., one individuable purely by its non-semantic properties) and (b) plays a fixed representational role in a symbol system.

An inner state that makes the same semantic contribution to all the larger states in which it figures thus counts as a "context-free atom." (Fodor's innate Language of Thought posits a set of innate representational atoms in just this sense.) Correlatively, consider the very idea of a microfeature. This can, it seems, be nothing but the idea of a context-free atom of the kind just described. What else *could* it be? At this point, nothing except the conservative, semi-classical vision of a symbolic economy, albeit one involving "smaller" symbols, seems to make sense.

Why is it then, that we seem to find it so hard to imagine "bottomless" context-sensitivity? The reason, as hinted above, is that the vocabulary in which the question is posed is itself deeply classical, and leads us to make a crucial error. The mistake is to try to understand an essentially dynamic approach in essentially static terms. This becomes clear if we recall the definition of "context-free" just given and ask what is the "it" that is either supposed to make, or fail to make, "the same semantic contribution to any larger representational state of which it is a part"? In familiar classical approaches, the answer is clear: namely, a syntactically defined item that is taken as the persisting vehicle of some single content. In such approaches, these items may be severally brought together in ways that reflect any contextual nuances. (See the discussion of "concatenative compositionality" in van Gelder, 1990, and Clark, 1993, Ch. 6). But the items themselves are unaffected by this bringing together, and contribute what they do largely independently of their current "neighbors." In connectionist approaches of the kind detailed above, however, such items simply do not exist. Symbols, insofar as we can use such talk at all, exist in these systems only as a present *response* to an (endogenously or exogenously originating) input. And these transient constructs internally reflect the local context in their very structure. What we would need to justify the claim that there are context-free atoms acting as the representational base-line of such systems is to discover syntactic structures that persist unaltered and that carry a fixed content. The only contenders for such persistent content-bearers are the weights that moderate the various context-sensitive activity patterns. But a given weight, or set of weights, cannot be identified with any fixed content in these superpositional systems. For each weight contributes to *multiple* representational abilities. It makes no sense to ask what this or that weight means. It is only relative to some specific inputs that the weights give rise to activation patterns that can be seen as expressing this or that (context-involving) content.

The mistake, then, is to suppose that connectionist systems use symbols *in the familiar sense* and then to ask whether some of them are best seen as a base-line of atomic representations. Context-sensitivity does not have to bottom out in context-free symbolic primitives because there are no *symbols* here at all, taken independently of some context. Symbols, in short, do not exist in these systems except as context-sensitive responses to inputs. No wonder it is context-sensitivity all the way!

It is worth commenting that one possible source of confusion here is the common (and laudable) use of techniques such as cluster analysis (see, e.g., Sejnowski & Rosenberg, 1987) to generate a static symbolic description of a network's knowledge. Such analyses in effect transform a body of episodes of actual context-reflecting processing into a static (context-transcending) symbolic description. Such analysis is indeed vital; but we should not then be misled into thinking that the symbols thus discerned somehow exist as long-term syntactic items in the network. They do not.

To sum up, it is indeed possible to view superpositional distributed connectionist systems as context-sensitive all the way down. For the fixed resources out of which the various representations (activity patterns) are non-arbitrarily constructed are just the weights that (1) encode *multiple* items and types of knowledge and (2) do so in a way that yields, on any given occasion, only a highly context-reflecting occurrent representation. When Smolensky claims that in his approach there are no symbols that figure both in mental processes and mental representations he may thus be taken as meaning that the idea of a symbol as both a bearer of some fixed content and a persistent inner item over which computational processes can be defined is inapplicable to these models. Now we know why.

Strong Representational Change

Several of the themes we have been developing come together to yield the final property to be explained: the potential of connectionist approaches to model what I call strong representational change. The notion of strong representational change is best understood by contrast with (did you guess?) one of weak representational change. An example of weak representational change is found in Jerry Fodor's discussions of nativism and concept-learning (see, e.g., Fodor, 1975, 1986). Fodor suggests that concept-learning is best understood as a rational process involving hypothesis generation and testing. "Rational" here is contrasted with other ways in which our conceptual repertoire might be affected by experience such as "being hit on the head . . . having your cortex surgically re-wired, etc." (Fodor, 1986, p. 275). Fodor does not say in more positive terms what makes a process rational, but it seems safe to assume that the idea is that in rational concept-learning the perceived *contents* of the training data are the cause of the acquisition or grasp of the new concept, as against other (non-rational) routes that bypass the training data (e.g., being hit on the head), or depend on it in some weaker way (e.g., as soon as you see anything at all, you find the concept "mother" has been triggered in your mind). (The content was not *given* in the training data, but merely required some nudge from the environment to bring it into play.) Fodor's key claim is that the only model we have of a rational process of concept learning is one in which we use evidence given in training cases to assess various internally constructed hypotheses concerning the content of some new concept. Thus, for example, if you have to learn the meaning of "FLURG," you construct a hypothesis of the form "Something is an instance of 'FLURG' if and only if *P*," where *P* is filled in in some way suggested by your experiences to date. Then you check your hypothesis by seeing if future cases fit it. Once you have found a way of unpacking *P* that fits all the cases you are encountering, you have completed the process of rational content assignment.

In short:

> Concept learning involves . . . the formulation and confirmation of hypotheses about the identity of the concept being learned. (Fodor, 1986, p. 267)

But the sense in which such a model is properly termed "weak" is now manifest. As Fodor puts it:

> On this view, learning the concept FLURG is learning that a certain hypothesis is true; viz., the hypothesis that the concept FLURG is the concept of something that is green or square. But learning that the hypothesis is true is in turn said to be a matter of entertaining that hypothesis and comparing it with the data. But that hypothesis itself *contains* the concept GREEN OR SQUARE. Now, surely, if a concept is available to you for hypothesis formation, then you have that concept. And if you have to have a concept in order to perform the concept-learning task, then what goes on in the concept-learning task cannot be the learning of that concept. So the concept-learning task is not a task in which concepts are learned. (Fodor, 1986, p. 269)

Concept learning thus understood involves only *weak* representational change insofar as it does not result in any increase in the *representational power* of the system. For to learn a "new" concept at all, on the generate and test model, you must *already* possess the representational resources to express its content. The potential repertoire of the system is fixed by the repertoire available for expressing hypotheses. And this (for obvious reasons) cannot *itself* have been acquired by the generate and test method, so (Fodor concludes) it must be innate. The mind is thus viewed on the model of a body of static, evolutionarily determined text subject only to minor re-combinatorial episodes of weak re-organization. More on this in a moment. First, we must guard against an all-too-easy misreading of Fodor's claim.

Fodor's image of rational representational change is *weak* insofar as it depicts the products of such change as necessarily falling *within* the expressive scope of the original representational base. It is weak, I repeat, insofar as it depicts rational representational change as limited by a *preceding* and *representational* base. Notice, then, how this goes beyond the mere fact that the evolution of a system is constrained by its original potential. The latter is trivially true. As Christiansen and Chater (1992) note,

> potential of *any* sort cannot increase . . . for a system to learn or do anything, it must necessarily have had the potential to learn or do it in the first place. (Christiansen and Chater, 1992, p. 245)

But this is not Fodor's point. The point is that concept-learning (according to Fodor) can only involve either (or both) of two processes, namely, (1) the *triggering* of an innate representational atom and/or (2) the deployment of such atoms in generate and test style learning.

Either way, the basic representational power of the system remains unaltered—its representational scope is *fixed* by the innate representational repertoire. It is this fixing of expressive scope in virtue of an innate *representational* repertoire that is the distinctive, and non-trivial, feature of Fordor's story.

Fodor thus depicts all rational processes of representational change as (a) involving an innate representational base, (b) exploiting the training environment only as either a source of triggering experiences (for the base) or as a source and test-bed for conjectures (for the rest), and hence (c) as being fundamentally conservative, since representational power never really increases. This is the image of mind as preexisting text. Connectionist approaches invite us, I believe, to think again. By shifting much more weight onto the training environment, the connectionist is able to treat learning as essentially rational (according to our earlier definition) yet as capable of genuinely transforming the representational capacities of the system. To see how, let's briefly contrast the connectionist stories concerning (a) and (b) above.

Regarding the putative innate representational base, we should note first that many connectionist models acquire domain knowledge without the benefit of any such resource. These are those models that begin with a set of random connection weights and learn about a domain "from scratch." Famous examples include NETtalk (Sejnowski & Rosenberg, 1987) and the past tense learning network(Rumelhart & McClelland, 1986a). It is true that the initial choices of architecture, number of units and layers, etc. amount to the building in of a little knowledge; but it is not at all akin to the provision of an innate base of atomic symbols ready for combination into hypotheses to be tested against the evidence. The only real candidate for such a base is not the overall architecture but the connection weights. And there is a sense (though not, as we'll see, a very interesting one) in which the weights do participate in a generate and test procedure, namely,

> the hypotheses are embedded in the weights of the network, the test is the measure of network performance (such as sum-squared error), and the procedure for generating new hypotheses, given the successes or failures of past hypotheses, is given by the learning algorithm. (Christiansen & Chater, 1992, p. 244)

The point, however, is that the initial weights (assuming a random starting point) are not usefully seen as constituting a set of representational elements (ask yourself what such weights represent!); and, *a fortiori*, the subsequent learning of the network is not usefully understood as constrained by the representational limitations of an initial "language."

The random-start case is a useful existence proof of the ability of some systems to engage in something like rational concept acquisition without an innate representational base. For they do not acquire knowledge by accident (not like the "bang on the head" case) nor by simple maturation or external re-wiring. Instead, what they learn is a consequence of the *contents* of the training cases, and the process is thus a rational one as we (and I think, Fodor) are using the term.

The existence proof is useful, but we should not be carried away into thinking that connectionism must buy into a "tabula rasa" model of knowledge acquisition. Such a model would be implausible on well-documented empirical grounds (see, e.g., Baillargeon,1987; Spelke, 1991). The precise way in which such knowledge (e.g., about physics, faces, and language) may be built in remains an open question. But one obvious option is for evolution to preset some or all of the weights so as to embody some initial domain of knowledge. Even so, notice that we no longer need hold that this knowledge limits future learning in anything like the way Fodor imagines. For it does not constitute a set of representational resources in terms of which any target knowledge must already be expressible if it is ever to be acquired. If some target knowledge lies outside the expressive repertoire of the inbuilt representations, it may still be acquired (supposing the system's basic resources are adequate and local minima are avoided) in the same way as it would be in the random-start case. Correlatively, if some of the inbuilt knowledge turned out to be false or misleading in the domain as evidenced by later training data, the system can undo the pre-set weightings and try again (see Rumelhart & McClelland, 1986b, p. 141). Finally, we might also note that the *distributed* nature of the method of encoding knowledge (i.e., encoding it across a whole set of weights) allows us to make sense of *partial innate knowledge* namely, a system might, as Bates and Elman put it, encode "90% or 10% of any innate idea" (Bates & Elman, 1992, p. 17). This ability to model degrees of innateness of ideas is hailed by Bates and Elman as potentially constituting "connectionism"'s greatest contribution to developmental cognitive neuroscience" (op. cit., p. 17). Whether this is so or not, it certainly does seem to be the case that connectionist approaches offer a rich and varied apparatus for dealing with questions of innate knowledge and the developmental trajectory of learning, and do so without the implausible and overly constraining hypothesis of an innate symbol system whose representational resources are fixed once and for all.

The second point of contrast concerns the role of the environment and the training data. The effects of these are severely downplayed in Fodor's approach. The various inputs to the system may trigger dormant innate representations, or they may contribute to the choice of hypotheses concerning target meanings, or they may be used to test such hypotheses. But in all cases, the training data subserve a fully fledged preexisting and unchanging representational economy—an economy that from the outset possesses its complete representational repertoire and is organized as a systematic symbol system. In connectionist approaches, by contrast, the training environment, for good or ill, is a major determinant of both the knowledge and the *processing* profile acquired by a network. To give just one example, many classical models depict the use of rule-systems in various linguistic tasks (e.g., pronunciation, past-tense formation) as dependent on a distinct processing resource that is innately given (though it does not emerge until a certain point in the maturational cycle). Connectionist approaches, however, can display the emergence of a rule-system as a product of training in a single network. (For a nice discussion, see Marchman, 1992.) A further interesting prospect is that much of what is currently attributed to innate modular structure in the brain might in fact be achieved by a process of highly input-sensitive learning that results in degrees of *functional modularity*, that is, sets of units "that are powerfully interconnected among themselves and relatively weakly connected to units outside the set" (Rumelhart & McClelland, 1986b, p. 141). The effects of localized damage to the hardware (the brain) might as a result vary according to variation in the "modularizing" learning experiences of the individual. And damage to the same area at different *times* during the course of learning might produce quite different effects. (See also Marchman, 1992; Bates & Elman, 1992.)

More crucial than all this, however, is the way continued training can bring about *qualitative changes* in the performance of a network. It is here that the connectionist emphasis on process becomes visible. We touched on one example of this above, namely, the case in which a single connectionist mechanism that begins as a rote memorizer of past tenses can, as a result of continued training, begin quite suddenly to exhibit knowledge of the rule[2] for the regular past tense (while still retaining knowledge of irregular cases, subject to an initial U-curve effect—see, e.g., Plunkett & Marchman, 1991). It is worth stressing the suddenness of the transition, since it is direct evidence of the ability of these systems to model *stages* in a dynamic process of cognitive development. Thus we read that:

> Performance in the network reveals a *critical mass* effect. . . . That is to say, the transition from a state of rote representation to a state of systematic representation (involves) . . . a sudden transition between the two modes of representation when the training set reaches a given critical size. (Plunkett & Sinha, 1991, pp. 30–31)

In a similar vein, another treatment notes that:

> In trying to achieve stability across a large number of superimposed, distributed patterns (a) network may hit on a solution that was "hidden" in bits and pieces of the data; the solution may be transformed and generalized across the system as a whole resulting in what must be viewed as a qualitative shift. (Bates & Elman, 1992, p. 15)

More detailed examples of such qualitative shifts are presented in Clark (1993). For now, we need only note that the *root* of this ability lies in the deep interpenetration of knowledge and processing characteristics that is fundamental to connectionist approaches. Processing in these systems involves the use of connection weights to create (or re-create) patterns of activation yielding desired outputs. But these weights just *are* the network's store of knowledge. And new knowledge has to be stored superpositionally, that is, by amending existing weights. Changes in the knowledge base and in the processing characteristics thus go hand in hand. As McClelland, Rumelhart, and Hinton put it:

> The representation of the knowledge is set up in such a way that the knowledge necessarily influences the course of processing. Using knowledge in processing is no longer a matter of finding the relevant information in memory and bringing it to bear: It is part and parcel of the processing itself. (McClelland, Rumelhart, & Hinton, 1986, p. 32)

Text (knowledge) and process (the use and alteration of knowledge) are thus inextricably intertwined. And critical mass effects are thus to be *expected*; for we are dealing with systems that are highly data-driven and whose processing profiles are an immediate consequence of their state of superposed stored knowledge. A few extra items of data can push such a system out of error-inducing local minima and completely re-structure its output profile. The great promise of connectionism in this respect is thus to offer a detailed understanding of how the combination of this distinctive mode of processing/storage and the stream of environmental inputs can yield the kind of qualitative increments in performance that have so long been at the heart of psychological theories of cognitive development.

Conclusions: From Text to Process

The radical connectionist vision is pretty much the antithesis of Fodor's vision. Fodor depicts the mind as a static representational storehouse, subserved by innate modules, and evincing qualitative representational change only if it occurs as part of a fixed maturational program. Connectionist approaches allow (rather subtle kinds of) innate knowledge but do not restrict learning to manipulations of a fixed representational base. They allow us to explore modularization as a knowledge-driven process, and they encourage the investigation of strong representational change as a product of genuine learning. What Fodor sees as the starting point of our mental life (a store of context-free, recombinable representational atoms) the connectionist sees as (at best) the end-point[3] of a sustained developmental process. Each approach has its attendant attractions and problems. But the feature-complex distinctive of the radical connectionist vision has at last fully emerged. It comprises the ability to build representational fluidity (see the section on intrinsic context-sensitivity) and representational change (above) into the heart of a practicable approach to cognitive modeling. Where the classicist thinks of mind as essentially static, recombinable *text*, the connectionist thinks of it as highly fluid environmentally coupled dynamic *process*. This shift of emphasis, I believe, constitutes connectionism's most fundamental contribution to shaping the future of cognitive science and the philosophy of mind.

Notes

This chapter draws heavily on material from Chapter 2 of Clark (1993). Thanks to the MIT Press for permission to use that material here.

1. For introductions see Rumelhart, McClelland, and the PDP Research Group (1986) vols. I and II and (with a more philosophical slant) Clark (1989) and/or Churchland (1989).

2. Many readers will be aware of the long debate concerning the original past tense model proposed in Rumelhart and McClelland (1986a). The model was heavily criticized in Pinker and Prince (1988). The model treated here is a different version developed as a response to that critique. See Plunkett and Sinha (1991), pp. 22–33.

3. I first came upon this characterization of the difference in Cussins (1990) and subsequently in a personal (e.mail) communication from Jeff Elman.

References

Baillargeon, R. R. 1987. "Young Infants Reasoning about the Physical and Spatial Properties of a Hidden Object." *Cognitive Development*, 2, pp. 170–200.

Bates, E. and Elman, J. 1992. *Connectionism and the Study of Change*. Technical Report 9202, Center for Research in Language, University of California, San Diego.

Christiansen, M. and Chater, N. 1992. "Connectionism, Learning and Meaning." *Connection Science*, 4 (3–4), pp. 227–52.

Churchland, P. M. 1989. *The Neurocomputational Perspective*. Cambridge: MIT Press/Bradford.

Clark, A. 1989. *Microcognition: Philosophy, Cognitive Science and Parallel Distributed Processing*. Cambridge: MIT Press/Bradford.

Clark, A. 1993. *Associative Engines: Connectionism, Concepts and Representational Change*. Cambridge: MIT Press/Bradford.

Cussins, A. 1990. "The Connectionist Construction of Concepts," in M. Boden (ed.), *The Philosophy of Artificial Intelligence*, pp. 368–440. New York: Oxford University Press.

Elman, J. 1991a. "Representation and Structure in Connectionist Models," in Altmann (ed.), *Cognitive Models of Speech Processing*. Cambridge: MIT Press.

Elman, J. 1991b. "Distributed Representations, Simple Recurrent Networks and Grammatical Structure." *Machine Learning*, 7, pp. 195–225.

Fodor, J. 1975. *The Language of Thought*. New York: Crowell.

Fodor, J. 1986. "The Present Status of the Innateness Controversy," in J. Fodor (ed.), *Representations: Philosophical Essays on the Foundations of Cognitive Science*, pp. 257–316. Brighton, Sussex: Harvester Press.

Marchman, V. 1992. *Language Learning in Children and Neural Networks: Plasticity, Capacity and the Critical Period*. Technical Report 9201, Center for Research in Language, University of California, San Diego.

McClelland, J. and Kawamoto, A. 1986. "Mechanisms of Sentence Processing: Assigning Roles to Constituents," in McClelland, Rumelhart, and PDP Research Group (eds.), *Parallel Distributed Processing: Explorations in the Microstructure of Cognition,* vol. II, pp. 272–326. Cambridge: MIT Press/Bradford.

McClelland, J., Rumelhart, D., and Hinton, G. 1986. "The Appeal of Parallel Distributed Processing," in McClelland, Rumelhart, and PDP Research Group (eds.), *Parallel Distributed Processing: Explorations* in the *Microstructure of Cognition,* vol. II, p. 344. Cambridge: MIT Press/Bradford.

Pinker, S. and Prince, A. 1988. "On Language and Connectionism. Analysis of Parallel Distributed Processing." *Cognition*, 28, pp. 73–193.

Plunkett, K. and Sinha, C. 1991. "Connectionism and Developmental Theory." *Psykologisk Skriftserie Aarhus,* vol. 16, no. 1, pp. 1–77.

Plunkett, K. and Marchman, V. 1991. "U-shaped Learning and Frequency Effects in Multi-layered Perception: Implications for Child Language Acquisition." *Cognition*, 38, pp. 1–60.

Rumelhart, D. and McClelland, J. 1986a. "On Learning the Past Tenses of English Verbs," in *Parallel Distributed Processing: Explorations* in the *Microstructure of Cognition,* vol. II, pp. 216–71. Cambridge: MIT Press/Bradford.

Rumelhart, D. and McClelland, J. 1986b. "PDP Models and General Issues in Cognitive Science," in *Parallel Distributed Processing: Explorations in the Microstructure of Cognition,* vol. I, pp. 110–46. Cambridge: MIT Press/Bradford.

Rumelhart, D., Smolensky, P., McClelland, J. and Hinton, G. 1986. "Schemata and Sequential Thought Processes in PDP Models," in *Parallel Distributed Processing: Explorations* in the *Microstructure of Cognition,* vol. II, pp. 7–58. Cambridge: MIT Press/Bradford.

Sejnowski, T. and Rosenberg, C. 1987. "Parallel Networks That Learn to Pronounce English Text." *Complex Systems*, 1, pp. 145–68.

Smolensky, P. 1988. "On the Proper Treatment of Connectionism," in *Behavioral and Brain Sciences,* vol. II.

Smolensky, P. 1991. "Connectionism, Constituency and the Language of Thought," in B. Lower and G. Rey (eds.), *Jerry Fodor and His Critics*, pp. 201–29. Oxford: Basil Blackwell.

Spelke, E. 1991. "Physical Knowledge in Infancy: Reflections on Piaget's Theory," in S. Cary and R. Gelman (eds.), *Epigenesis of theMind: Essays in Biology and Knowledge*. Hillsdale, NJ: Erlbaum.

van Gelder, T. 1990. "Compositionality: A Connectionist Variation on a Classical Theme." *Cognitive Science*, 14, pp. 355–84.

van Gelder, T. 1991. "What Is the 'D' in 'PDP'? A Survey of the Concept of Distribution" in R.W. Ramsey, S. Stich, and D. Rumelhart (eds.), *Philosophy and Connectionist Theory,* pp. 33–59. Hillsdale, NJ: Erlbaum.

13

William Bechtel

Embodied Connectionism

Cognition Disembodied

Among the characteristics of the cognitive revolution, there are two that require careful consideration. The first is a tendency to think of cognitive activities as activities occurring exclusively within the mind/brain. The mind/brain might receive information from outside itself (stimuli) and generate outputs into the external world (behaviors), but cognitive activity consists in processing information within the system itself. My central objective in this chapter is to argue against this feature of the cognitive revolution. The tendency to localize cognitive activity inclusively within the mind/brain is encouraged by a second feature of the cognitive revolution, the tendency to focus principally on "higher" cognitive activities such as reasoning and problem solving. These are activities in which language frequently plays a central role. Generally, the problems themselves are posed linguistically. Moreover, solving the problems is often thought to require manipulation of linguistic symbols.

These two features of the cognitive revolution provided mutual support for those working within what is broadly construed as the symbolic approach to modeling cognition. What this approach assumed was that in order to account for higher cognitive activities, information had to be represented within the mind/brain in symbol strings and processed by performing manipulations of these symbols. In attempting to articulate the conceptual commitments of cognitive scientists adopting this approach, Fodor (1975) characterized the mental symbols as constituting *a language of thought*. What was crucial for Fodor in treating these mental symbols as constituting a language was that they were composed according to syntactical principles and that the cognitive processes that operated on these symbols were specified in terms of this syntax. One result of attempting to explain higher mental processes as involving formal operations on syntactically structured strings was further to support the first feature of the cognitive revolution. Only if all information on which the mind/brain had to rely was represented within the system could the system function by applying formal rules to syntactically structured representations. Thus, it is not surprising that Fodor (1980) also endorses a program of *methodological solipsism* for cognitive science.

The reemergence of connectionism as a modeling framework in cognitive science is particularly problematic in light of these features of the cognitive revolution. Connectionism seems far more suitable to modeling lower cognitive processes such as recognizing and categorizing input patterns or controlling motor activity than to modeling higher cognitive activities. These seem particularly challenging for a connectionist since they do not employ a language-like representational system. Fodor has contended that the assumption of a language-like representational system is not just a feature of one approach to cognitive modeling but is a necessary aspect of any adequate account of cognitive processing. In his initial presentation of the language of thought hypothesis Fodor (1975) presented a number of cognitive activities, including perceptual identification and learning natural languages, which he claimed could only be explained by positing a language of thought. More recently (1987; Fodor & Pylyshyn, 1988), he has argued that there are fundamental general features of thought that require a language of thought. The feature of Fodor's that has attracted the most discussion is systematicity. Since connectionists cannot explain the systematicity of thought, insofar as it does not posit language-like representations, Fodor argues that connectionism is not viable as an approach to cognitive modeling.

In claiming that thought is systematic, Fodor is claiming that the capacity to think each thought is not developed independently. Rather, any system that is capable of thinking a particular thought, for example, that the florist loves Mary, will be able to think a variety of related thoughts, such as that Mary loves the florist. This systematicity, Fodor contends, is readily explained if we assume a system of mental representations in which symbols representing individual entities or activities can be composed in accordance with syntactical rules. Without a combinatorial syntax governing the construction of mental representations, however, Fodor contends that such systematicity would be difficult to explain. According to Fodor, connectionist networks do not inherently provide a combinatorial syntax. Rather, each representation in a connectionist network is discrete. While the activation of one representation may tend to excite other representations in a network, such co-activations will not manifest the sort of systematicity Fodor is concerned with. One reason for this is that when multiple representations are active in a connectionist network, there is no way to keep track of the appropriate relations between them. For example, in Figure 13.1 there are units active for red, blue, circle, and square. But, since there is no syntax governing composition, there is no way to determine whether this is supposed to represent a red circle and blue square or a blue circle and a red square.

Connectionists have explored a variety of strategies for answering Fodor. Many of these have adopted a strategy articulated by van Gelder (1990). While classical symbolic models rely on explicit, structural compositionality in which, as in natural language, one can directly identify the component representations in the compound representations, van Gelder argues that connectionist networks may exhibit only implicit, functional compositionality. When compositionality is implicit and functional, one might be able to perform operations on the representations that are

red   blue   green   brown   square   triangle   circle

Figure 13.1. An illustration of a problem facing connectionist networks insofar as they do not employ a combinatorial syntax and semantics.

dependent on their constituents, but one would not be able to identify components of the compound representations that represent different constituents. An example of a system that generates functional compositionality is Pollack's (1990) RAAM network, in which compressed compound representations of tree structures are constructed from which the whole tree can be regenerated. What distinguishes RAAM representations of a tree structure from more traditional symbolic representations is that within the compressed compound representation one cannot identify components standing for the parts. Nonetheless, these representations can be employed not only to regenerate the original representations, but also to generate other compressed compound representations, such as a compressed representation of an active sentence from the compressed representation of a passive sentence (Blank, Meeden, & Marshall, 1992; Chalmers, 1990). (An alternative approach for achieving functional compositionality is Smolensky's use of tensor product operations to build compound representations in which particular components of a compound representation are bound to each other in such a manner that they can be later extracted. See Smolensky, 1990, and Dolan, 1989.)

These developments within the connectionist tradition are exciting. But insofar as they employ syntactic principles, albeit only functionally, to build compounds, they engender a response that Fodor and Pylyshyn (1988) already anticipated. Fodor and Pylyshyn did not completely repudiate connectionism. They allowed that it might provide an account of the implementation of a classical symbolic architecture. But they contended that *cognitive* modeling was modeling at the level of the symbolic architecture, not at the level of its implementation. Connectionists who have developed implementations of classical principles typically reject this response, contending that connectionist implementations of classical architectures are cognitive insofar as they explain some features of performance, such as the ability to generalize to similar cases, that should count as cognitive. But this does point out a limitation of this approach to answering Fodor and Pylyshyn: it minimizes what might be the distinctive contributions of connectionist modeling. Consequently, it is worth considering whether there is another way of responding to Fodor and Pylyshyn. Later in this chapter I will sketch such an alternative. What is distinctive about this alternative is that it rejects the idea that cognitive activities take place totally within the cognitive system. Rather, I will suggest that many cognitive activities, especially higher ones, require interaction between a cognitive system and features of the world in which it is embodied. In particular, I will argue that, to the degree that cognition is systematic, it is due to the ability of an embodied cognitive system to use symbols in the environment that afford syntactical structuring. But first I want to reflect further on the feature of cognitivism to localize cognitive activity in the head.

Such localization is not a peculiar feature of cognitivism, but is a symptom of a common explanatory strategy employed in science.

Localization as a Feature of Developing Mechanistic Explanations

One common goal in scientific explanation is to develop an account of the mechanism that is responsible for the phenomenon of interest. One assumes that the phenomenon is generated by a mechanism, that is, a machine that consists of differentiable components that do various different tasks and through their interaction generate the phenomenon. Accordingly, in developing an explanation one tries to determine what the parts of the responsible system are, what these parts do, and how they interact. Underlying the quest for such mechanistic explanations are some fundamental assumptions about nature that are widely shared. One of the most fundamental is that nature is what Herbert Simon (1980) refers to as *nearly decomposable*. The idea is that entities in nature behave as they do largely as a result of their internal composition so that one can separate parts of a larger system from each other without substantially affecting their operation. This assumption is grounded on a conception of how the natural world is put together. In Simon's conception, components are put together into stable sub-assemblies, and these are in turn put together into larger assemblies. The strongest forces are used to put the lowest level sub-assemblies together, and progressively higher levels are put together with forces of progressively weaker strength (Simon, 1973).

When one tries to explain a mechanism, one is doing the opposite of what someone does in trying to build the mechanism: one tries to take it apart, at least conceptually, rather than putting it together. When the phenomenon we are interested in is natural, as opposed to the product of human design, we do not have access to the blueprints and the challenge is to discover what are the parts and how they are put together. A crucial first step in developing a mechanistic explanation is to differentiate the system responsible for the phenomenon from its environment. Richardson and I (Bechtel & Richardson, 1993) term such a system the *locus of control* for the phenomenon (Figure 13.2). The transmissions across the boundary of the system are viewed as relatively insignificant. They may provide stimuli to the system or conduct outputs from the system, but the system itself is construed as operating by internal principles. The phenomena to be explained are the outputs the system generates when given the inputs. For example, biological respiration was recognized by the beginning of the nineteenth century as utilizing oxygen and foodstuffs taken into the body to release energy and to generate carbon dioxide and water. For much of the nineteenth century there were disagreements over what system in the body was responsible for this activity. Initially there was support for the claim that respiration occurred in the lungs, but toward the middle of the century the bloodstream and the tissue cells became the two chief candidates. Finally, Pflüger (1872) provided a convincing theoretical case for the tissue cells as the locus of control for respiration. (For details, see Bechtel & Richardson, 1993.)

After the first step is taken, the second step is to try to explain how the system that is judged to be the locus of control performs its function. Frequently, the strat-

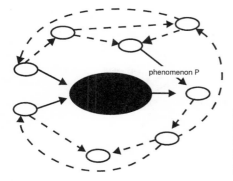

Figure 13.2. Identifying a system as the locus of control for a phenomenon. The system interacts with other systems, but phenomenon P is considered to be primarily the product of internal activities in the shaded system.

egy followed here is a replication of the strategy in establishing a locus of control. Researchers seek to identify a component within the system that can be construed as responsible for this activity of the system (Figure 13.3).Thus, after respiration was localized within the cell, researchers began to posit components within cells that were responsible for respiration. As the idea of biological catalyst or enzyme gained popularity, many investigators pursued the proposal that there was an enzyme in the cell responsible for respiration. In the early years of the twentieth century two competing proposals took center stage: Otto Warburg proposed that there was an oxygen-activating catalyst in the cell that sufficed for causing respiration, while Heinrich Wieland proposed that there was a dehydrogenase that removed hydrogen atoms from substrates and prepared them to combine with molecular oxygen. Richardson and I refer to this attempt to identify a single factor that is responsible for the phenomenon in question as *direct localization*. Sometimes direct localization is correct: there is a single component within the system responsible for the activity in question. But an important feature to note is that, when this is so, the explanatory task has not yet been engaged. To understand how a machine accomplishes a particular effect it is necessary to determine how components, each performing different simpler tasks, interact to generate the phenomenon. If the direct localization is correct, it is necessary to go to a lower level, at which one takes this component apart, to differentiate sub-components that interact, to generate the phenomenon.

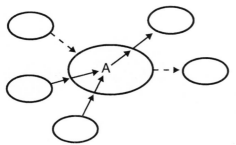

Figure 13.3. Direct localization. A component within the system is credited with responsibility for the system exhibiting a certain behavior.

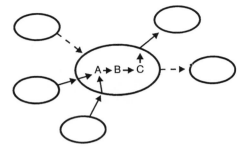

Figure 13.4. Complex localization. A number of components operating in a linear series are assumed to generate the behavior of the system that is of interest.

In many historical cases, the decomposition was accomplished without going to yet a lower level. Researchers discovered that the factor in which the phenomenon was directly localized was not itself sufficient to produce the phenomena. For example, by the 1920s it began to be clear to most researchers that both Warburg and Wieland were partly correct: there is an enzyme that reacts directly with molecular oxygen but there are also dehydrogenases that react with substrates. Neither alone, however, was sufficient. As often happens, as this idea began to be accepted, researchers discovered a variety of other factors that figured in the overall process of respiration and began to propose what Richardson and I term a *complex localization*. As researchers first begin to contemplate a complex structure within the system, they generally begin by positing a linear structure (Figure 13.4).

But often it turns out that there are feedback loops and other interactions between the components that make the system into a *complex integrated system* (Figure 13.5).

In *Discovering Complexity*, Richardson and I focused on cases in which initial direct localizations are supplanted by models of complex integrated systems. While we discuss at length how the decision to identify something as the locus of control for a phenomenon is controversial, we do not explore the possibility that the initial posit might also have to be revised. This is what I want to explore in the case of cognition. Cognition was one of the cases Richardson and I used to try to show how controversial identifying a locus of control can be. The battle between cognitivism and behaviorism was a battle over whether the locus of control of cognitive performance was within the mind/brain or resided in the system external to the mind/brain. The success of the cognitive revolution seemed to settle the issue in favor of the

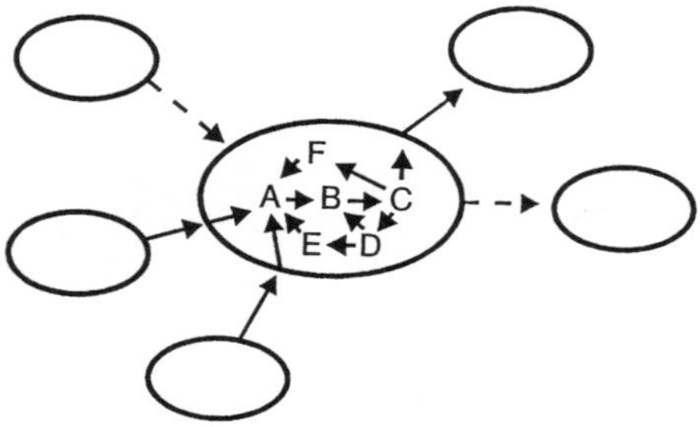

Figure 13.5. An integrated system. The components within the system responsible for its behavior interact through a variety of feedback loops, compromising decomposability.

mind/brain being the locus of control. But what I want to suggest is that just as the recognition of complex interactions often forces researchers to abandon the view that components within a system are nearly decomposable and to attend carefully not just to what components do but to how they interact, in the case of cognition we may need to abandon the view that the mind/brain can be theoretically isolated from an organism's body and environment in such a manner that we minimize the interactions between it and the environment. Rather, we may need to recognize quite complex interactions between the mind/brain and the environment in generating behavior and come to think of cognition as the product of complex interactions between the mind/brain and its environment.

Embodying Connectionist Networks

One way to look at the proposal that I am making is that it reduces the demands on the cognitive system from those imposed by the symbolic tradition. If we construe the cognitive system as a system isolatable from the cognizer's body and its environment, then it seems as if we must view the cognitive system as representing the features of the external world and operating only on those representations. This makes sense when we also view the cognitive system as primarily an inference system: an inference is the generation of a new representation from those previously contemplated. But a connectionist network is quite different from a symbolic inference system. A connectionist network is a model of a dynamical system in which physical components are conceived to be causally interacting with each other. Although in most connectionist modeling, as in symbolic modeling, one thinks of the system as largely isolated from its environment, only receiving inputs and sending back its outputs, this approach is hardly necessary. The causal activities contemplated within the system are much like those occurring outside the system or at the system interface. To see how this can reduce the demands on the internal system, it may help to begin by thinking about modeling the cognitively most basic activities, or even activities not thought to be cognitive. For example, some of the most basic activities the brain performs are those regulating functions elsewhere in the body, such as heart rate. If we take seriously the idea that the brain is *regulating* these activities, we realize that it does not have to employ a representation of the system it is regulating. It needs a source of information about the system it is regulating (that is, it needs to be causally affected by it in the appropriate manner) and a procedure for altering the behavior of the system. It need not calculate how a change will affect the regulated system; it can let the affected system respond, and then respond to it. If evolution has properly prepared it, the brain can be an effective regulator without building up an internal model.

When we turn from regulating physiological functions to regulating behaviors, we can see how this approach might significantly reduce the demands often placed on the mind/brain in the course of cognitive modeling. The traditional way of envisioning the mind/brain as operating is as possessing a detailed representation of the system's body and the surrounding environment and using this both to determine a course of action and to make predictions about how the world will appear after the course of action. Then new input is received and the mind/brain checks to make sure that the accruing conditions correspond to those that were predicted. The alternative

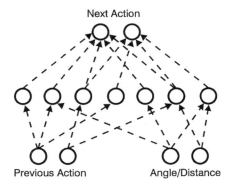

Figure 13.6. The network used in Nolfi et al.'s (1990) first simulation.

approach is to construe the system as interacting with the environment and responding to the changes in the environment over time. This does not deny that the interactions with the environment are goal directed. There can be goals in terms of which the system assesses the activity. But the system need not form a representation of its body and the world in which it is acting and calculate, using these representations, how to move its body so as to secure its goal. It need not even form a representation of the goal.

A simple connectionist simulation will suggest how this might be accomplished. Nolfi, Elman, and Parisi (1990) developed a connectionist network that was designed to move hypothetical organisms through a simple environment. Nolfi et al. used this network to explore how selection and learning could interact with one another in determining the behavior of these organisms. At first, though, they simply offered a model of selection changing behavior. The brains of the organisms were modeled by feedforward networks consisting of two sets of two input units, seven hidden units, and two output units (Figure 13.6).

The two sets of inputs encoded the previous action the hypothetical organism took and the angle and distance of the nearest food source. The output units specified the action to be taken: halting, moving left, right, or forward. The organisms were permitted to move through a variety of simple environments, consisting of 10 x 10 grids. Some cells were designated as food sites and any time an organism reached one of these sites it was envisioned to receive food. In each generation, 100 organisms were permitted to make 5,000 moves through a series of such environments. In the first generation, all the weights for the 100 organisms were chosen randomly. A typical trajectory of an organism at this stage is shown in Figure 13.7.

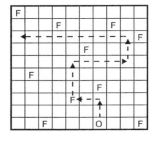

Figure 13.7. A typical trajectory through an environment of an organism in the first generation.

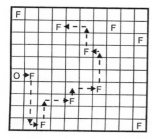

Figure 13.8. A typical trajectory through an environment of an organism in the fiftieth generation.

Evolution was construed to act by selecting the 20 organisms that reached the most food squares and creating five slight variants of each to constitute the next generation. After 50 generations, the organisms evolved so as to find many more food squares. Figure 13.8 shows the path taken by a typical organism in the fiftieth generation.

As a result of simulated evolution, these organisms have become much more efficient at finding food squares. Their behavior seems to be directed toward that goal. Pretty clearly, however, they do not have a representation of that goal. This I do not find problematic; in fact, I would want to argue that such a representation is not necessary. But what is problematic at this stage is that the organisms cannot alter their behavior so as to improve their ability to obtain this goal. To see how this limitation might be overcome, we need to consider how Nolfi et al. added learning to their model. To do this, they added a second set of output units to the networks (Figure 13.9).

On these units the networks were to predict what the sensory input (angle and distance of nearest food site) would be after the movement. This is the same pattern as would be given to the network on the next cycle as part of its input.[1] The error between the prediction and what was to be presented on the next cycle as input was used to change the weights leading to these units and from the input units to the hidden units. (The weights from the hidden units to the motor output units were fixed by evolution and not subject to learning.) This learning procedure was applied during

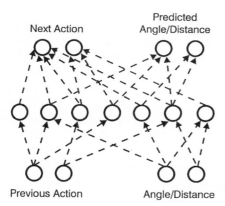

Next Action

Predicted
Angle/Distance

Previous Action

Angle/Distance

Figure 13.9. The network used to add learning to natural selection in simulation of Nolfi et al. (1990).

the life cycle of each organism, and organisms were selected for reproduction in the same manner as in the preceding simulation. When it was time to reproduce, the new organisms were based on the *original* weights of the preceding generation, not those acquired by learning (hence, there was no inheritance of acquired characteristics). These weights were randomly altered to a small degree, and a new generation was created that could begin learning.

Including learning in the simulation not only led to organisms that visited more food squares, but also to subsequent generations exhibiting improved performance even though there was no inheritance of acquired characteristics. Nolfi et al. offered as an explanation for this that the process of learning provided a way of determining which organisms were most likely to produce more successful offspring by random mutations to their weights. An organism that gains from learning is one that has a set of initial weights that, on the basis of small changes during learning, will produce even better results. Then variation and selective retention are also able to produce more fit organisms. This is an interesting suggestion as to how learning and evolution might interact. For my purposes, however, there is another feature of this simulation that is of more interest: evolution has seemed to provide these organisms with goals that learning is then able to promote. But this does not require that the organism form a representation of its goal and the environment and perform calculations using those representations; the seven hidden units in this network are hardly able to accomplish something so sophisticated. Rather, what the network is able to do is develop and improve *procedures* for interacting with its environment by moving its body and responding to the information then available to it. This is accomplished, admittedly, by training the network to predict the future inputs it will receive from its environments. But this ability is based on learning to recognize patterns in current input information (specifying previous action and current environmental input), not from operating on an internal model of the environment. The network has reached this level of performance through cooperative interaction between itself and its environment, which involves a sequence of movements and predictions, followed by new inputs. As a result, the network has adapted to the task imposed on it in a specific range of environments and has learned what procedures are required to accomplish that task.

Nolfi et al.'s simulation is extremely limited. The information the organisms receive is not like what would be provided by a real visual system, and the outputs it generates are not what would be required to guide a real organism. While recognizing this, we should also consider how one would go about complicating the system. One would not just direct one's attention to the network, although almost certainly much more elaborate networks than feedforward networks will be required to generate appropriate behavior (Bechtel, 1993). One would focus equally if not more on the organism's body and environment. That is, one would focus on the sorts of information that would be available to the organism about its environment from its senses, and how this information would be provided to the mind/brain as well as the kinds of motion that the organism can control and the ways in which this control must be exercised. Intelligent behavior, on this model, will not be just a matter of a powerful mind/brain, but of a mind/brain appropriately linked to its environment so as to extract relevant information from the environment and control motion in the environment. This may not require the cognitive system to build up complete representations of the environment, but only procedures for interacting with it.

Embodied Networks and Higher Cognitive Activities

One might respond to a simulation such as Nolfi et al.'s by claiming that simple procedural knowledge of how to interact with an environment might conceivably account for the behaviors of some simple organisms, but that it cannot account for truly cognitive performance in which features such as systematicity make their appearance. For such performance one must move beyond the simple procedural knowledge that might be modeled by connectionist networks and posit a cognitive system that stores and manipulates representations. What I want to propose is that the internal system required for performing higher cognitive tasks need not be *fundamentally* different from that which can acquire procedures for guiding an embodied system through an environment. What is different is that the system is able to make use of specific items in its environment to represent information for the system. By learning to manipulate these external symbols the system acquires abilities it did not previously possess.

The idea that a cognitive system might be enhanced by use of external symbols was suggested by Rumelhart, Smolensky, McClelland, & Hinton (1986; see also Clark, 1989), who used arithmetic problems to illustrate this idea. In solving all but the simplest arithmetic problems, we rely on writing symbols in canonical fashion in an external medium, and then use those enscribings as input for further processing. For example, if we want to multiply the following two three-digit numbers, we begin by writing them in canonical form:

$$343$$
$$\underline{822}$$

Writing the problem in this way permits us to decompose it into several simple component tasks, each of which can be solved mentally by applying procedural knowledge of how to multiply two one-digit numbers. This does not require internal computation; rather, through schooling we have simply learned the procedures. As soon as we *recognize* one of the simple problems, the answer comes to mind. The first task in the above problem is 2×3, whose answer we write in the appropriate position:

$$343$$
$$\underline{822}$$
$$6$$

This representation then directs us to the next task, that of multiplying 2×4. What we have learned is a routine for dealing with a complex problem in a step-by-step manner, with each step requiring limited cognitive effort. The point I want to emphasize is that a problem that would be quite difficult if external symbols were not available is rendered much simpler with these symbols. (Margolis, 1987, also suggests that cognitive problems might be solved by repeated processes of pattern recognition, although he does not develop his analysis within a connectionist framework.) External symbols might not only provide a way to decompose and solve complex problems, but also a way to account for the apparent systematicity of cognition. The external symbols (which include not just written symbols but spoken words, manual signs, or privately rehearsed words[2]) afford syntactical structuring whereby complex structures are composed of simpler structures. Such syntax is what Fodor claims explains systematicity. The cognitive system must be able to operate on these symbols while respecting their syntax. But this only requires the sort of capacity illustrated in the arithmetic example: the system must be able to recognize patterns in the symbols and respond to them appropriately, often by altering the pattern of external symbols. Of course, to develop a connectionist system that can actually exhibit these abilities is a complex task. A suggestion as to how one might proceed is found in the network developed by St. John and McClelland (1990) that learned to develop case role representations from a variety of natural language sentences. Rather than pursuing natural language here (I have explored it further in Bechtel, in press), I will take up another cognitive task in which systematic behavior is exhibited—natural deduction in sentential logic. Natural deduction is nearly the epitome of the process of formal symbol manipulation posited in the symbolic approach. In natural deduction one applies rules to strings of symbols without regard to their semantics so as to generate new strings of symbols. The idea that something like this might be going on in the head provided an inspiration for thinking of the cognitive system as a formal symbol manipulator. If, however, we focus on how humans actually acquire ability in natural deduction, we might think of the activity differently, as one involving interaction with external symbols.

Before even learning natural deduction, students must learn the simple inference principles such as *modus ponens*. When one observes students who have not previously learned these principles trying to learn them, one observes that they are learning to recognize new patterns. It is not sufficient in introductory logic courses just to state the inference principles and contrast them with invalid inference principles such as affirming the consequence. Often a great deal of practice is required for students to learn to employ the valid principles and avoid the invalid principles reliably. In teaching introductory logic, where the focus was on arguments from natural language, I developed simple computer-aided instruction exercises that required students to identify the forms of arguments or complete enthymemes. When new students work on these exercises, it is common for them to write out templates of the different valid and invalid argument forms and compare new arguments explicitly with these templates. After a period of time this ability becomes automatic and they cease to rely on their templates. Inspired by what I observed in the classroom, I developed connectionist networks that could evaluate argument forms and complete enthymemes (Bechtel & Abrahamsen, 1991).

Inference principles such as *modus ponens* are the basic tools for natural deduction, but natural deduction involves more than individual applications of these principles. Rather, these principles must be deployed in an appropriate manner so as to lead from the premises that are given to the desired conclusion. How is this accomplished? One possibility is that one might employ a set of formal rules that specify the circumstances when it is appropriate to make a particular inference. One can in fact develop such rules and use them in programs designed to generate natural deductions. Students often seem to desire such rules. When I am teaching natural deduction and do a derivation as an example, students often query me as to why I execute a certain step. I respond by trying to articulate a strategy that I am following, but I strongly suspect that I am confabulating. One piece of evidence for confabulation is that students might later observe another situation in which the strategy I articulated would seem to be applicable, but I instead do something differently. It seems more likely that expertise in natural deduction is comparable to the sorts of expertise Dreyfus and Dreyfus (1986) describe. According to Dreyfus and Dreyfus, rules figure only at the stage of competent performance; experts rely instead on what they call *intuition*. Intuition is not intended to constitute a mysterious ability, but rather the ability to recognize directly that a particular situation is comparable to one experienced previously, and to use the solution to the previous situation as a template for the solution to this problem. By the time one teaches natural deduction, one typically has a great deal of experience developing derivations, and on this basis is able to recognize immediately that a particular problem is comparable to a previous case and to employ the strategy that worked for it.

Intuition is the sort of ability that seems naturally modeled in connectionist networks. But how are we to capture the systematic structure that appears in natural deductions with connectionist networks? What I contend is that from being exposed to the systematic structure that is exhibited in natural deductions, as they are exhibited in symbols encountered in the external world, networks and mind/brains can learn to respect the structure found in those derivations and to conform to that structure in generating new derivations. To demonstrate that such an approach might be sufficient, I have developed a connectionist network that has learned to construct a variety of natural deductions using a limited set of inference principles: *v*-intro, *v*-elim, &-intro, &-elim, and ⊃-elim. (The description of the network provided here is abbreviated; for a more detailed description of this network, see Bechtel, 1994). For training, 14 schemas of natural deductions were employed (Table 13.1, on the next page). These schemas involved derivations using either three or four premises and either three, four, or five inferential steps. A line in the derivation could consist of one or two atomic sentences, possibly negated, and one two-place connective. In constructing the problem set, three permutations of the order of the first three premises were employed for all 14 derivations patterns: 1, 2, 3; 2, 1, 3; and 3, 1, 2.[3]

Twenty-four derivations were constructed from each of these schemas using the sentential constants A, B, C, D in all possible permutations to replace the sentential variables p, q, r, and s. I will refer to a particular set of derivations in which an assignment of sentential constants was made to the sentential variables as a *substitution set*. Derivations constructed from three-fourths of the substitution sets were used as the training set; accordingly, the training set consisted of 756 derivations. Derivations from the remaining one-fourth of the substitution sets (chosen to be representative of the overall distribution) were reserved for testing generalization. These totaled 252.

A feedforward connectionist network with one layer of hidden units was used for this task (Figure 13.10). The input layer consisted of 104 units that received activations of 0 or 1. The input pattern presented the target conclusion, the premises, and any inferential steps that had been completed. The input units fed into a layer of 20

Table 13.1. The 14 Derivation Schemas Used with the Natural Deduction Networks

	I			II	
1. $p \vee q$		:pr	1. $p \vee q$		:pr
2. $q \supset r$		:pr	2. $q \supset r$		:pr
3. $\sim p$		:pr	3. $\sim p$		:pr
4. q		:1,3 v-elim	4. q		:1,3 v-elim
5. r		:2,4 $\supset$-elim	5. r		:2,4 $\supset$-elim
6. $r \vee s$		:5 v-intro	6. r & q		:5,4 &-intro

	III			IV	
1. $p \supset q$		:pr	1. $p \supset q$		:pr
2. $q \supset r$		:pr	2. $q \supset r$		:pr
3. p		:pr	3. p		:pr
4. q		:1,3 $\supset$-elim	4. q		:1,3 $\supset$-elim
5. r		:2,4 $\supset$-elim	5. r		:2,4 $\supset$-elim
6. $r \vee s$		:5 v-intro	6. r & q		:5,4 &-intro

	V			VI	
1. $p \vee q$		:pr	1. $p \supset q$		:pr
2. $q \supset r$		:pr	2. $q \supset r$		:pr
3. $\sim p$ & s		:pr	3. p & q		:pr
4. $\sim p$		:3 &-elim	4. p		:3 &-elim
5. q		:1,4 v-elim	5. q		:1,4 $\supset$-elim
6. r		:2,5 $\supset$-elim	6. r		:2,5 $\supset$-elim

	VII			VIII	
1. p & q		:pr	1. p & q		:pr
2. $p \supset \sim r$		:pr	2. $p \supset r$		:pr
3. $r \vee s$		:pr	3. $q \supset s$		:pr
4. p		:1 &-elim	4. p		:1 &-elim
5. $\sim r$		:2,4 $\supset$-elim	5. r		:2,4 $\supset$-elim
6. s		:3,5 v-elim	6. q		:1 &-elim
7. q		:1 &-elim	7. s		:3,6 $\supset$-elim
8. s & q		:6,7 &-intro	8. r & s		:5,7 & intro

	IX			X	
1. $\sim p$ & q		:pr	1. p & q		:pr
2. $p \vee r$		:pr	2. $p \vee \sim r$		:pr
3. $r \supset s$		:pr	3. $r \vee s$		:pr
4. $\sim p$		:1 &-elim	4. $\sim p$		:1 &-elim
5. r		:2,4 v-elim	5. $\sim r$		:2,4 v-elim
6. s		:3,5 $\supset$-elim	6. s		:3,5 v-elim
7. q		:1 &-elim	7. q		:1 &-elim
8. s & q		:6,7 &-intro	8. s & q		:6,7 & intro

	XI			XII	
1. $p \supset q$		:pr	1. $p \vee \sim q$		:pr
2. $q \supset r$		:pr	2. $q \vee \sim r$		:pr
3. $r \supset s$		:pr	3. $r \vee s$		:pr
4. p		:pr	4. $\sim p$		:pr
5. q		:1,4 $\supset$-elim	5. $\sim q$		:1,4 v-elim
6. r		:2,5 $\supset$-elim	6. $\sim r$		:2,5 v-elim
7. s		:3,6 $\supset$-elim	7. s		:3,6 v-elim
8. s & q		:7,5 &-intro	8. s & $\sim r$		:7,6 & intro

Table 13.1. (*continued*)

XIII		XIV	
1. $p \lor q$	:pr	1. $p \supset {\sim}q$	:pr
2. $q \supset r$	:pr	2. $q \lor {\sim}r$	:pr
3. $r \supset s$	:pr	3. $r \lor s$	:pr
4. ${\sim}p$	:pr	4. p	:pr
5. q	:1,4 v-elim	5. ${\sim}q$	:1,4 $\supset$-elim
6. r	:2,5 $\supset$-elim	6. ${\sim}r$	:2,5 v-elim
7. s	:3,6 $\supset$-elim	7. s	:3,6 v-elim

hidden units, which in turn fed into an output layer consisting of 13 units. On these the network was trained to produce an encoding of the next step in the derivation. On the input and output layers 13 units sufficed to encode each line of the derivation in such a manner that a unit having an activation of 1 or close to 1 represented the presence of a negation, sentential constant, or connective assigned to that unit. Activation of unit 1 served to negate the first sentential constant; activation of unit 9 negated the second sentential constant. Units 2-5 and 10-13 specified the first and second sentential constants; the first unit in each of these clusters signified D, the second C, etc. Units 6-8 encoded the connective (unit 6 represented &, unit 7 $\supset$, and unit 8 v). Figure 13.11 shows the manner in which the proposition ${\sim}A$ & B was encoded. If the line consisted simply of an atomic letter or its negation, the first eight units were assigned an activation of 0.

To present a complete derivation problem to the network, the appropriate activation pattern was presented to the three or four sets of units encoding the premises

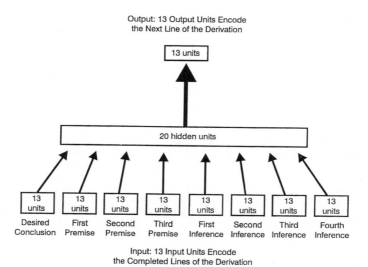

Figure 13.10. Network trained to construct natural deductions from three premises. Units in groups for lines not yet completed are left blank.

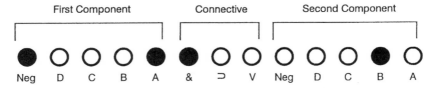

Figure 13.11. Encoding schema used to encode one line in a derivation on 13 units. Each unit represents a negation, a sentential constant, or a connective. The units active in this case represent the statement ~A & B.

as well as the set of units encoding the desired conclusion, and the remaining sets of units were left off. The network then was required to construct the proper next line on the output units (i.e., it had to make the first inference). Then, regardless of what pattern was generated by the network on the output units, the desired pattern (the correct first inference) became part of the input on the next step. That is, the network was supplied with the premises, desired conclusion, and correct first inference. From this input the network's job was to generate the second inference. This process was repeated until the network had constructed all of the steps of the derivation. (Note that the only representation of the previously completed steps is on the input layer, which corresponds to the visual input a human might have who has written the problem and the steps completed so far on paper. The network, like a human, is going through a series of interactions with external symbols.)

To train each derivation, the network was presented with each of the steps in the appropriate sequence. For each step, the network produced its answer on the output units, was informed of the correct answer, and, in accord with the backpropagation learning algorithm, used the difference between its answer and the correct answer to revise the weights on the connections in the network. An epoch of training consisted of one pass through each of the derivations in the training set. After only a few epochs of training the network began to get most of the steps correct. After 500 training epochs the network had mastered the training set.

The 252 derivations reserved for testing generalization involved 936 total inferences. The network was judged to have made the inference correctly if all units that had values of 1 in the target had activations greater than 0.5 and units that were 0 in the target had activations less than 0.5.[4] The network was correct on 767 of these inferences (81.9%). All of the errors the network made involved the sentential constants. There were no errors involving either the negation operators or the main logical connectives. The errors involving the sentential constants fell into three categories: either no unit representing a sentential constant was activated above 0.5, both the correct and one incorrect unit were activated above 0.5, or an incorrect unit was activated above 0.5.Thus, even when the network responded incorrectly, its answers were interpretable as either failures to respond, uncertainty as to the correct response, or simple errors.

In those cases in which either no unit for a sentential constant was activated above 0.5 or more than one was, one can examine which unit had the highest activation. In 80 of the 169 cases in which errors were made, the correct unit was the one with the greatest activation. There is a clean-up technique that can be used to raise the activation of the winning unit and suppress the activation of all of the other units in a cluster (this involves adding excitatory connections from a unit to itself and inhibitory connections to all other units in the cluster). If such a technique had been employed, the percentage correct would have risen to 90.5%. The network does seem to have learned to respect the systematicity of natural deduction.

While I cannot present the details here (see Bechtel, 1994), an analysis of the errors made by the network reveals that many of them are also quite systematic. Very frequently, if there were two or more problems in the same substitution set that required the same type of inference, if the network made an error on one, it would often make the same error on several or even all of the instances. Moreover, in many cases the same errors would be made even when the premises, on which the inferences depended, were presented in a different order. It appears that the network has failed to learn the correct generalization of specific inference patterns. Such a pattern of errors, together with the network's highly accurate performance in other cases, supports the claim that this network has learned to generalize patterns of inference. It is not treating each case separately.

In this first experiment, only *A, B, C,* or *D,* not their negations, were employed as substitutions for the sentential constants *p, q, r,* and *s.* In a second simulation, the previous set of derivation sets was supplemented with three additional sets in which the substitution for either *q,* for *p* and *r,* or for *q, r,* and *s* were negated. (Thus, in the first case, if *A* originally was to be substituted for *q,* ~*A* would be substituted instead; double negations were automatically removed.) This introduction of additional substitution sets required the network to be extremely sensitive to the proper placement of the negation signs. Because of the increased complexity of the stimulus set, the number of hidden units was increased to 40. In a simulation in which one-fourth of the derivations were selected for the test set in such a manner that if a substitution set in which one assignment of sentential constants to sentential variables was chosen for the test set, then none of the other substitution sets differing only in assignments of negations was chosen, the network, after 100 epochs of training, made no errors on either the training or the generalization set. In a second variation, the opposite principle was employed: If a substitution set in which one assignment of sentential constants to sentential letters was chosen for the test set, the other three substitution sets differing in assignments of negations were also chosen. Again, after 100 epochs of training, the network was perfect on the complete training set. On the test for generalization, the network was correct on 2,112 out of 2,496 inferential steps it was required to make (84.6% correct). In this variation of the experiment, five errors arose when the network failed to supply the correct negation sign. The rest of the errors, as in the experiment without the extra negations, involved activating the unit for an incorrect sentential constant, failing to activate any unit for an atomic sentence above 0.5, or activating the units for two atomic sentences. (When the more lax criterion requiring only that the unit representing the correct atomic sentence be the most active in the group was employed, the network was correct on 2,336 inferences, yielding 93.6% correct.) Moreover, the errors that the network made were systematic, not random. If an error was made on one inference in one substitution set, it was likely to be made on other problems in the substitution set that required the same inference and on the problems in the other three substitution sets that differed only in placement of negation signs.

In one last test of this network, I explored how well it would perform on the novel derivation problems shown in Table 13.2. These schemas are modeled on those used in the training, but introduce variations. For example, schema XV follows the general pattern of schemas I and II, but employs two disjuncts in the first two premises rather than one or two conditionals. Once again, three permutations of the first three premises were employed. Four substitution sets were constructed, substituting for *p, q, r,* and *s* either *B, A, D, C; A, ~D, B, C; ~D, B, ~A, C;* or *B, ~D, A, ~C.* There were a total of 252 inferential steps in these derivations, of which the network was correct on 198 (78.6% correct). Once again, there was systematicity in the errors the network produced.

These explorations reveal a surprising ability of a simple connectionist network to construct natural deductions. From being trained on a variety of problems, the network has generalized the ability to construct a range of natural deductions. It has learned the systematicity exhibited in the training set. But it does not develop an internal representation of this structure. All the network does at any given stage is complete the next inferential step. Of course these networks have not demonstrated the full range of capacities required for natural deduction. The inference rules used by these networks are not complete. For a natural deduction system to be complete, one must make use of indirect proofs such as those found in *reductio* arguments. Moreover, each of the lines employed in these derivations involved at most two atomic propositions and a logical connective. Real natural deduction systems allow for embedding of compound propositions within the components of a proposition, as in ((*A* & *B*) v *C*). Further, only four sentential constants were used. As a result, the network may not have learned the general principle that whenever a sentence of the form $p \supset q$ appeared as a previous line and the sentential constant that substitutes for *p* appears on another line, then the sentential constant that substitutes for *q* could be derived as a new line; thus, even if the encoding system allowed for encoding a new sentential constant *E*, it is doubtful that the network could now generalize to a new derivation using it. Whether it is possible to overcome these limitations with a simple feedforward network such as I have been using is unclear and can only be determined by further investigation.

What I want to stress is that the network has achieved the level of proficiency it has, not by learning to carry out a complex set of inferences internally, but by interacting with external symbols in which the problems and partial solutions are represented. The network is not itself a symbolic inference engine; it is only when it is supplemented with external symbols that it then can manipulate that it obtains this level of performance. At the beginning of the previous section I suggested that recognizing the embodiment and environmental embeddedness of a cognitive system might reduce the demands on the cognitive system. This is due to a division of labor. The division of labor envisaged here can be illustrated with a Turing machine. There are two crucial components to a Turing machine. The first is a finite state device that is capable of reading the symbol on a particular square of a tape and following rules that specify, depending on the symbol read and the particular state the device is in, either the writing of a new symbol, moving right or left, or stopping. It is by supple-

Table 13.2. Six New Derivation Schemas Used to Test Generalization of the Natural Deduction Network

XV			XVI		
1. $p \vee \sim q$	:pr		1. $p \supset \sim q$	:pr	
2. $q \vee r$	:pr		2. $q \vee r$	:pr	
3. $\sim p$	:pr		3. p	:pr	
4. $\sim q$	:1,3 v-elim		4. $\sim q$	:1,3 ⊃-elim	
5. r	:2,4 v-elim		5. r	:2,4 v-elim	
6. $r \vee s$	:5 v-intro		6. $r \,\&\, \sim q$	:5,4 &-intro	

XVII			XVIII		
1. $p \supset \sim q$	:pr		1. $p \,\&\, q$	:pr	
2. $q \vee r$	:pr		2. $p \supset r$	:pr	
3. $p \,\&\, q$	:pr		3. $r \supset s$	:pr	
4. p	:3 &-elim		4. p	:1, &-elim	
5. $\sim q$	:1,4 ⊃-elim		5. r	:2,4 ⊃-elim	
6. r	:2,5 v-elim		6. s	:3,5 ⊃-elim	
			7. q	:1 &-elim	
			8. $s \,\&\, q$	:6,7 &-intro	

XIX			XX		
1. $p \vee \sim q$	:pr		1. $p \supset q$	:pr	
2. $q \vee r$	:pr		2. $q \supset \sim r$	:pr	
3. $r \supset s$	:pr		3. $r \vee s$	:pr	
4. $\sim p$	:pr		4. p	:pr	
5. $\sim q$	:1,4 v-elim		5. q	:1,4 ⊃-elim	
6. r	:2,5 v-elim		6. $\sim r$	:2,5 ⊃-elim	
7. s	:3,6 ⊃-elim		7. s	:3,6 v-elim	
8. $s \,\&\, \sim q$	:7,5 &-intro				

menting this finite state device with a potentially infinite tape on which symbols are stored that the Turing machine achieves its capabilities, including exhibiting systematicity. Some theorists have construed the mind itself as having the power of a Turing machine. What I am proposing is that the cognitive system may only have to correspond to the finite state device of a Turing machine as long as it has the capacity to read and write external symbols in order to exhibit systematicity. It is by being embodied and existing in an environment in which such symbols can be stored that a cognitive system of this sort can obtain the powers of a Turing machine.

Conclusion

The assumption that cognition occurs within the mind/brain and the focus on higher cognitive activities fit well with the symbolic program in cognitive science. But the reemergence of connectionism can lead us to reexamine these features of the cognitive revolution. Connectionism is naturally suited to modeling lower cognitive activities that rely primarily on pattern recognition and responses to the patterns recognized. To perform these activities, the mind/brain does not have to represent the external world and calculate a response; rather it can adapt to that world by developing appropriate procedures. I described a network that might guide a simple organism in such interactions with an environment. I suggested that the way to develop this simple model further is to maintain the focus on the interaction of the network with the body and environment of the organism. Focus on lower cognitive tasks that enable organisms to function in their environment is likely to be the most fruitful way to further develop the connectionist approach to cognition. But for many cognitive scientists, cognition consists fundamentally of higher tasks in which, for example, systematicity is exhibited. In the last section I turned to such tasks and proposed that the enhanced cognitive performance is once again not due to developments exclusively within the mind/brain, but to developments in the interaction of the mind/brain and the environment. By employing external symbols, a cognitive system might bootstrap itself to higher levels of performance. I illustrated this with the example of a connectionist network that learns, in conjunction with external representations of natural derivations, to construct such derivations.

I contended that it is a natural step as humans attempt to explain phenomena in nature to demarcate a system from its environment, and to treat that system as the locus of control for the phenomenon. But sometimes that strategy is mistaken: the phenomenon is the product of the interactions of the identified system and its environments. This, I am suggesting, may be the case with cognition. If it is, then we must reassess an important aspect of the cognitive revolution.

Notes

1. The fact that this information will be available to the organism after it acts separates this stimulation from many stimulations that use feedback networks trained by backpropagation. With backpropagation the network has to be told what correct performance would have been, and this is not usually information that would be available to the system.

2. Vygotsky (1962) argued that the capacity to produce private speech by talking to oneself is a further development of the ability to engage in public speech. It may well involve many of the same activities as are involved in the production of public speech except that the activities of the vocal tract are suppressed. Responding to such private speech may also involve many of the same activities that are involved in responding to the public speech of others. I am construing such private speech using natural language as a product of the cognitive system and hence external to it.

3. This was done to ensure that the network was not relying on the order of the premises and any local cues that it might provide. Otherwise, rather than learning a principle of logic, the network might simply learn that in certain circumstances it should copy over what appears on the final five units of the third line of the derivation. John Nolt, for example, in commenting on a paper in which a network using just the first six argument forms from Table 1 was presented, generated a set of three rules, two of which used position cues, that were sufficient to construct all the derivations.

4. In fact, in this and the other tests reported below, using this relatively lax criterion for correctness only marginally improved the percentage correct. Most of the units that should have been on had activations well above 0.9 and those that should have been off had activations well below 0.1.

References

Bechtel, W. (in press). "What Knowledge Must be in the Head in Order to Acquire Knowledge?" In B. Velichkovsky and D. M. Rumbaugh (eds.), *Uniquely Human: Origins and Destiny of Language*. Hillsdale, NJ: Erlbaum.

Bechtel, W. 1993. "Currents in Connectionism." *Minds and Machines*, *3*, pp. 125–53.

Bechtel, W. 1994. "Natural Deduction in Connectionist Systems." *Synthese*, 101, pp. 433–63.

Bechtel, W., and Abrahamsen, A. A. 1991. *Connectionism and the Mind*. Oxford: Basil Blackwell.

Bechtel, W., and Richardson, R. C. 1993. *Discovering Complexity: Decomposition and Localization as Scientific Research Strategies*. Princeton, NJ: Princeton University Press.

Blank, D. S., Meeden, I., and Marshall, J. B. 1992. "Exploring the Symbolic/Subsymbolic Continuum: A Case Study of RAAM," in J. Dinsmore (ed.), *Closing the Gap: Symbolism vs. Connectionism*. Hillsdale, NJ: Erlbaum.

Chalmers, D. J. 1990. "Mapping Part-whole Hierarchies into Connectionist Networks." *Artificial Intelligence*, *46*, pp. 47–75.

Clark, A. 1989. *Microcognition: Philosophy, Cognitive Science, and Parallel Distributed Processing*. Cambridge: MIT Press.

Dolan, C. P. 1989. *Tensor Manipulation Networks: Connectionist and Symbolic Approaches to Comprehension, Learning, and Planning*. Los Angeles. A. I. Lab. Report, University of California.

Dreyfus, H. L., and Dreyfus, S. E. 1986. *Mind over Machine: The Power of Human Intuition and Expertise in the Era of the Computer*. New York: Free Press.

Fodor, J. A. 1975. *The Language of Thought*. New York: Crowell.

Fodor, J. A. 1980. "Methodological Solipsism Considered as a Research Strategy in Cognitive Psychology." *The Behavioral and Brain Sciences*, *3*, pp. 63–109.

Fodor, J. A. 1987. *Psychosemantics: The Problem of Meaning in the Philosophy of Mind*. Cambridge: MIT Press.

Fodor, J. A., and Pylyshyn, Z. W. 1988. "Connectionism and Cognitive Architecture: A Critical Analysis." *Cognition*, *28*, pp. 3–71.

Margolis, H. 1987. *Patterns, Thinking, and Cognition: A Theory of Judgment*. Chicago: University of Chicago Press.

Nolfi, S., Elman, J. L., and Parisi, D. 1990. *Learning and Evolution in Neural Networks*, No. 9019. San Diego: University of California. Center for Research in Language.

Pflüger, E. 1872. "*Über die Diffusion des Sauerstoffs, den Ort und die Gesetze der Oxydationsprocesse im thierischen Organismus.*" Pflügers Archiv für die gesamte Physiologie des Menschen und der Tiere, 6, pp. 43–64.

Pollack, J. 1990. "Recursive Distributed Representations." *Artificial Intelligence*, *46*, 77–105.

Rumelhart, D. E., Smolensky, P., McClelland, J. L., and Hinton, G. E. 1986. "Schemas and Sequential Thought Processes in PDP Models," in J. L. McClelland, D. E. Rumelhart, and T. P. R. Group (eds.), *Parallel Distributed Processing: Explorations in the Microstructure of Cognition*, Vol. 2: *Psychological and Biological Models*. Cambridge: MIT Press.

St. John, M. F., and McClelland, J. E. 1990. "Learning and Applying Contextual Constraints in Sentence Comprehension." *Artificial Intelligence*, *46*, pp. 217–57.

Simon, H. A. 1973. "The Organization of Complex Systems," in H. H. Pattee (ed.), *Hierarchy Theory: The Challenge of Complex Systems*, pp. 1–27. New York: Braziller.

Simon, H. A. 1980. *The Sciences of the Artificial*, 2nd ed. Cambridge: MIT Press.

Smolensky, P. 1990. "Tensor Product Variable Binding and the Representation of Symbolic Structures in Connectionist Systems." *Artificial Intelligence*, *46*, pp. 159–216.

van Gelder, T. 1990. "Compositionality: A Connectionist Variation on a Classical Theme." *Cognitive Science*, *14*, pp. 355–84.

Vygotsky, L. S. 1962. *Language and Thought*. Cambridge: MIT Press.

14

Sidney J. Segalowitz & Daniel Bernstein

Neural Networks and Neuroscience

What Are Connectionist Simulations Good for?

Introduction

Around the turn of the century, the Sears mail order catalogue advertised some articles as "electric," despite their having no electric properties, presumably to lend an air of modernity, high technology, and scientific magic. Dr. Scott's Electric Hair Brush, the Electric Fabric Cleanser, the New Electric Washing Machine (at $2.98, it contained a rotary lever but no motor), and the Radium Shoulder Brace ($0.36) were thoroughly modern for the 1908 catalogue from Sears, Roebuck & Co., but the descriptions were connotative rather than denotative (Schroeder, 1971). We are now in an age when "neural" has such an aura. Those of us who study psycholinguistics or brain functions of language often have found ourselves on mailing lists for books and workshops in "neurolinguistic programming," which is not related to the fields of neuroscience, linguistics, or programming. In this short chapter, we explain why we similarly see so-called neural networks, the result of PDP modeling, as not directly reflecting anything neural except in a superficial way. We propose to cover some general issues about the adequacy of Parallel Distributed Processing modeling (also referred to as PDP models, neural networks, and connectionist models) as an aid to cognitive neuroscience. By stepping back from specific implementations of PDP models, we will consider what such neural networks are capable of offering cognitive neuroscience, as well as their limitations in this endeavor.

We come to this discussion as people interested in tying together what we know about brain function with what we know about cognitive function. We have to ask ourselves what role the exercise of PDP modeling has in this enterprise. Please note that we are not referring to the exercise of simulating the neural innervation of the heart during heart fibrillations, or the exercise of getting a robot to perform a certain task well. (See Widrow, Rumelhart, & Lehr, 1994, for a recent review of the variety of uses available.) These are laudable goals, but have little to do with understanding the neurological networks involved in human cognitive function. What we wish to consider is perhaps a caricature of the connectionist model, but one that still captures its logical status within neuroscience. In a way, we hope that our argument is eventually proven incorrect, for the goals of connectionists are lofty (McClelland, 1989). In addition, computer time is a cheaper and more convenient resource than laboratory work. We fear nonetheless that these goals are not attainable for this form of simulation any more than for other forms. Psychology has a long history of attempts (and successes) at creating mathematical models for behavior. Hull's model of rote learning (Hull, Hovland, Ross, Hall, Perkins, & Fitch, 1940) is perhaps the epitome of such simulations; but others involving even more complex behaviors also exist, for example, Rapoport's (1962) outline of various models of social interaction. The interesting aspect of this tradition is that, while sometimes useful for heuristic purposes, this modeling has not led to new insights about or theories of behavior.

However, an added risk that is built in to PDP modeling that did not exist for the previous model builders is that the terminology is highly seductive. Referring to the information units as "neurons" is a terrific advertising gimmick, but does not in itself mean that the system is modeled after how the brain works, as is frequently claimed in science magazines (e.g., Cross, 1994). Even in scholarly contributions, this is implied rather strongly. Consider the following quotation describing the sorts of problems that neural networks are especially good at solving in engineering fields: "These tasks often involve ambiguity, such as that inherent in handwritten character recognition. Problems of this sort are difficult to tackle with conventional methods [i.e., non-PDP models] because the metrics used by the brain to compare patterns may not be very closely related to those chosen by an engineer designing a recognition system" (Widrow et al., 1994). This suggests that, compared to traditional models, PDP models are more closely related to actual neural function, and not just to mental function. Even if PDP models did represent some aspects of true brain function, we argue that the only role PDP modeling has in furthering our knowledge of cognitive neuroscience is by suggesting experimentally testable hypotheses—that is, they can add ideas but not empirical data.[1]

We say this knowing that Plaut and Shallice (1993) recently have presented a lucid description of how a connectionist model can be used to generate ideas about certain behavioral syndromes arising from brain damage. It may prove useful as a heuristic for developing hypotheses about mental models; and we have no problem with this. Further, a PDP model may test assumptions about the potential or feasibility of a particular cognitive architecture with respect to the limits of prediction. However, the basic issue we wish to address is (1) whether a connectionist model ever can be part of a test of how humans actually work; and (2) whether a connectionist model ever tells us anything new about brain organization or function. We think the answer to both is "No." The question then remains how a PDP simulation might be useful to cognitive neuroscientists.

What Is a Simulation Good for?

The purpose of theories and models is to help move forward one's understanding of the phenomenon at hand—knowing what the critical input variables are, how the output relates to them, and what sort of information transformation goes on in between. Simulations are useful in helping us do this by forcing us to make the model explicit. The argument we will develop here is that a PDP network is not a simulation of either a cognitive, a neuropsychological, or a neurological model: It is rather an associationistic black box with highly adaptive weight-adjustment going on inside, which does not necessarily evaluate our ideas about the critical input factors, processes, or relations to output. Consider, for example, a reading model that has numerous input variables representing graphemic factors. What if one or two of the input variables end up being weighted less than the others? Does this mean that human brains hardly make use of the information encoded in those graphemic cues? It would seem odd to conclude this (or anything else about humans) from the model, given that there are many possible variations on the particular PDP model that may have yielded other results, and there conceivably are other PDP models that may yield identical results.

Can Either a Successful or Unsuccessful PDP Model Tell Us Anything Directly About the Brain?

A successful network (say, one linking graphemic input to word comprehension) does show us that the inputs can lead to the outputs. This is an existence proof, but this itself does not convince us that humans actually operate in this particular way. In order to do this, one must employ standard experimental cognitive psychology, whereby the *processes* that are inherent in the model are substantiated as fitting human information processing. What a successful PDP model (or any other mathematical simulation) does tell us is that our thinking can be coded systematically and is logically consistent. But it is not clear how this can be a substitute for studying the organism of interest—humans.

What if we are not successful in developing a PDP model of, say, reading? The failure can be due to a variety of reasons, and yet does not logically mean that we chose incorrect input variables to represent how the brain deals with the task. There are a host of factors that could account for it. A PDP model inevitably incorporates parameters that are not critical to the data set being simulated, but may influence the viability of the result. Also, it inevitably omits other parameters that may be critical, without the knowledge of the experimenter. For example, some finite set of valid input-to-output cases must be decided on; some decision must be made with respect to whether the weightings are nominal or continuous; the number and size of hidden layers must be chosen; some but not all aspects of the stimulus input characteristics must be coded; and so on.

We now want to consider another example, closer to our own field of research—understanding the factors affecting electrophysiological measures of attention during altered levels of arousal. We could build a PDP model with inputs representing reticular formation activation level, level of serotonin, amount of ambient noise, number of hours since last sleep episode, current state of hunger, the loudness of the stimulus to be attended to, the stimulus' probability of occurrence, etc., while trying to predict the P300 amplitude or some other component of the event-related potential (e.g., Segalowitz, Ogilvie, & Simons, 1990). If we were successful, the resultant model would not tell us what the mechanisms were or even how the input variables were related to the output in human functioning. It might give us an interpretable possibility that we might want to test in humans, but most of the time the results would be uninterpretable (see below). Perhaps more important, if the model was unsuccessful, we still could not conclude that, say, reticular formation activation was irrelevant to attentional processes.

A parable to illustrate this point: Consider the odd situation of finding an organism that had a central nervous system totally unrelated to ours (perhaps it has been kidnapped by an expedition to another planet). Say we judge this organism to have evolved a well-developed perceptual and physiological system to support locomotion in three-dimensional space, just like us. Considering that the information inputs for the alien and for us are identical when we are beside each other (gravity, texture, slope of the ground), we could wonder which of our central nervous systems it is that a prospective PDP model actually simulates. Surely we have no way to judge; and surely it actually could be similating neither. Successfully simulating movement is an interesting exercise; but it is not an adequate substitute for bench neuroscience.

What Is a PDP Model a Model of?

We find McCloskey's (1991) perspective especially illuminating about how a PDP back-propagation model works: We give it inputs, specify correct outputs, and then watch it grow. That is, the simulation appears on its own once we have set the original parameters. The result must be highly dependent on the type of inputs used, the weighting processes, and the number and configuration of layers chosen. If this were not the case (i.e., if inputs, layers and weighting really did not matter), then all models would reduce to each other and we no longer would be testing any particular theory.

Given this, what does the PDP simulation evaluate? Does it evaluate the set of inputs, the set of hidden layers, or the weighting strategy? All of these involve relatively arbitrary choices at some level. For example, the weighting strategy could be discrete rather than continuous. It could be on a logarithmic scale instead of a linear one. It could be built on Boolean principles in order to mimic human reasoning or it could avoid such principles (and we would hope that any particular PDP model would be specific enough not to be characterized by both at the same time). It could be constructed to have localized positive feedback, in order to mimic the notion of arousal activation within a limited domain, such as in semantic activation. Some decision must be made as to how many hidden units should be included to capture properly the structure of the human brain (although we could argue that we are really only looking for simulations of the human mind). Some decisions must be made as to the nature of the inputs. Inevitably, a PDP model incorporates aspects that are not critical to the theory supposedly being tested, and also inevitably omits other aspects.

The fundamental issue here is the question of what we are trying to simulate. It appears that we start off with a set of hypothesized cognitive inputs (which are certainly not brain inputs since they are already highly abstracted, e.g., they are phonemes and not acoustic primitives, or letters and not retinal points of light). We know what the results should be, and we wait for the model to tell us what to have in between. This surely is a simulation of *the data set* and not of a processing theory. We choose the inputs and we choose the outputs; and if it works we have chosen well. But why would we want to test with an existence proof whether or not a known and verified set of data is possible? There may be good philosophical reasons; but they are outside the domain of empirical neuroscience.

In other words, a PDP simulation in reality simulates a data set that derives from experiments done to test a theory. It does not really test the theory that we have in mind. This is presumably the purpose of a simulation, and yet seems to get lost in the exercise (McCloskey, 1992).

What Would a Real Brain Simulation Look Like?

What would happen if we really did simulate the brain on a real scale? Would this help cognitive neuroscience? Let's say we successfully simulated a person by connecting a million input units to a billion hidden elements, which themselves were connected (each with several tens of thousands of connections) to other hidden layers, and so on until the information accessed the output motor system with its tens of millions of neurons. The result would be a (very) large set of weightings linking the inputs to the first hidden layer, and the hidden layer with other hidden layers, etc. How do we go about finding out what the mind is like from this? Or for that matter how the brain is organized? We know, for example, that inputs are not actually connected in such a linear and hierachical way to outputs in the human brain (Kolb & Whishaw, 1990); yet this is not a necessary restriction in PDP modeling. Perhaps such an investigation would tell us more about the nature of PDP modeling than the nature of the human brain. Perhaps such modeling can tell us something about the human mind; but that remains to be seen. (See, e.g., such debates as in Farah, 1994.) The bottom line is that a neural network with a hundred connections is extremely difficult to interpret; and there is no known way to fathom the theoretical significance of thousands or millions of connections.

The PDP Model as a Level of Analysis Problem

Smolensky (1988) argues that one could see PDP models as bridging the gap between the highly reductionist level of *neural* information processing, and the abstract *symbolic* (or conceptual) level of cognitive information processing. The second (symbolic processing) is often referred to as a computational approach; and PDP modeling, according to Smolensky, forms the amalgam needed to build a truly integrative cognitive neuroscience by supplying the *subsymbolic* interface. This appears to be an attempt to use PDP simulations as a bridge between the mind and the brain. The cognitive output of the mind is shown to be a systematic result of the computations involved at the subsymbolic level. Does this work? Since PDP models recently have been used to disavow the cognitive structurings suggested from studies in experimental psychology, (e.g., Cohen, Romero, & Farah, 1992; Farah, O'Reilly, & Vecera, 1993; Seidenberg & McClelland, 1989), it does not seem to account for these structurings. Once again, such PDP modeling certainly can be used as existence proofs to show that some particular structuring is feasible, if people actually worked that way—for example, that separate processes for implicit and explicit recognition are not needed for a model of face perception. (See Farah, O'Reilly, & Vecera, 1993.) But the symbolic neurons are not actual neurons and cannot tell us anything new about flesh-and-blood neurons. It is unclear to us, therefore, how in principle this subsymbolic level can bridge the gap, except by capturing what already is known from other studies.

There are more modest goals of PDP modeling that do not try to capture neural functioning, goals related to the logical adequacy of ideas. For example, Kimberg and Farah (1993) show that the special sorts of difficulty experienced by patients with frontal lobe damage on a variety of tasks can be simulated in a PDP model that illustrates a general weakening of associations between items in memory rather than the loss of a "central executive." The debate about logical difficulties with the notion of a neural central executive is not new (e.g., Kupfermann & Weiss, 1978) nor is the notion that the frontal lobe is dependent on good input and maintenance of communication with the rest of the brain (Goldberg & Bilder, 1987). Nevertheless, there are cognitive neuropsychologists who certainly seem to suggest that the notion of a central executive should be taken literally (e.g., Shallice, 1982, 1988) rather than as a metaphor for the kinds of errors associated with frontal-lobe damage; and so Kimberg and Farah's cautionary remarks are appropriate. However, as they point out (p. 423), the particular architecture of the model is not a necessary component of the argument. This is good, since the structure of the PDP model can be used for many purposes, some clearly unacceptable. (See Young, 1994.)

The point is that to test a brain hypothesis, one needs evidence, not a simulation. It seems to us, then, that the debates raised by PDP models are not debates about *neural* functioning, but about *mental* functioning. In other words, subsymbolic models compete with the conceptual, computational level of modeling (Olson & Caramazza, 1991). However, we fail to see how subsymbols can be mapped onto the neurological level in a way that reflects real brain structure. A PDP model may at times suggest what we can look for, but never what the neural structure must be. On the contrary, it is neuroscience based on evidence that can dictate what the input should be in a PDP model. (See, e.g., Pollen, Gaska, & Jacobsen, 1989, who outline constraints on models from physiological work.) A PDP model can never dictate parameters or structure to a neuroscience model that is based on evidence.

Conclusion

PDP models are associationist networks that are a considerable improvement over paradigms, such as Markov chaining, that previously were the strongest modeling available. However, calling them neural networks does not make them into models of brain function. We see PDP models as at most valid attempts to suggest cognitive architectures for mental functions, but not as a substitute for human investigation either at the cognitive or neuronal levels.

Notes

1. There is a similarity between mathematical modeling of behavior and thought experiments, of which even the epitome (Einstein's theory of special relativity and Galileo's revision of Aristotle's view of gravity) were only hypotheses waiting for empirical verification (Pickering, 1994).

References

Cohen, J., Romero, R.D., and Farah, M.J. 1992. "Disengaging from the Disengage Mechanism: A Re-interpretation of Attentional Deficits Following Parietal Damage." Paper presented at the *International Neuropsychological Society*, February 5–8, 1992, San Diego.

Cross, M. 1994. "The Smart Way to Record Fading Patents." *New Scientist*, February 19, p. 18.

Farah, M.J. 1994. "Neuropsychological Inference with an Interactive Brain: A Critique of the 'Locality' Assumption." *Behavioral and Brain Sciences*, 17, 43–104.

Farah, M.J., O'Reilly, R.C., and Vecera, S.P. 1993. "Dissociated Overt and Covert Recognition as an Emergent Property of a Lesioned Neural Network." *Psychological Review*, 100, pp. 571–88.

Goldberg, E. and Bilder, R. 1987. "Frontal Lobes and Hierarchic Organization of Neurocognitive Control," in E. Perecman (ed.), *The Frontal Lobes Revisited*, pp. 159–87. New York: JRBN Press.

Hull, C.L., Hovland, C.I., Ross, R.T., Hall, M., Perkins, D.T., and Fitch, F.B. 1940. *Mathematico-Deductive Theory of Rote Learning: A Study in Scientific Methodology*. Westport, CN: Greenwood Press.

Kimberg, D.Y. and Farah, M.J. 1993. "A Unified Account of Cognitive Impairments Following Frontal Lobe Damage: The Role of Working Memory in Complex, Organized Behavior." *Journal of Experimental Psychology: General* 122, pp. 411–28.

Kolb, B. and Whishaw, I.Q. 1990. *Fundamentals of Human Neuropsychology*, 3rd ed. New York: W.H. Freeman.

Kupfermann, I. and Weiss, K.R. 1978. "The Command Neuron Concept." *Behavioral and Brain Sciences*, 1, pp. 3–39.

McClelland, J.L. 1989. "Parallel Distributed Processing: Implications for Cognition and Development," in R.G.M. Morris (ed)., *Parallel Distributed Processing: Implications for Psychology and Neurobiology*, pp. 9–45. Oxford: Clarendon Press.

McCloskey, M. 1991. "Networks and Theories: The Place of Connectionism in Cognitive Science." *Psychological Science*, 2, pp. 387–95.

McCloskey, M. 1992. "Connectionist Modeling of Cognitive Deficits: What Can We Learn from Damaged Networks?" Paper presented at TENNET meeting, May 12–14, 1992, Montreal.

Olson, A. and Caramazza, A. 1991. "The Role of Cognitive Theory in Neuropsychological Research," in F. Boller and J. Grafman (eds.), *Handbook of Neuropsychology,* vol. 5, pp. 287–309. Amsterdam: Elsevier.

Pickering, Alan D. 1994. "Neural Nets Cannot Live by Thought (Experiments) Alone." Psycoloquy 94.5.35.pattern-recognition.6.pickering. Washington, DC: American Psychological Association (internet discussion publication).

Plaut, D.C. and Shallice, T. 1993. "Perseverative and Semantic Influences on Visual Object Naming Errors in Optic Aphasia: A Connectionist Account." *Journal of Cognitive Neuroscience*, 5, pp. 89–117.

Pollen, D.A., Gaska, J.P., and Jacobson, L.D. 1989. "Physiological Constraints on Models of Visual Cortical Function," in R.M.J. Cotterill (ed.), *Models of Brain Function*, pp. 115–135. Cambridge: Cambridge University Press.

Rapoport, A. 1962. "Mathematical Models of Social Interaction," in R.D. Luce, R.R. Bush, and E. Galanter (eds.), *Handbook of Mathematical Psychology,* vol. II, pp. 493–579. New York: Wiley.

Schroeder, J.J. Jr. (ed.) 1971. *Sears, Roebuck & Co: 1908 Catalogue No. 117.* Northfield, IL: Digest Books.

Segalowitz, S.J., Ogilvie, R.D, and Simons, I.A. 1990. "An ERP State Measure of Arousal Based on Behavioral Criteria," in J. Horne (ed.), *Sleep '90*, pp. 23–25. Proceedings of the European Sleep Research Society. Stuttgart: Gustav Fischer Verlag.

Seidenberg, M.S. and McClelland, J.L. 1989. "A Distributed, Developmental Model of Word Recognition and Naming." *Psychological Review*, 96, pp. 523–68.

Shallice, T. 1982. "Specific Impairments in Planning," in D.E. Broadbent and L. Weiskrantz (eds.), *The Neuropsychology of Cognitive Function*, pp. 199–209. London: The Royal Society.

Shallice, T. 1988. *From Neuropsychology to Mental Structure*. Cambridge: Cambridge University Press.

Smolensky, P. 1988. "On the Proper Treatment of Connectionism." *Behavioral and Brain Sciences*, 11, pp. 1–74.

Widrow, B. Rumelhart, D.E., and Lehr, M.A. 1994. "Neural Networks: Applications in Industry, Business and Science." *Communications of the ACM* (March, 1994), pp. 93–105.

Young, A.W. 1994. "What Counts as Local?" *Behavioral and Brain Sciences*, 17, pp. 88–89.

15

Itiel E. Dror & Marcelo Dascal

Can Wittgenstein Help Free the Mind from Rules?

The Philosophical Foundations of Connectionism

The question whether or not the construct "rule" is essential for cognition is one of the main divisions between connectionist and rival approaches in cognitive science. In this chapter, we consider the philosophical significance of this division, and its implications for cognitive research, in the light of several possible interpretations of Wittgenstein's paradox of following a rule. The conclusion is that the rejection of rules by connectionism makes it philosophically incompatible with the symbolic rule-governed approach; nevertheless, the rejection of rules does not necessarily lead, on its own, to a single way of conceptualizing the mind and its place in nature. Wittgenstein's notions of "form of life" and "language games" are used as an aid for understanding the philosophical foundations of connectionism.

> This was our paradox: no course of action could be determined by a rule, because every course of action can be made out to accord with the rule. The answer was: if everything can be made out to accord with the rule, then it can also be made out to conflict with it. And so there would be neither accord nor conflict here. (Wittgenstein, 1953)

1. The "New Science of the Mind" now has two entrenched and well-developed paradigms fighting for supremacy. As candidates for pure and applied scientific theories, both paradigms seek to gather support from their ability to provide coherent formalized models of mental activities and from their successful implementations. Furthermore, both are believed to provide a set of principles capable of explaining all cognitive phenomena, that is, a unified theory of cognition and eventually of the mind. (See Anderson, 1983; Newell, 1990, for the symbolic approach; Grossberg, 1982; Rumelhart & McClelland, 1986, for connectionism.)

Although philosophical issues are often mentioned in the debate, they are usually overshadowed by the quest for "empirical support." (See Dror & Young, 1994, for a discussion of how this quest for empirical support can affect the development of cognitive science.) However, the interpretation of such support depends on divergent philosophical assumptions. Consequently, progress in the debate can only be achieved if, in conjunction with the empirical and practical development of each paradigm, their philosophical assumptions and orientations are carefully examined. In this chapter, we examine one cluster of such assumptions, namely, those pertaining to the role of "rules" in cognition. More specifically, we inquire what are the philosophical implications of the connectionist attempts to provide a "rule-free" account of cognition.

2. The so-called "cognitive revolution" in psychology brought about a rehabilitation of mentalism, in the wake of the alleged inability of behaviorism to account for higher cognitive processes. Once reinstated, the mentalistic outlook legitimized the use of several concepts that had been ruled out by behavioristic strictures. Among them was the idea that cognition is the exercise of a set of competences, which are best described as the mastery and application of rules. Thus, Newell, Shaw, and Simon (1963) investigated the rules people use for reasoning and problem solving; Chomsky (1965, 1980, 1986) conceived the grammar of a language as the set of rules that every competent speaker internalizes; Atkinson and Shiffrin (1968) considered memory as a content-addressed, rule-based system; visual recognition was explored through paradigms of template-matching, feature analysis, and structural descriptions (Palmer, 1975; Winston, 1975).

The development of the modern computer provided both a useful metaphor and support for the view that rules lie at the center of cognitive processes. The software-hardware distinction, interpreted as parallel to the mind-brain distinction, permitted the symbolic approach to emphasize the non-reductionistic character of the new approach (e.g., Putnam's (1967) notion of "functionalism"). The mind was conceived as equivalent to a set of software, rather than to its neuronal underpinnings. Like computer programs, it was said to operate by following symbolic rules. In so-called "traditional AI," computers were used to model such processes as well as to make use of such models. Newell and Simon (1972) implemented the means-ends problem-solving rule in their General Problem Solver; Winograd (1972) applied Chomsky's theory of grammar to simulate language understanding in SHRDLU; structural descriptions have been used in computer learning and in computer vision (Ballard & Brown, 1982), to name a few examples.

The underlying ideology of this approach is that a *physical symbol system* has the necessary and sufficient means for general intelligence (Newell & Simon, 1976). The physical symbol system contains elements that are put together through symbolic structures to form expressions. Rules — which are themselves expressions — create, modify, reproduce, or destroy other expressions based on their symbolic structure and the elements they contain.

3. The rise of connectionism was prompted by dissatisfaction with the meager practical achievements of the symbolic approach, if compared with its pretensions, as well as by advances in exploring the neural networks of the brain. (For a discussion on the relationship between the biology of the brain and cognitive computations, see Dror and Gallogly, 1996.) The insistence on using rules loomed large as a possible cause for the failures of the symbolic approach. Rules were relatively easy

to formulate; but they rendered systems too rigid to be able to capture the specific properties of cognitive processes. The use of rules resulted in failure to exhibit a number of processes, such as pattern recognition, pattern completion, automatic generalization, and graceful degradation, that are important for models of cognitive processes. (See Rumelhart & McClelland, 1986.)

With the development of an alternative way of simulating and accounting for the computational processes underlying cognitive abilities, connectionists also developed an alternative conceptual framework for those cognitive processes that previously seemed to be naturally conceptualized in terms of rules. (See Dror & Young, 1994.) Based on the idea of flow of activations between massively interconnected simple-units, a framework was established where rules—as explanatory constructs—became altogether unnecessary. The idea of parallel distributed processing would now unmask rules as fictitious theoretical constructs. In this respect, Wittgenstein's denunciation of the paradox underlying the use of the notion of "following a rule," and its possible interpretation as rejecting the explanatory value of rules altogether, could be a powerful philosophical ally to this framework.

4. Wittgenstein shows that explanations in terms of rule following must be completely mistaken. For example, how can one ascertain that a rule has been followed, say, in constructing a given sequence of numbers? Suppose you are given a sequence of numbers, <2,4>; What is the rule for generating it and its continuation? None—Wittgenstein would say—because an infinite number of rules could generate the given sequence (e.g.: *multiply the preceding number by 2*; *add 2 to the preceding number*; etc.). One might suggest that the next number will determine which rule has been used: if 6, then the rule is *add two*; if 8, then it is *multiply by 2*. Yet, no matter how many exemplars of the sequence are given, as long as the list is finite—as it necessarily is—there will be indefinitely many rules that are able to generate the sequence. Thus, it is impossible to determine which is the rule that is being followed; and, in fact, one can raise the question whether a rule is being followed at all. Suppose we enlarge the sequence of numbers from <2,4> to <1,2,4,8>. Is its rule now *multiply the preceding number by 2*, or *add the two preceding numbers and add to the sum the position of the number in the sequence minus 2*, since both yield the same sequence?

The appeal to rule following, therefore, seems to have lost its explanatory value, since indefinitely many rules can be made to conform to any course of events. Thus, we can never determine which rules underlie a phenomenon; and this leads us to question their existence altogether. This conceptual argument should bolster connectionism's resolve to provide an account of cognition that does not rely at all on the problematic construct "rule."

5. Regarding the role of rules, the contrast—both practical and conceptual—between the connectionist and the symbolic approaches can be best illustrated by considering an example. THEO and TheoNet are two models that have been proposed to account for processes of reasoning used by experts who forecast solar flares. This kind of reasoning involves the use of informal reasoning within a frame of constraints, and the use of partial and inaccurate information. Both models

receive information on recent activity of the sun (number of flares, size, distribution, and so on). THEO is a rule-based expert system (Shaw, 1989), whereas TheoNet is a three-layer connectionist network (Bradshaw et al. 1989).

THEO processes the data through the knowledge encoded in it, in the form of symbolic rules. For example, a rule in the knowledge base of the expert system may state: "If more than one Zürich class flare has occurred in the past 24 hours and its size was larger than 5, then there is a .85 probability that an M-class flare will occur in the next 24 hours." In other words, the system relies on a symbolic meaningful connection between entities such as Zürich class flare, its size, and the probability of an M-class flare; and all this is encoded in the system in the form of rules. Thus, the rules of the system are the underlying entities that capture its knowledge and action. In contrast, TheoNet does not have such rules. TheoNet processes the data through a set of connections that go through a hidden layer of units before the final output vector is produced. Each connection in the network has a weight associated with it (which the network learns through training). The input data triggers a simple flow of activation between the units. The system has no underlying symbolic rules.

Accounting for processes of reasoning used by experts that forecast solar flares can rely either on the connectionist or on the symbolic rule-governed approaches. THEO and TheoNet perform equally well, and both perform as well as skilled human forecasters. Alternative accounts, such as these, based on the two competing approaches, have been proposed for a variety of cognitive processes (e.g., models for word identification; for a review see Rueckl & Dror, 1994). Such "matches," however, have not been conclusive, and have not proven the empirical supremacy of either approach.

6. But what is the deeper significance, if any, of this contrast between the two approaches? Is it a symptom of their radical incompatibility or can they be reconciled in some higher theoretical synthesis? Are they guided by and oriented toward radically different visions of mind and cognition; or are they mere "notational variants" of the same philosophical attitude toward the mind?

Replies to these questions vary broadly. Some reject the incompatibility thesis on the grounds that one of the approaches (connectionism) is not in fact an alternative explanation for cognitive processes at the psychological level (Fodor & Pylyshyn, 1988). Others accept both approaches, viewing them as complementary, insofar as they account for different cognitive processes (Estes, 1988; Schneider & Detweiler, 1987). Still others see the two approaches as different levels of description of the same process, and consider connectionism to be either the proper implementation of the symbolic approach (Broadbent, 1985), or an intermediate level between the symbolic and the neural levels (Smolensky, 1988).

One might argue, however, that the two approaches are not so easily reconcilable because (1) they are indeed conceptually incompatible, and (2) they represent completely different conceptions of cognition and mind. On this view, the divergence concerning the role of rules is crucial because it uncovers the deeper differences in philosophical outlook. The issue is not just whether rules are implicitly represented in connectionist systems through some form of hidden representations (Hinton,

1986), or whether the connectionist approach merely introduces a sub-symbolic level (Smolensky, 1988). Such interpretations of connectionism may be remnants of the "old" paradigm forced on the "new" one—a very common phenomenon when revolutionary ideas are introduced (Dror & Young, 1994; see J. Dror, 1994, 1995, for how the acceptance of new ideas is constrained by previous beliefs). It is not uncommon to find concepts derived from the symbolic approach embedded in connectionist systems (e.g., Miikkulainen, 1993). Another example is Pollack's (1988) attempt to employ a commonly used computer structure—a stack—in a connectionist network. Other examples are provided by the new generation of "hybrid" models that combine modules of symbolic rules with modules of connectionist networks within a single operating system. Such systems may prove to be technologically efficient, as they seek to exploit the best in each approach.

But the occasional success of such eclectic implementations is not, by itself, proof of explanatory compatibility and conceptual harmony between the two approaches, for it may be due to the plasticity in the practice of the programmers. From a theoretical and philosophical point of view, one should focus rather on a—presumably ideal—notion of "pure connectionism" and examine where it may lead us. In particular, what is the significance of its claim to be, in sharp contrast to "pure symbolism," entirely "rule-free"?

7. The implications of such an entirely "rule-free" connectionist interpretation could be far-reaching; a different conception of cognition is likely to emerge. With the rejection of the need for "rule" modes of explanation, the last bastion of dualism is conquered; not only is the mental no longer a separate ontological domain, but it is no longer a distinct epistemological domain, with its special mode of explanation, either. The mind is thereby completely and finally naturalized, as it never had been before. In this way, a remarkable Ockhamian ontological economy is achieved, along with a no less valuable epistemological economy obtained through the simplification of the explanatory machinery of science.

8. Wittgenstein's paradox—one could argue—would lend further support to this conceptualization. The trouble is that, construed as above, Wittgenstein's argument undermines not only the appeal to rules but also the appeal to any other form of data-based generalization. For it shows that an indefinite number of hypotheses are equally possible candidates to capture *the* regularity underlying *any* given set of cases—the well-known fact that data underdetermine theory. To be sure, this would provide a uniform account of regularities across all domains, but only at the cost of making such an account equally problematic throughout. It is doubtful that connectionism would be willing to give up the reliability of generalizations of all sorts.

The games "interpreting Wittgenstein" and "interpreting Wittgenstein's rule following argument" are among the toughest in town. Despite the danger, it may be worthwhile, for the purposes of understanding the philosophical foundations of connectionism, to make a move. Wittgenstein's argument against rules can be interpreted as showing the theoretical uselessness of this concept, since its alleged work can be performed by resorting to more economical concepts such as "activity" or "training." This is particularly clear in his remarks on language. Wittgenstein rejects the notion that language is taught ostensively—that is, "[T]he teaching of language is not explanation, but training." Language is not a set of rules—that is, "(L)anguage is part of an activity, or of a form of life"; "Explanation is never completed . . . and never shall!" Wittgenstein is interpreted as rejecting "rules" as the underlying concept of cognitive processes and mental operations, whereby rules operate on mental symbols. That is, Wittgenstein is attacking the very conception of the symbolic approach as pointless. "Bingo!" would exclaim the connectionist; "Our enemy's foe is our friend; thank you Mr. Wittgenstein!"

9.   But the connectionist should beware of his newly acquired friend. For one thing, Wittgenstein is not necessarily ruling out the notion of rules altogether; he is not substituting the "rule" approach with a "rule-free" approach. Wittgenstein rather suggests conceptualizing regularities as "social practices." After all, one might attempt to account for such practices in terms of the acquisition of rules, which are then put to use. Furthermore, Wittgenstein stresses that such practices do not necessarily share a common denominator to which they all can be reduced (see Gert, 1995). Human activity, and presumably the mind, are to be conceived as an irreducible plurality of (language) games grounded in a variety of "forms of life."

"Bingo again!" the connectionist euphorically might exclaim. "After all, aren't we also talking about a plurality of processors working in parallel, with no single 'executive' that controls all the moves in the game or even all the games that are played in the mind? The 'practices' you talk about, the 'language-games,' are the 'form of life' of the connectionist network playing 'games' of recurrent patterns, interactive flow of activations, and modifications of weights. Wittgenstein's social network of interactions is really a wonderful model for the inner structure and workings of the connectionist framework and of the mind. Thanks again, Mr. Wittgenstein!"

10.   But now, in view of these notions of Wittgenstein, can connectionism provide insight into the workings of cognition? If we no longer are seeking to reveal an "inner essence" of symbolic rules that allegedly govern cognition, what is it that we are seeking? Connectionism's reply is that cognition arises through information processing that occurs by parallel distributed processing of simple activations (Wittgenstein's practices of language games and forms of life). Connectionism can aid in understanding the mechanisms and operations of this processing in a variety of ways. Connectionism can examine computationally how complex cognitive abilities decompose into different subsystems, or what the principles and characteristics are by which information is processed (for a discussion on the ways connectionism can be used for studying high-level cognition, see I. Dror, 1994). For example, if we want to explore which subsystems are involved in a cognitive ability, we can use the "split networks" technique or modular architecture as means of decomposing a cognitive ability into its subsystems. These techniques determine which processes are computationally distinct and are thus likely to be carried out by different subsystems. Jacobs et al. (1991) used modular architecture to explore the ways in which visual images are decoded. Their investigation showed that visual processing is divided into distinct subsystems that process "what" and "where" information.

Another example for using connectionism to explore cognition is by lesioning connectionist models and examining how they break down. The changes in performance due to artificial damage to specific connectionist models are then compared to changes in performance in people with brain damage. This technique has been used to explore the operations involved in reading and their breakdown in dyslexia. (For a summary of these models, see Rueckl & Dror, 1994.)

11. The above examples show how connectionism can account for mental phenomena that previously had been accounted for by the symbolic rule-governed approach. However, Wittgenstein's notions also can open new horizons in exploring cognition. Given that the mind is not governed by symbolic rules, we can investigate phenomena that we observe but have no understanding, so far, of how they work.

One way for conducting such studies is to examine the performance of a connectionist network while the representation of information is systematically manipulated. This technique of manipulating representations examines the effectiveness of different processing schemes to achieve a given cognitive ability. Given that we know what information is initially available, we can examine the flow of information processing backwards—from the cognitive performance back to the initial information. We begin by constructing a connectionist model that is trained to perform a certain cognitive task. Instead of giving the network the initial input, we take this initial input and experiment on it; we process it in a variety of ways, each of which produces a different representation of the initial input. We then feed each representation to the network, and examine which one(s) enable(s) the network to perform the cognitive task in question. This way we are able to ascertain the information processing necessary for a given task. The systematic manipulation of representations provides many important insights into the information processing involved in cognitive abilities.

Dror, Zagaeski, and Moss (1995) used this technique to examine the information processing involved in three-dimensional object recognition independent of orientation, based on sonar. They trained a three-layer feed-forward network that had 248 units to recognize three-dimensional shapes. Then they examined the network's ability to generalize and recognize the shapes when they were presented in orientations that were not used in the training. The manipulation of the representation included two formats: One was a time domain representation that was created by digitizing the echo waveform (it had the maximum temporal resolution and no frequency resolution); the other was a frequency domain representation that was created by the power spectrum (it had the maximum frequency resolution and no temporal resolution). The network that used the pure time domain representation failed to perform the task, whereas the network that used the pure frequency domain representation was able to generalize and correctly recognize the shapes even when they were presented in novel orientations. Thus, the performance of the connectionist networks suggests that one needs to process frequency domain information for object recognition by sonar. In addition, a cross-correlation between the emitted sonar signal and the returning echoes was created as input to the network. The network was unable to perform the task using cross-correlational representations. Given the success of the network using only information contained in the actual echo, the connectionist network demonstrated that one does not need to examine the echoes relative to the emitted sound, but rather that the echo by itself encodes the shape information of the object.

12. At this point, we have a better idea of what is at stake in trying to understand the philosophical orientation of connectionism's view of mind, in the light of Wittgenstein's view of rules. To be against the symbolic approach *per se* does not yet determine whether one's naturalization of the mind should go inwards (to the domain of activities within the brain) or outwards (to the domain of activities within the social perspective), or perhaps both ways. Freud laid the foundation of psychology with the assumption that the *inner* powers can explain behavior. However, Freud's means to achieve access to the inner powers were rejected as unscientific. Methodological behaviorism, claiming the scientific impenetrability of the inner powers, resorted to the outer powers. The cognitive revolution, in turn, proposed to go into the inner powers through the notion of executive-driven symbolic manipulation according to rules. Wittgenstein renounces this latter notion and therefore rejects it as a means of explaining the inner powers. Similar to behaviorists, Wittgenstein turns to the outer powers. Connectionism provides a means to account for the inner powers without resorting to the false and illusory construct of rules. Seen in this light, the relationship between Wittgenstein and connectionism is not just one of "complementarity" (Mills, 1994; for a critique see Dascal, 1995); the former providing a phenomenological description of the phenomena that the latter attempts to explain. Rather, Wittgenstein should be seen as someone who clears the ground for new modes of explanation of cognitive phenomena through the rejection of a familiar, commonsense, and yet unreliable theoretical construct.

Acknowledgments

We thank William K. Estes, Daniel L. Schacter, Jay G. Rueckl, James Intriligator, and Ofra Rechter for their comments on an earlier version of this chapter. Please send all correspondence concerning this chapter to: Itiel Dror, Psychology Department, Benton Hall, Miami University, Oxford, OH 45056. USA

References

Anderson, J.R. 1983. *The Architecture of Cognition*. Cambridge: Harvard University Press.
Atkinson, R.C., and Shiffrin, R.M. 1968. "Human Memory: A Proposed System and Its Control Processes," in K.W. Spence and J.T. Spence (eds.), *The Psychology of Learning and Motivation*, Vol. 2. New York: Academic Press.
Ballard, D.H. and Brown, C.M. 1982. *Computer Vision*. Englewood Cliffs, NJ: Prentice-Hall.
Bradshaw, G., Fozzard, R., and Ceci, L. 1989. "A Connectionist Expert System that Actually Works," in D. Touretzke (ed.), *Advances in Neural Information Processing Systems*. Morgan Kaufmann.
Broadbent, D. 1985. "A Question of Levels: Comment on McClelland and Rumelhart." *Journal of Experimental Psychology: General, 114*, pp. 189–192.

Chomsky, N. 1965. *Aspects of the Theory of Syntax.* Cambridge: MIT Press.

Chomsky, N. 1980. *Rules and Representations.* New York: Columbia University Press.

Chomsky, N. 1986. *Knowledge of Language: Its Nature, Origin, and Use.* New York: Praeger.

Dascal, M. 1995. "Cognitive Science in the Philosophers' Mill." *Pragmatics and Cognition, 3,* pp. 133–145.

Dror, I. E. 1994. "Neural Network Models as Tools for Understanding High-level Cognition: Developing Paradigms for Cognitive Interpretation of Neural Network Models," in M. C. Mozer, P. Smolensky, D. S. Touretzky, J. L. Elman, and A. S. Weigend (eds.), *Proceedings of the 1993 Connectionist Models Summer School,* pp. 87–94. Hillsdale, NJ: Erlbaum.

Dror, I. E. and Gallogly, D. P. 1996. *Computational Analyses in Cognitive Neuroscience: In Defense of Biological Implausibility.* Unpublished manuscript.

Dror, I. E. and Young, M. J. 1994. "The Role of Neural Networks in Cognitive Science: Evolution or Revolution?" *Psycoloquy, 5* (79).

Dror, I. E., Zagaeski, M., and Moss, C. F. 1995. "Three-dimensional Target Recognition via Sonar: A Neural Network Model." *Neural Networks, 8,* pp. 143–54.

Dror, J. M. 1994. "Using the Cultural Pentagon to Analyze the Reception of Cultural Objects." *Sociological Abstracts, 42,* pp. 429–430.

Dror, J. M. 1995. "Developing the Archival Template and the Cultural Pentagon: The Forces Influencing the Reception of Corneille's *Le Cid.*" *Journal of Culture and Society, 3,* 27–65.

Estes, W.K. 1988. "Toward a Framework for Combining Connectionist and Symbol-Processing Models." *Journal of Memory and Language, 27,* pp. 196–212.

Fodor, J.A. and Pylyshyn, Z.W. 1988. "Connectionism and Cognitive Architecture: A Critical Analysis." *Cognition, 28.*

Gert, H. J. 1995. *Resemblances in Wittgenstein and Connectionism.* Unpublished manuscript.

Grossberg, S. 1982. *Studies of Mind and Brain: Neural Principles of Learning, Perception, Development, Cognition, and Motor Control.* The Hague: Reidel.

Hinton, G.E. 1986. "Learning Distributed Representation of Concepts." *Proceedings of the Eighth Annual Conference of the Cognitive Science Society.*

Jacobs, R. A., Jordan, M. I., and Barto, A. G. 1991. "Task Decomposition Through Competition in a Modular Connectionist Architecture: The What and Where Vision Tasks." *Cognitive Science, 15,* pp. 219–250.

Miikkulainen, R. 1993. *Subsymbolic Natural Language Processing: An Integrated Model of Scripts, Lexicon, and Memory.* Cambridge: MIT Press.

Mills, S. 1994. "Wittgenstein and Connectionism: A Significant Complementarity?" in C. Hookway and D. Peterson (eds.), *Philosophy and Cognitive Science.* Cambridge: Cambridge University Press.

Newell, A. 1990. *Unified Theories of Cognition.* Cambridge: Harvard University Press.

Newell, A., Shaw, J.C., and Simon, H.A. 1963. "Empirical Explorations with the Logic Theory Machine: A Case Study in Heuristics," in E.A. Feigenbaum and J. Feldman (eds.), *Computers and Thought.* New York: McGraw Hill.

Newell, A. and Simon, H.A. 1972. *Human Problem Solving.* Englewood Cliffs, NJ: Prentice-Hall.

Newell, A. and Simon, H.A. 1976. "Computer Science as Empirical Inquiry: Symbols and Search." *Communication of the Association for Computing Machinery,* Vol. 19, No. 3.

Palmer, S.E. 1975. "Visual Perceptions and World Knowledge: Notes on a Model of Sensory-Cognitive Interactions," in D.A. Norman and D.E. Rumelhart (eds.), *Explorations in Cognition.* New York: Freeman.

Pollack, J. 1988. "Recursive Auto-Associative Memory: Devising Compositional Distributed

Representation," in *Proceedings of the Tenth Annual Conference of the Cognitive Science Society*.

Putnam, H. 1967. "The Mental Life of Some Machines," in H. Putnam, *Philosophical Papers*, vol. 2 (1975). Cambridge: Cambridge University Press.

Rueckl, J. G. and Dror, I. E. 1994. "The Effect of Orthographic-Semantic Systematicity on the Acquisition of New Words," in C. Umilta and M. Moscovitch (eds.), *Attention and Performance, XV*, pp. 571–88. Hillsdale, NJ: Erlbaum.

Rumelhart, D.E., and McClelland, J.L. (eds.). 1986. *Parallel Distributed Processing,* Vol. 1: *Foundations*. Cambridge: MIT Press.

Schneider, W., and Detweiler, M. 1987. "A Connectionist/Control Architecture for Working Memory," in G.H. Bower (ed.), *The Psychology of Learning and Motivation*. New York: Academic Press.

Shaw, D. 1989. "Theophrastus." *Proceedings of the Fourth Annual Rocky Mountain Conference on Artificial Intelligence*, 7–17.

Smolensky, P. 1988. "On the Proper Treatment of Connectionism." *Behavioral and Brain Sciences, 11*, 1–74.

Winograd, T. 1972. *Understanding Natural Language*. New York: Academic Press.

Winston, P.H. 1975. "Learning Structural Descriptions from Examples," in P.H. Winston (ed.), *The Psychology of Computer Vision*. New York: McGraw-Hill.

Wittgenstein, L. 1953. *Philosophical Investigations*. Oxford: Basil Blackwell.

16

Timothy van Gelder

The Dynamical Alternative

Introduction

What is cognition? Contemporary orthodoxy maintains that it is computation: the mind is a special kind of computer, and cognitive processes are the internal manipulation of symbolic representations. This broad idea has dominated the philosophy and the rhetoric of cognitive science—and even, to a large extent, its practice—ever since the field emerged from the postwar cybernetic meleé. Many arguments have been advanced in its favor, and perhaps equally many against it. For a number of years, the situation has been one of stalemate, for generally neither computationalists nor their opponents have found each other's arguments convincing. There is no doubt, however, that for a long time computationalists maintained a strategic advantage, for their opponents were unable to provide an alternative conception that could match computationalism in empirical scope and theoretical well-foundedness. Computationalists could thus always fall back on what has become known as the "what else could it be?" argument.

In recent years, however, this strategic advantage has been diminished, if not eliminated. Within cognitive science an alternative vision of the nature of cognition is emerging, a worthy challenger to computational incumbency. That alternative is not, as many have supposed, connectionism (as such), but rather the dynamical conception of cognition. In a nutshell, if cognition is not computation, it might be the behavior of dynamical systems. This chapter aims to provide an informal introduction to the dynamical conception. It begins with a somewhat unusual detour, via the early industrial revolution in England, circa 1788.

The Governing Problem

A central engineering challenge for the industrial revolution was to find a source of power that was reliable, smooth, and uniform. In the latter half of the eighteenth century, this had become the problem of translating the oscillating action of the steam piston into the rotating motion of a flywheel. In one of history's most significant technological achievements, Scottish engineer James Watt designed and patented a gearing system for a rotative engine. Steam power no longer was limited to pumping; it could be applied to any machinery that could be driven by a flywheel. The cotton industry was particularly eager to replace its horses and water-wheels with the new engines. However, high-quality spinning and weaving required that the source of power be highly uniform, that is, there should be little or no variation in the speed of revolution of the main driving flywheel. This is a problem, since the speed of the fly-wheel is affected both by the pressure of the steam from the boilers, and by the total workload being placed on the engine, and these are constantly fluctuating.

It was clear enough how the speed of the flywheel had to be regulated. In the pipe carrying steam from the boiler to the piston was a throttle valve. The pressure in the piston, and thus the speed of the wheel, could be adjusted by turning this valve. To keep engine speed uniform, the throttle valve would have to be turned, at just the right time and by just the right amount, to cope with changes in boiler pres-sure and workload. How was this to be done? The most obvious solution was to employ a human mechanic to turn the valve as necessary. However, this had a num-ber of drawbacks: mechanics required wages, and were often unable to react suffi-ciently swiftly and accurately. The industrial revolution thus confronted a second engineering challenge: design a device that can *automatically* adjust the throttle valve so as to maintain uniform speed of the flywheel despite changes in steam pres-sure or workload. Such a device is known as a *governor*.

Difficult engineering problems often are best approached by breaking the over-all task down into simpler sub-tasks, continuing the process of decomposition until one can see how to construct devices that can directly implement the various com-ponent tasks. In the case of the governing problem, the relevant decomposition seems clear. A change need only be made to the throttle valve if the flywheel is not currently running at the correct speed. Therefore, the first sub-task must be to mea-sure the speed of the wheel; and the second sub-task must be to calculate whether there is any discrepancy between the desired speed and the actual speed. If there is no discrepancy, no change is needed, for the moment at least. If there *is* a discrep-ancy, then the governor must determine by how much the throttle valve should be adjusted to bring the speed of the wheel to the desired level. This will depend, of course, on the current steam pressure, and so the governor must measure the current steam pressure and then on that basis calculate how much to adjust the valve. Finally, of course, the valve must be adjusted. This overall sequence of sub-tasks must be carried out as often as necessary to keep the speed of the wheel sufficiently close to the desired speed.

A device that can solve the governing problem would have to repeatedly carry out these various sub-tasks in the correct order, and so we can think of it as obey-ing the following algorithm:

1. Measure the speed of the flywheel.
2. Compare the actual speed against the desired speed.
3. If there is no discrepancy, return to step 1; otherwise
   a. Measure the current steam pressure.

b. Calculate the desired alteration in steam pressure.

c. Calculate the necessary throttle valve adjustment.

4. Make the throttle valve adjustment.

Return to step 1.

There must be some physical device capable of actually carrying out each of these sub-tasks, and so we can think of the governor as incorporating a tachometer (for measuring the speed of the wheel); a device for calculating the speed discrepancy; a steam pressure meter; a device for calculating the throttle valve adjustment; a throttle valve adjuster; and some kind of central executive to handle sequencing of operations. This conceptual breakdown of the components of the governor may even correspond to its actual breakdown; that is, each of these components may be implemented by a distinct, dedicated physical device. The engineering problem would then reduce to the (presumably much simpler) problem of constructing the various components and hooking them together so that the whole system functions in a coherent fashion.

As obvious as this approach now seems, it was not the way the governing problem was actually solved. For one thing, it presupposes devices that can swiftly perform some quite complex calculations, and although some simple calculating devices had been invented in the seventeenth century, there was certainly nothing available in the late eighteenth century that could have met the demands of a practical governor.

The real solution, adapted by Watt from existing windmill technology, was much more direct and elegant. It consisted of a vertical spindle geared into the main flywheel so that it rotated at a speed directly dependent on that of the flywheel itself (see Figure 16.1). Attached to the spindle by hinges were two arms, and on the end of each arm was a metal ball. As the spindle turned, centrifugal force drove the balls outward and hence upward. By a clever arrangement, this arm motion was linked

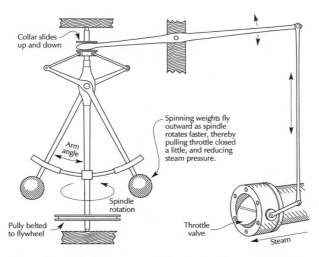

Figure 16.1. The Watt centrifugal governor for controlling the speed of a steam engine (Farey, 1827). From van Gelder, T.J., "Dynamics and Cognition," in J. Haugeland, ed., (1997) *Mind Design II: Philosophy, Psychology, and Artificial Intelligence*, rev. ed. Cambridge: MIT Press.

directly to the throttle valve. The result was that as the speed of the main wheel increased, the arms raised, closing the valve and restricting the flow of steam; as the speed decreased, the arms fell, opening the valve and allowing more steam to flow. The engine adopted a constant speed, maintained with extraordinary swiftness and smoothness in the presence of large fluctuations in pressure and load.

It is worth emphasizing how remarkably well the centrifugal governor actually performed its task. This device was not just an engineering hack employed because computer technology was unavailable. In 1858 *Scientific American* claimed that an American variant of the basic centrifugal governor, "if not absolutely perfect in its action, is so nearly so, as to leave in our opinion nothing further to be desired."

But why should any of this be of any interest in the philosophy of cognitive science? The answer may become apparent as we examine a little more closely some of the differences between the two governors.

Two Kinds of Governors

The two governors described in the previous section are patently different in construction, yet both solve the same control problem, and we can assume (for purposes of this discussion) that they both solve it sufficiently well. Does it follow that, deep down, they are really the same kind of device, despite superficial differences in construction? Or are they deeply different, despite their similarity in overt performance?

It is natural to think of the first governor as a *computational* device; one that, as part of its operation, *computes* some result, namely, the desired change in throttle valve angle. Closer attention reveals that there is in fact a complex group of properties here, a group whose elements are worth teasing apart.

Perhaps the most central of the computational governor's distinctive properties is its dependence on *representation*. Every aspect of its operation, as outlined above, deals with representations in some manner or other. The very first thing it does is measure its environment (the engine) to obtain a symbolic representation of current engine speed. It then performs a series of operations on this and other representations, resulting in an output representation, a symbolic specification of the alteration to be made in the throttle valve; this representation then causes the valve-adjusting mechanism to make the corresponding change. This is why it is appropriately described as *computational* (now in a somewhat narrower sense): it literally computes the desired change in throttle valve by manipulating symbols according to a schedule of rules. Those symbols, in the context of the device and its situation, have meaning, and the success of the governor in its task is owed to its symbol manipulations being in systematic accord with those meanings. The manipulations are discrete operations that necessarily occur in a determinate *sequence*; for example, the appropriate change in the throttle valve can only be calculated after the discrepancy, if any, between current and desired speeds has been calculated. At the highest level, the whole device operates in a *cyclic* fashion: It first measures (or "perceives") its environment; it then internally computes an appropriate change in throttle valve; it then effects this change ("acts" on its environment). After the change has been made and given time to affect engine speed, the governor runs through the whole cycle again . . . and again. . . . Finally, notice that the governor is *homuncular* in construction. Homuncularity is a special kind of breakdown of a system into parts or components, each of which is responsible for a particular sub-task. Homuncular components are ones that, like departments or committees within bureaucracies, interact by communication (i.e., by passing meaningful messages). Obviously, the representational and computational nature of the governor is essential to its homuncular construction; if the system as a whole did not operate by manipulating representations, it would not be possible for its components to interact by communication.

These properties—representation, computation, sequential and cyclic operation, and homuncularity—form a mutually interdependent cluster; a device with any one of them will standardly possess others. Now, the Watt centrifugal governor does not exhibit this cluster of properties as a whole, nor any one of them individually. As obvious as this may seem, it deserves a little detailed discussion and argument, since it often meets resistance, and some useful insights can be gained along the way.

Since manipulable representations lie at the heart of the computational picture, the non-representational nature of the centrifugal governor is a good place to start. There is a common and initially quite attractive intuition to the effect that the angle at which the arms are swinging is a representation of the current speed of the engine, and it is because the arms are related in this way to engine speed that the governor is able to control that speed. However, this intuition is misleading; arm angle and engine speed are of course intimately related, but their relationship is not representational. There are a number of powerful arguments favoring this conclusion. They are not based on any unduly restrictive definition of the notion of representation; they go through on pretty much any reasonable characterization, based around a core idea of some state of a system that, by virtue of some general representational scheme, stands in for some further state of affairs, thereby enabling the system to behave appropriately with respect to that state of affairs (Haugeland, 1991).

A useful criterion for the presence of representations—a reliable way of telling whether a system contains them or not—is to ask whether there is any explanatory utility in describing the system in representational terms. If you really can make substantially more sense of how a system works by concretely describing various identifiable parts or aspects of it as representations in the above sense, that's the best evidence you could have that the system really does contain representations. Conversely, if describing the system as representational lets you explain nothing over and above what you could explain before, why on earth suppose it to be so? Note that very often representational descriptions *do* yield substantial explanatory benefits. This is certainly true for pocket calculators, and mainstream cognitive science is premised on the idea that humans and animals are like that also. However, a noteworthy fact about standard explanations of how the centrifugal governor works is that they never talk about representations. This was true for the informal description given above, which apparently suffices for most readers; more important, it has been true of the much more detailed descriptions offered by those who have actually been in the business of constructing centrifugal governors or analyzing their behavior. Thus, for example, a mechanics manual for construction of governors from the middle of the last century, Maxwell's original dynamical analysis (see below), and contemporary mathematical treatments all describe the arm angle and its role in the operation of the governor in non-representational terms. The reason, one might reasonably conclude, is that the governor contains no representations.

The temptation to treat the arm angle as a representation comes from the informal observation that there is some kind of correlation between arm angle and engine speed; when the engine rotates at a certain speed, the arms will swing at a given angle. Now, supposing for the moment that this is an appropriate way to describe their relationship, it would not follow that the arm angle is a representation. One of the few points of general agreement in the philosophy of cognitive science is that mere correlation does not make something a representation. Virtually everything is correlated, fortuitously or otherwise, with something else; to describe every correlation as representation is to trivialize representation. For the arm angle to count, in the context of the governing system alone, as a representation, we would have to be told *what else* about it justifies the claim that it is a representation.

However, to talk of some kind of correlation between arm angle and engine speed is grossly inadequate; and once this is properly understood, there is simply no incentive to search for this extra ingredient. For a start, notice that the correlation at issue only obtains when the total system has reached its stable equilibrium point, and is immediately disturbed whenever there is some sudden change in, for example, the workload on the engine. At such times, the speed of the engine quickly drops for a short period, while the angle of the arms adjusts only at the relatively slow pace dictated by gravitational acceleration. Yet, even as the arms are falling, more steam is entering the piston, and hence the device is already working; indeed, these are exactly the times when it is *most crucial* that the governor work effectively. Consequently, no simple correlation between arm angle and engine speed can be the basis of the operation of the governor.

The fourth and deepest reason for supposing that the centrifugal governor is not representational is that, when we fully understand the relationship between engine speed and arm angle, we see that the notion of representation is just the wrong sort of conceptual tool to apply. There is no doubt that, at all times, the arm angle is in some interesting way related to the speed of the engine. This is the insight that leads people to suppose that the arm angle is a representation. Yet appropriately close examination of this dependence shows exactly why the relationship *cannot* be one of representation. For notice that, because the arms are directly linked to the throttle valve, the angle of the arms is at all times determining the amount of steam entering the piston, and hence, at all times, the speed of the engine depends in some interesting way on the angle of the arms. In short, at all times, arm angle and engine speed are both determined by, and determining, each other's behavior. As we will see below, there is nothing mysterious about this relationship; it is quite amenable to mathematical description. Yet, it is much more subtle and complex than the standard concept of representation can handle, even when construed as broadly as in the Haugeland definition. In order to describe the relationship between arm angle and engine speed we need a *more* powerful conceptual framework than mere talk of representations. That framework is the mathematical language of dynamics, and, in that language, the two quantities are said to be *coupled*. The real problem with describing the governor as a representational device, then, is that the relation of representing—something standing in for some other state of affairs—is too simple to capture the actual interaction between the governor and the engine.

If the centrifugal governor is not representational, then it cannot be computational, at least in the specific sense that its processing cannot be a matter of the rule-governed manipulation of symbolic representations. Its non-computational nature can also be established another way. Not only are there no representations to be manipulated, there are no distinct manipulatings that might count as computational operations. There are no discrete, identifiable steps in which one representation gets transformed into another. Rather, the system's entire operation is smooth and continuous; there is no possibility of non-arbitrarily dividing its changes over time into distinct manipulatings, and no point in trying to do so. From this, it follows that the centrifugal governor is not *sequential* and not *cyclic* in its operation in anything like the manner of the computational governor. Since there are no distinct processing steps, there can be no sequence in which those steps occur. There is never any one operation that must occur before another one can take place. Consequently, there is nothing cyclical about its operation. The device has, to be sure, an "input" end (where the spindle is driven by the engine) and an "output" end (the connection to the throttle valve). However the centrifugal governor does not follow a cycle where it first takes a measurement, then computes a throttle valve change, then makes that adjustment, then takes a measurement, etc. Rather, input, internal activity, and output are all happening continuously and at the very same time, much as a radio is producing music at the very same time as its antenna is receiving signals.

The fact that the centrifugal governor is not sequential or cyclic in any respect points to yet another deep difference between the two kinds of governor. There is an important sense in which *time does not matter* in the operation of the computational governor. There is of course the minimal constraint that the device must control the engine speed adequately, and so individual operations within the device must be sufficiently fast. There is also the constraint that internal operations must happen in the right sequence. Beyond these, however, there is nothing that dictates *when* each internal operation takes place, *how long* it takes to carry it out, and *how much time* elapses between each operation. There are only pragmatic implementation considerations: which algorithms to use, what kind of hardware is needed to run the algorithms, and so forth. The timing of the internal operations is thus essentially *arbitrary* relative to that of any wider course of events. It is as if the wheel said to the governing system: "Go away and figure out how much to change the valve to keep me spinning at 100 rpm. I don't care how you do it, how many steps you take, or how long you take over each step, as long as you report back within (say) 10 ms."

In the centrifugal governor, by contrast, there is simply nothing that is temporally unconstrained in this way. There are no occurrences whose timing is arbitrary relative to the operation of the engine. All behavior in the centrifugal governor happens in the very same real-time frame as change in the speed of the flywheel. We can sum up the point this way: The two kinds of governor differ fundamentally in their *temporality*, and the temporality of the centrifugal governor is essentially that of the engine itself.

Finally, it need hardly be labored that the centrifugal governor is not a homuncular system. It has parts, to be sure, and its overall behavior is the direct result of the organized interaction of those parts. The difference is that those parts are not modules interacting by communication; they are not like little bureaucratic agents passing representations among themselves as the system achieves the overall task.

Conceptual Frameworks

In the previous section I argued that the differences in nature between the two governors run much more deeply than the obvious differences in mechanical construction. Not surprisingly, these differences in nature are reflected in the kind of conceptual tools that we must bring to bear if we wish to understand the operation of these devices. That is, the two different governors require very different conceptual frameworks in order to understand how it is that they function *as governors*, that is, how they manage to control their environment.

In the case of the computational governor, the behavior is captured in all relevant detail by an algorithm, and the general conceptual framework we are bringing to bear is that of mainstream computer science. Computer scientists are typically concerned with what you can achieve by stringing together, in an appropriate order, some set of basic operations: either how best to string them together to achieve some particular goal (programming theory of algorithms), or what is achievable in principle in this manner (computation theory). So we understand the computational governor as a device capable of carrying out some set of basic operations (measurings, subtractings, etc.), and whose sophisticated overall behavior results from nothing more than the complex sequencing of these basic operations. Note that there is a direct correspondence between elements of the governor (the basic processing steps it goes through) and elements of the algorithm that describes its operation (the basic instructions).

The Watt centrifugal governor, by contrast, cannot be understood this way at all. There is nothing in that device for any algorithm to lock onto. Very different conceptual tools have always been applied to this device. The terms in which it was described above, and indeed by Watt and his peers, were straightforwardly mechanical: rotations, spindles, levers, displacements, forces. Last century, more precise and powerful descriptions became available, but these also have nothing to do with computer science. In 1868 the physicist James Clerk Maxwell made a pioneering extension of the mathematical tools of *dynamics* to regulating and governing devices (Maxwell, 1868). The general approach he established has been standard ever since. Though familiar to physicists and control engineers, it is less so to most cognitive scientists and philosophers of mind, and hence is worth describing in a little detail.

The key feature of the governor's behavior is the angle at which the arms are hanging, for this angle determines how much the throttle valve is opened or closed. Therefore, in order to understand the behavior of the governor, we need to understand the basic principles governing how arm angle changes over time. Obviously, the arm angle depends on the speed of the engine; hence we need to understand change in arm angle as a function of engine speed. Now, there is a whole branch of mathematics specialized for describing rates of change: namely, the calculus and differential equations. A differential equation just is a specification of the rate of change at some moment of time of some variable, as a function of the values at that time of other variables and parameters. If we assume for the moment that the centrifugal governor is disconnected from the throttle value, then it turns out to be governed by a particularly succinct differential equation that tells us the *acceleration* (rate of change in the *change* in arm angle) as a function of the current arm angle and the current rate of change of arm angle.[1] This equation effectively determines the *shape* of the change in both arm angle and change in arm angle. The particular shape of the change depends on the current engine speed. As far as this equation is concerned, engine speed is assumed to stay fixed, or, if it changes, that change is assumed to be independent of change in the governor. When a factor influences variables in a system, but is not itself influenced by them, it is said to be a *parameter*. Parameters are said to *fix the dynamics* of the system.

This differential equation is perfectly general and highly succinct: it is a way of describing how the governor behaves for any arm angle and engine speed. This generality and succinctness comes at a price, however. If we happen to know what the current arm angle is, how fast it is changing, and what the engine speed is, then from this equation the *only* thing we can figure out is the current instantaneous acceleration. If we want to know at what angle the arms will be in a half-second, for example, we need to find a *solution* to the general equation—that is, another equation that tells us what values the arm angle takes as a function of time, which "satisfies" the differential equation. There are of course any number of such solutions, corresponding to all the different behavioral trajectories that the governor might exhibit, but these solutions often have important general properties in common; thus, as long as the parameters stay within certain bounds, the arms will always eventually settle into a particular angle of equilibrium for that engine speed; that angle is known as a *point attractor*.

Thus far I have been discussing the governor without taking into account its effect on the engine, and thereby indirectly on itself. Here, the situation gets a little more complicated, but the same mathematical tools apply. Suppose we think of the steam engine itself as a dynamical system governed by a set of differential equations. One of these equations will tell us the change in engine speed as a function of a number of other variables and parameters. One of those parameters will be the current setting of the throttle valve—but that setting depends directly on the current angle of the arms in the governor! We can thus think of the arm angle as a parameter of the engine system, just as engine speed is a parameter of the governor system. Equivalently, and more simply, we can think of the governor and steam engine as belonging to a single dynamical system in which both arm angle and engine speed are state variables. Two systems related in this way are said to be *coupled*. The relationship is a particularly interesting and subtle one. Changing a parameter of a dynamical system changes its total dynamics (i.e., the way its state variables change their values depending on their current values, across the full range of values they may take). Thus, any change in engine speed, no matter how small, changes not the state of the governor directly, but rather the way the state of the governor changes, and any change in arm angle changes the way the state of the engine changes. Again, however, the overall system (coupled engine and governor) settles quickly into a point attractor, that is, engine speed and arm angle remain constant, which is exactly the desired situation. Indeed, the remarkable thing about this coupled system is that under a wide variety of conditions it always settles swiftly into states at which the engine is running at a particular speed.

In this discussion, two very broad, closely related sets of conceptual resources have (in a very modest way) been brought into play. The first is *dynamical modeling*, that branch of applied mathematics that attempts to describe change in real-world systems by describing the states of the system numerically and then writing equations that capture how these numerical states change over time. The second set of resources is *dynamical systems theory*, the general study of dynamical systems considered as abstract mathematical structures. Roughly speaking, dynamical modeling attempts to understand natural phenomena as the behavior of real-world realizations of abstract dynamical systems, whereas dynamical systems theory studies the abstract systems themselves. There is no sharp distinction between these two sets of resources, and for our purposes they can be lumped together under the general heading of *dynamics*.

Morals

This discussion of the governing task suggests a number of closely related lessons for cognitive science:

- First, various different kinds of systems, fundamentally different in nature and requiring very different conceptual tools for their understanding, can subserve sophisticated tasks—including interacting with a changing environment—that may initially appear to demand that the system have knowledge of, and reason about, its environment. The governing problem is one relatively simple example of such a task; it can be solved either by a computational system or by a non-computational dynamical system, the Watt centrifugal governor.
- Second, in any given case, our sense that a specific cognitive task *must* be subserved by a (generically) computational system *may* be due to deceptively compelling preconceptions about how systems solving complex tasks must work. Many people are oblivious to the possibility of a non-computational, dynamical solution to the governing problem, and so all too readily assume that it must be solved in a computational manner. Likewise, it may be that the basically computational shape of most mainstream models of cognition results not so much from the nature of cognition itself as it does from the shape of the conceptual equipment that cognitive scientists typically bring to the study of cognition.
- Third, cognitive systems may in fact be *dynamical* systems, and cognition the behavior of some (non-computational) dynamical system. Perhaps, that is, cognitive systems are more relevantly similar to the centrifugal governor than they are similar either to the computational governor, or to that more famous exemplar of the broad category of computational systems, the Turing machine.

The Dynamical Conception of Cognition

Do these suggestions have anything to do with reality? Is there any evidence in the day-to-day practice of cognitive science that cognitive systems may be dynamical systems?

As it turns out, all across cognitive science there are researchers conceptualizing cognitive processes in a thoroughly dynamical fashion. These researchers are bringing the tools of dynamical modeling and dynamical systems theory to bear on aspects of cognition ranging from "peripheral" functions such as perception and motor control through to central cognitive processes such as decision making; and from "low-level" aspects such as neural behavior through to "high-level" issues such as syntax and semantics.[2] This is not exactly a new phenomenon; dynamical approaches were prominent in the cybernetic era (1945–1960), and dynamical modeling has been a continuous undercurrent of cognitive science research ever since (e.g., Grossberg, 1982). However, recent years have seen rapid growth in dynamical work, due in part to the convergence of a number of key factors.

First, the last few decades have witnessed something of an explosion in the pure mathematics of dynamical systems theory, particularly the study of nonlinear systems that can produce highly complex behaviors. The conceptual resources that have become available form the foundations for the scientific study of natural and artificial complex systems. Second, the continuing exponential growth in computing power available to everyday scientists, together with the development of sophisticated software packages, has made possible detailed simulation and exploration of a wide variety of dynamical systems. Consequently, throughout the natural sciences rapid advances are being made in understanding phenomena that were previously beyond the reach of standard theoretical tools. The emergence of dynamical approaches to the study of cognition is one manifestation of this general scientific trend. Cognitive scientists now have at their disposal vastly more powerful resources than 30 or 40 years ago, with the result that a wide variety of aspects of cognition that at that time may have seemed could only be computational now have become natural topics for dynamical description and exploration.

Third, the relentless accumulation of knowledge in neuroscience has provided an increasingly rich and provocative source of insights into neural processing. This combined with other factors led to the rapid growth in the 1980s of what came to be known as *connectionist* research, the modeling of cognition using dynamical systems in the form of networks of idealized processing units. Connectionism has demonstrated that it is possible to produce dynamical models across the full spectrum of cognitive capacities. Recently, connectionist modeling has itself become increasingly dynamical in flavor; that is, connectionists are deploying the conceptual resources of dynamics in increasingly sophisticated ways in describing their models and thereby cognition itself. In this trend, connectionism is joining forces with, and providing inspiration for, others who are providing dynamical models that are *not* specifically connectionist in character.

Another key factor has been widespread disillusionment with traditional computational modeling, leading to the search for alternative frameworks for the study of cognition. Practical difficulties in producing models for particular aspects of cognition have come to be regarded by many as symptomatic of deep problems inherent in the computational approach itself. These deep problems directly reflect long-standing philosophical critiques of the assumptions about the nature of mind and cognition that have formed the background for mainstream computational cognitive science. Hubert Dreyfus and others have shown how the actual difficulties encountered by artificial intelligence and mainstream cognitive science are exactly those one would expect if the challenges launched at the traditional conception of mind by philosophers such as Heidegger, Ryle, and Merleau-Ponty were correct. Conversely, it is becoming increasingly clear that the alternative picture they provide, of mind as deeply temporal, embodied, and "pre-theoretical," is fundamentally compatible with the dynamical conception of cognitive processes as the emergent, self-organized behaviors of nonlinear dynamical systems.

Some of the dynamical models being produced are actually surprisingly similar to the centrifugal governor (see, for example,Townsend, 1992; Townsend & Busemeyer, 1989). Generally, however, the centrifugal governor stands to dynamical models in a similar relation as Turing machines stand to computational models of cognition. Nobody supposes that Watt governors or Turing machines themselves constitute good models of any aspect of cognition. Rather, they are particularly simple and clear exemplars of a broad class of systems defined by possession of a cluster of deep properties. The computational conception of cognition is founded on what might be called the computational hypothesis, which is roughly the claim that all cognitive systems are computational systems. [The most well-known version of this hypothesis is Newell and Simon's *Physical Symbol System Hypothesis* (Newell, 1980; Newell & Simon, 1976), but see Haugeland, 1978; 1981; 1985) for better formulations of the same idea.] We might intuitively express this hypothesis as the claim that, in order to exhibit cognitive functions, systems must be relevantly similar to Turing machines. Analogously, the core of the dynamical conception of cognition is the *dynamical hypothesis*, that cognitive systems are dynamical systems; or, intuitively, that in order to exhibit cognitive functions, systems must be relevantly similar to the Watt governor.

Of course, much work needs to be done to clarify the dynamical hypothesis in relation to the computational hypothesis. What, more precisely, are dynamical systems; and what makes them essentially different from computational systems? What is the "relevant similarity" to Turing machines or Watt governors? Are the computational and dynamical hypothesis competitors across all aspects of cognition, or do they have restricted proper domains? And so forth. Unfortunately, space does not allow detailed exploration of these issues here (see van Gelder & Port, 1995, for initial treatment of some of them.) It should be stressed, however, that both the computational and dynamical conceptions of cognition are characterized by clusters of commitments at four distinct levels. First, there is the level of the formal model: is it a computational system, or a dynamical system? Second, there is the level of the conceptual resources that form the general framework within which models are developed and understood: are they drawn from computer science or from the mathematics of dynamics? Third, there is the level of basic assumptions about the nature of cognitive systems: are such systems generically computational (representational, computational—i.e., based on discrete representation-transforming operations—discrete, homuncular, cyclical, atemporal) or are they more a matter of coupled coevolution of continuous quantities? Finally, as hinted above, there is a fourth level, that of the broad philosophy of mind that molds one's preconceptions about the nature of cognition, the best models, conceptual resources, etc. The two main choices are between a generically Cartesian conception (mind as fundamentally the inner, representational cause of intelligent behavior) and a generically post-Cartesian conception (mind as fundamentally a matter of embodied, skillful interaction with the world).

These differing clusters of commitments at the four levels are naturally compatible with each other: a choice of a computational model fits best with the use of the tools of computer science, a conception of cognition as representational, etc., and a broadly Cartesian picture of mind. However, it is possible to do a certain amount of mixing and matching of elements, and any given theorist's stance on some aspect of cognition may diverge from the basic computational or dynamical perspective in a number of respects. This is how best to understand much connectionist work, especially the well-known "PDP"-style research that modeled cognitive processes using layered distributed backpropagation networks (e.g., the famous past tense model of Rumelhart & McClelland, 1986). This kind of work was based on dynamical systems, but much of the theoretical superstructure that dictated the particular structure of these models and how they were interpreted as models of cognition was drawn from the computational worldview. We can thus see much connectionist work as a kind of hybrid, occupying a middle ground between thoroughly computational approaches to cognition on one hand and thoroughly dynamical approaches on the other. Whether this kind of intermediate stance will turn out in the long run to be no more than a temporary phase produced by shifting trends in cognitive science, or exactly what is needed to unlock the secrets of cognition, remains to be seen, though the increasingly dynamical orientation of most connectionist work suggests that the former is more likely.

More generally, we should reject the standard characterization of cognitive science as being dominated by two schools of thought, the mainstream computational approach and connectionism. This characterization distorts cognitive science in many ways, but for current purposes it should be seen as failing to recognize a theoretically more penetrating contrast between the computational approach and the *dynamical* approach. Much connectionist work is thoroughly dynamical (e.g., Port, Cummins, & McAuley, 1995) but much of it is really a kind of conceptual mongrel combining computational and dynamical elements. Recognizing a dynamical approach enables us to see important affinities between some connectionist work and a wide variety of non-connectionist dynamical research efforts. For interesting reasons, specifically connectionist dynamical modeling may well turn out to the most widespread and productive form that the dynamical approach takes; nevertheless, seeing this work as fundamentally dynamical in orientation, and secondarily connectionist in detail, gives us a deeper understanding of this research and how it contrasts with computational orthodoxy.

Over the last few decades, many general arguments have been raised in favor of the idea that cognitive processes must be basically computational in nature (e.g., Pylyshyn, 1984). Many of these arguments were quite persuasive, but every one was ultimately based on the premise that there was no other competing account of the nature of cognition that could provide remotely plausible explanations of the kind of phenomena targeted by the argument. Now, however, it is becoming increasingly clear what the alternative conception of the nature of cognition is, and conceptual innovations are leading to sketches of how those phenomena might find powerful explanations (e.g., Petitot, 1995; Pollack, 1991). In other words, the "what else could it be?" argument is being dealt with head-on, and consequently none of the standard general arguments for the computational need be regarded as conclusive; they provide interesting considerations supporting the computational approach, but the astute observer now maintains a "wait-and-see" attitude. Meanwhile, a wide variety of general considerations can be adduced to support the dynamical approach over and above its computational competitor. These considerations cannot be listed here (see van Gelder & Port, 1995), but it distorts matters only a little to say that most of them boil down to one fundamental insight: cognitive processes, like all other natural process, essentially unfold in real time. The most powerful accounts of cognition will describe *how* they unfold in time, and dynamics is the framework that science always has used to describe change in time. The science of cognition must *begin* with that insight, and build an account of everything that is distinctive about cognition in particular on that foundation.

Notes

This chapter is a highly truncated version of my "What Might Cognition Be, if not Computation?" *Journal of Philosophy*, 1995, *91*, 345–381, with additions and some changes.

  1. That equation is

$$\frac{d^2\theta}{dt2} = (n\omega)^2 \cos\theta \sin\theta - \frac{g}{l} \sin\theta - r \frac{d\theta}{dt}$$

where θ is the angle of arms, n is a gearing constant, ω is the speed of engine, g is a constant for gravity, l is the length of the arms, and r is a constant of friction at hinges (See Beltrami, 1987, p. 163).

  2. Rather than cite individual examples, I merely list here some overviews or collections that the interested reader can use as a bridge into the extensive realm of dynamical research on cognition. A representative sampling of current research is contained in Port and van Gelder (1995); this book contains guides to a much larger literature. An excellent illustration of the power and scope of dynamical research, in a neural network guise, is Grossberg (1988). Serra and Zanarini (1990) present an overview of a variety of dynamical systems approaches in artificial intelligence research. For the role of dynamics in developmental psychology, see Smith and Thelen (1993) and Thelen and Smith (1993).

References

Beltrami, E. 1987. *Mathematics for Dynamical Modeling*. Boston: Academic Press.

Farey, J. 1827. *A Treatise on the Steam Engine: Historical, Practical and Descriptive*. London: Longman, Rees, Orme, Brown and Green.

Grossberg, S. 1982. *Studies of Mind and Brain: Neural Principles of Learning, Perception, Development, Cognition, and Motor Control*. Boston: Dordrecht.

Grossberg, S. 1988. *Neural Networks and Natural Intelligence*. Cambridge: MIT Press.

Haugeland, J. 1978. "The Nature and Plausibility of Cognitivism." *Behavioral and Brain Sciences*, 1, 215–26.

Haugeland, J. 1981. "Semantic Engines: An Introduction to Mind Design," in Haugeland J. (ed.), *Mind Design*. Cambridge: MIT Press.

Haugeland, J. 1985. *Artificial Intelligence: The Very Idea.* Cambridge: MIT Press.

Haugeland, J. 1991. "Representational Genera," in W. Ramsey, S. P. Stich, and D. E. Rumel-hart (eds.), *Philosophy and Connectionist Theory.* Hillsdale, NJ: Erlbaum.

Maxwell, J. C. 1868. "On Governors." *Proceedings of the Royal Society*, 16, pp. 270–83.

Newell, A. 1980. "Physical Symbol Systems." *Cognitive Science*, 4, pp. 135–83.

Newell, A., and Simon, H. 1976 "Computer Science as Empirical Enquiry: Symbols and Search." *Communications of the Association for Computing Machinery*, 19, pp. 113–26.

Petitot, J. 1995. "Morphodynamics and Attractor Syntax," in R. Port and T. van Gelder (eds.), *Mind as Motion: Explorations in the Dynamics of Cognition.* Cambridge: MIT Press.

Pollack, J. B. 1991. "The Induction of Dynamical Recognizers." *Machine Learning*, 7, pp. 227–52.

Port, R., Cummins, F., and McAuley, D. 1995. "Modeling Auditory Recognition Using Attractor Dynamics," in R. Port and T. van Gelder (eds.), *Mind as Motion: Explorations in the Dynamics of Cognition.* Cambridge: MIT Press/Bradford.

Port, R., and van Gelder, T. (eds.). 1995. *Mind as Motion: Explorations in the Dynamics of Cognition.* Cambridge: MIT Press.

Pylyshyn, Z. W. 1984. *Computation and Cognition: Toward a Foundation for Cognitive Science.* Cambridge: MIT Press/Bradford.

Rumelhart, D. E., and McClelland, J. L. 1986. "On Learning the Past Tenses of English Verbs," in J. L. McClelland, D. E. Rumelhart, and The PDP Research Group (eds.), *Parallel Distributed Processing: Explorations in the Microstructure of Cognition*, Vol. 2: *Psychological and Biological Models.* Cambridge: MIT Press.

Serra, R., and Zanarini, G. 1990. *Complex Systems and Cognitive Processes.* Berlin: Springer-Verlag.

Smith, L. B., and Thelen, E. 1993. *Dynamic Systems in Development: Applications.* Cambridge: MIT Press.

Thelen, E., and Smith, L. B. 1993. *A Dynamics Systems Approach to the Development of Cognition and Action.* Cambridge: MIT Press.

Townsend, J. T. 1992. "Don't be Fazed by PHASER: Beginning Exploration of a Cyclical Motivational System." *Behavior Research Methods, Instruments Computers*, 24, pp. 219–27.

Townsend, J. T., and Busemeyer, J. R. 1989. "Approach-Avoidance: Return to Dynamic Decision Behavior," in H. Izawa (ed.), *Current Issues in Cognitive Processes.* Hillsdale, NJ.: Erlbaum.

van Gelder, T. J., and Port, R. 1995. "It's About Time: An Overview of the Dynamical Approach to Cognition," in R. Port and T. van Gelder (eds.), *Mind as Motion: Explorations in the Dynamics of Cognition.* Cambridge: MIT Press.

Appendix

The editors and the author, Tim van Gelder, decided to include the following series of questions and answers at this point in the text, on the grounds that it would provide readers with a valuable resource for understanding the wider significance of the preceding chapter. As part of the editing process, David Johnson taught a graduate and undergraduate class on a pre-publication version of this text in the fall and winter of 1995. The class had a particularly heated discussion about the philosophical implications of van Gelder's position. Johnson then invited one of the leaders in the debate to e-mail van Gelder to see if he would elaborate and expand his views on some of these topics. What appears here is the product of several exchanges and revisions. (The wording and the example in the last question—#5—were proposed by one of the students.)

1. Is the mind the same as the brain according to your account? If not, then is cognition something that is accomplished by the brain, or by the mind?

Answer: This is a quite general issue. The dominant perspective in cognitive science and contemporary Anglo-American philosophy of mind is that the mind is the brain, or at least the functioning of the brain. For example, cognitive science is often taken to be the science of the mind. There are good philosophical reasons to dispute this identification, however. If cognition is the domain of inner states and processes that are causally responsible for our sophisticated behaviors (including "internal" behaviors like thinking to oneself), then in the normal case there is much more to having a mind than having cognition. Mind is a complex ontological structure of which cognition is only one essential component. Rather than thinking of mind as the inner engine of behavior, we should think of cognition as the inner engine of mind.

2. Do you propose the dynamical approach as an exclusive approach to the mind? (I.e., do you take it to be a replacement or a supplement to the computational approach?)

Answer: At this stage of the game, the dynamical approach is best understood as purporting to provide the best account of every aspect of cognition (not mind in general), and, in that sense, to completely replace the computational approach to cognition. Put another way, the dynamical hypothesis is that all natural cognitive systems are dynamical systems rather than computational systems. Of course, it is an open empirical question to what extent either the dynamical or the computational hypothesis is true.

3. Do we have beliefs in the folk psychological sense? If we do, how can this be reconciled with a conception of cognition modeled on the non-representational Watt governor?

Answer: Two issues need to be clarified here. The first is the ontology of belief. A common thread running from Descartes through Fodor and Churchland is that beliefs are inner entities that are causally responsible for our behavior. Descartes thinks they must be aspects of a special mental substance, Fodor thinks they must be functionally characterized brain states, and Churchland thinks—looking at the brain—that there aren't any. However, a perspective that distinguishes mind from cognition, in the sense hinted at above, can allow beliefs to depend on, but be ontologically separate from, inner causal mechanisms. Think of Dennett's intentional stance attributions, for example. Laying out an adequate ontology of belief is not possible here; much of the relevant work has been done by Robert Brandom in *Making It Explicit* (1995). Suffice it to say that it is indeed possible to elaborate a conception of beliefs as representational entities that does not require the existence of beliefs as inner belief-like representational entities that are the crucial causal determinants of behavior.

The second issue is the role of representations in the dynamical approach. The Watt governor is certainly a nonrepresentational device, and the claim of the chapter is that, according to the dynamical approach, cognitive systems are more relevantly similar to the Watt governor than to a representational/computational governor or to a Turing machine. However, it must be stressed that the Watt governor is not itself being repesented as in any sense a serious model of cognitive mechanisms. Indeed, it is just as implausible a model of natural cognition as the Turing machine. Rather, the Watt governor functions as a kind of landmark indicating a whole class of systems—dynamical systems—within which adequate models of cognition might be found. Simplicity is its strength because the general reader can understand it relatively easily and gain an intuitive feeling for dynamical systems and dynamics. Simplicity is also its weakness, for there is very much more to cognition than simple coupling of one system variable to one external variable. Cognitive systems in general (though not in every aspect) exhibit much more disconnection from their environments and worlds than the Watt governor. Indeed, as Brian Smith has stressed, disconnection sets up the problem that intentionality steps in to solve. Thus, it is no surprise that serious dynamical models of cognition very often do build in representations of some kind. Three points should be noted, however. First, those representations have turned out to be radically different in form than the kinds of representations that figure in standard computational stories. Second, there are so many other differences between dynamical models and standard computational models that representational dynamical models should not be put in the same category simply on the basis of using representations. Third, one of the most exciting and challenging streams of dynamical research on cognition is, precisely, the development of genuinely nonrepresentational models of aspects of real cognition.

4. If minds are like Watt governors, how can there be such things as intentions, action, and moral responsibility?

Answer: Again, the problem here is that far too much pressure is being put on the Watt governor. That simple system cannot deliver everything you are asking of it. Systems capable of exhibiting these kinds of phenomena must be much more complex. For a thoroughly dynamical conception of some aspects of intention and action, see Chapter 5 of Scott Kelso's book *Dynamic Patterns* (1995). Also, remember that the dynamical approach is at best an account of cognition, and phenomena like moral responsibility are only instituted in much richer contexts (e.g., those in which there is God, or relevant social structures, etc.).

5. You claim: "In short, at all times, arm angle and engine speed are both determined by and determining each other's behavior" and that this relation is "much more complicated than the standard conception of representation can handle." But this kind of mutual determination also can be found in folk psychology. For example, my belief that it will rain and my desire to stay dry "cause" me to bring my umbrella with me to school, which in turn changes the beliefs I had about the location of my umbrella. The new belief that my umbrella is no longer hanging on the rack and my desire to hang my sun hat somewhere account for the fact that I hang my sun hat on

the rack where the umbrella used to be. . . . This seems to be a "differential" relationship, in a very crude sense of the word. Yet it also seems to be able to make sense of what I think and do, including the possibility of my having false beliefs.

Answer: Coupling is a very specific mathematically describable relationship. What you're finding here is a vague analogy in the relation of beliefs and desires to actions and the world. There are crucial differences, however. The situation you described involved a kind of cyclical influence: beliefs affect the world, which then affects beliefs, and so on. But genuine coupling involves simultaneous mutual determining of the shape of change. Fortunately, that relationship does *not* obtain between beliefs and what they are about. For example, I can believe that Canada is cold but this has no effect at all on the way Canada's weather is changing.

PART FOUR

David Martel Johnson

THE ECOLOGICAL ALTERNATIVE

Knowledge as Sensitivity to Objectively Existing Facts

"Compared to war, all other forms of human endeavor shrink to insignificance!" These words, spoken by the main character in a movie about the trials and triumphs of General George Patton in World War II, are a reminder that armed conflicts not only have effects on historical movements, political thinking, and so on. At least sometimes, they also have been (indirect) sources of intellectual achievements far removed from the practical issues of power and ideology that led to the conflicts in the first place. For example, there is a grain of truth in the idea that two theoretical frameworks for thinking about the mind emerged out of World War II — frameworks that psychologists, philosophers, linguists, and other thinkers have been extending and quarreling about ever since.

We already surveyed the most famous of these frameworks in Part One of this book, under the title "Good Old Fashioned Cognitive Science." To see how it was connected with World War II, consider the following point. The principal code used by the German military in that war was generated, and changed systematically through time, by a secret type of information-processing machine. Accordingly, Allied code-breakers (prominently including Alan Turing) were forced to guess at the principles of operation of this device, and then to try to build a workable version of the same machine themselves. From the perspective of hindsight, this was a crucial step toward the creation, in the period immediately following the war, of effective digital computers — so-called (because of John von Neumann's contributions to their development) "von Neumann machines." (See Kozaczuk, 1984, p. 96.) Furthermore, on a more abstract level, this and similar technical achievements inspired "traditional" cognitive scientists to claim that natural intelligence had to be analogous with the controlling procedures of such artificial thinking machines. For example, one implication of comparing or identifying the mind with a digital computer was that all information available to an intelligent human or other animal had to take the form of "internal representations," analogous to (say) the series of marks, perforations, or magnetized sections on a tape. (One especially clear expression of the latter idea is Jerry Fodor's principle of "methodological solipsism" as set forth in 1980.)

However, other theorists deny that this is the best—or even a coherent—way to conceive of human thought, knowledge, and action. At least some of these people were influenced by the work of the late psychologist J. J. Gibson. Again, it is fair to say that the time Gibson spent working as a psychologist for the U.S. Army during World War II was a crucial formative period for him. For example, one of his main assignments was to help redesign dashboard control panels, runway signals, etc. by organizing experiments to discover the visual cues pilots used to control their aircraft—in particular, when taking off and landing. The desperate military situation demanded quick, practically employable results. And this same pressure soon led Gibson to reject the previously standard psychological picture of an agent laboriously piecing together a internal picture of the world from separate, static mental representations obtained by perceiving. Instead, he found it was more effective to suppose that pilots could directly register certain pieces of information objectively located in their surroundings. For example, one means pilots used to do this—that is, "find" rather than "construct" information—was to take account of patterns present in the "visual or optical flow." (See Gibson, 1950, especially pp.120–31.)

This was the beginning of Gibson's so-called ecological approach to the mind. According to this viewpoint, an observer does not formulate hypotheses about what he or she sees, hears, etc. that presuppose certain internal mental structures and innate abilities. Instead, he takes account of external facts just by finding and "resonating to" them. (Cf. Crick, 1994, p. 75.) Gibson later extended his visual theory to the other senses as well (1966). And he refined it even further for the specific case of vision (1979).

Against this background, then, in the following two chapters Ulric Neisser and Edward Reed deploy a series of Gibson-like observations and arguments designed to show that theorists like Chomsky and Fodor are wrong to try to explain all mentality in terms of supposedly internal and innate forms of the mind or brain. One main theme of both these chapters is that we know from practical experience and experimentation that, for most purposes, minds are sufficiently sensitive and flexible to respond directly to whatever information they need in a given case.

References

Crick, F. 1994. *The Astonishing Hypothesis: the Scientific Search for the Soul.* London: Simon and Schuster.

Fodor, J. A. 1980. "Methodological Solipsism Considered as a Research Strategy in Cognitive Science." *The Behavioral and Brain Sciences*, Vol.3, No.1.

Gibson, J. J. 1950. *The Perception of the Visual World.* Boston: Houghton Mifflin.

Gibson, J. J. 1966. *The Senses Considered as Perceptual Systems.* Boston: Houghton Mifflin.

Gibson, J. J. 1979. *The Ecological Approach to Visual Perception.* Boston: Houghton Mifflin.

Kozaczuk, W. 1984. *Enigma: How the German Machine Cipher Was Broken, and How It Was Read by the Allies in World War Two.* University Publications of America, Inc.

Reed, E. S. 1988. *James J. Gibson and the Psychology of Perception.* New Haven: Yale University Press.

17

Ulric Neisser

The Future of Cognitive Science

An Ecological Analysis

As the twentieth century draws to a close, cognitive science finds itself confronted by two strangely different adversaries. On the one side, clearly visible and looming ever larger, is the rapidly advancing body of fundamental knowledge about the brain. In principle there is a wide range of possible relations between brain science and the study of cognition, but in fact advances in neuroscience may soon make some forms of cognitive science obsolete. On the other side is a broad assortment of cultural critics—not only the occasional post-modernist who rejects all scientific claims *a priori*, but the many humanist scholars who believe that scientific accounts of human life are necessarily hostile to value and meaning. Neither side has much use for psychological research: one because real discoveries can only be made in the brain, the other because no discovery of science can be humanly important.

I believe that both views are mistaken. Although neuroscience is indeed making rapid strides, its work is not and cannot be independent of psychological research at what I will call the *ecological* level of analysis. On the contrary, discoveries at that level provide the framework within which the basic questions of brain science must be posed. And although much psychological research has indeed been irrelevant (or worse) to what people really care about, there are also many findings of real significance. Thus, the strongest response to both lines of criticism may actually be the same: to show that cognitive psychology and related disciplines have in fact made important and ecologically valid discoveries. That is the chief aim of this chapter.

I am using the term "ecological" in a very broad sense, to encompass many different functional approaches to the study of perception, memory, language, and thought. While some of this work has been directly influenced by J.J. Gibson's (1979) ecological theory of perception, much of it draws on quite different intellectual traditions. There are cognitive linguists, developmental psychologists, visual scientists, students of everyday memory, cultural and cross-cultural psychologists, even psychometricians. It is a diverse group, with many internal disagreements and only a few attitudes in common. The most important of those attitudes is perhaps just an empirical bent: we all want to know what human beings actually do (or see, or know) in their normal commerce with their various environments.

Neuroscience and Cognitive Models

Such a definition may strike the reader as meaninglessly broad. Does it not encompass all of experimental psychology, or even all of empirical social science? In fact, it does not. The activity that dominates cognitive psychology today is not empirical exploration but something quite different: namely, the making and testing of hypothetical models. Ironically, the "hypothetico-deductive method" that was so strongly advocated by Hullian behaviorists half a century ago has become the stock-in-trade of their cognitivist successors. They argue that research should always begin with a theory: not just any theory, but a specific model of the internal processes that underlie the behavior of interest. That mental model is then tested as thoroughly as possible in carefully designed experimental paradigms. When it has been proven false (as it invariably is), a revised model is constructed so that the cycle can begin anew. The aim of such research is not to discover any secret of nature; it is to devise models that fit a certain range of laboratory data better than their competitors do.

The scope of model-making in today's cognitive science is remarkable. There are models of face recognition, of object recognition, of word recognition, of speech recognition; models of language understanding and language production; models of attention, of memory, of reading, of problem solving, of deductive reasoning. The methods used to test these models are highly sophisticated, and sometimes lead to interesting results. Nevertheless, new scientific developments threaten to undercut the entire enterprise. In the twenty-first century, when PET scans and MRI scans and other exotic techniques are revealing what really happens to information in the brain, who will still be interested in models like these?

Progress in neuroscience may soon enable us to understand much of the brain activity that underlies common forms of cognition. If this should happen, a good deal of serious intellectual housecleaning will be in order. The first step, according to some predictions (Churchland, 1988), will be the demise of so-called folk psychology. Everyday psychological concepts like "intention" and "emotion" will become obsolete: instead of talking about how they feel or what they think, people will talk about their synapses or their norepinephrin levels or their hippocampi. This outcome seems somewhat implausible to me, but I cannot help thinking that another and much more vulnerable target for elimination is close at hand. Information processing models of the classical kind, built on and tested by laboratory reaction-time experiments, may go the way of Ptolemaic epicycles. Contemporary modelers of attention and perception and memory are already scrambling to keep up with the latest neurological findings. Soon, I think, they will have to scramble harder still.

The Humanist Critique

Like most scientists, I have led a relatively sheltered intellectual life. There are some things I take for granted: that facts can be established, that arguments should be supported by evidence, that the increase of knowledge is intrinsically desirable, that scientists try to determine what is the case. For better or worse, these assumptions are no longer beyond question. From a post-modern perspective, science is just one ideology among many others — and an ugly, patriarchal, in-group-dominated ideology at that. And while scientists may claim that they are simply seeking the truth, every contemporary philosopher knows that truth itself is a fiction. Thus, chemistry and alchemy can be regarded as equally arbitrary cultural products, and no one has to believe what scientists may say.

The most remarkable aspect of this post-modern critique is that it is being made today, just when science is achieving such remarkable successes. We understand more about the structure of matter and the nature of the universe than ever before, and can expect to understand even more tomorrow. We have devised new kinds of materials, gone to the moon, established means of communication that outstrip the wildest dreams of earlier generations. It is not just the physical sciences that are advancing in this way. New techniques in biology have enormously increased our understanding of living cells and living organisms, of the immune system, of the transmission of life itself. We are also beginning to understand the basic structure of life on Earth: the distribution of species, the subtle interactions of ecological systems, the risks posed by technological development. What can have prompted such a sharp upsurge of skepticism about science at such a time?

There are many reasons for it, more than I can mention here and indeed more than I know. They range all the way from very valid concerns about technological pollution to simple jealousy of the success of a competing academic discipline. But a particularly important reason, obvious to everyone, is a kind of existential anxiety. Things are not what they used to be, and we don't like the way they are going. The world we live in, so strongly the product of modern science and technology, seems to have less and less space for subtleties of meaning and value. Medical technology keeps us alive whether we like it or not, progress in genetics seems to rob us of our choices, mind-altering drugs detach our feelings from our real situations, the future seems to change radically every year. In such a fix, a certain amount of anti-scientific feeling is not unreasonable.

Nor is it new. Humanist suspicion of science has not always been couched in such virulent language, but it goes back a long way. Often the target of that suspicion has not been science *per se*, but the implications of science for human values and human life. Biologists and psychologists may have learned a lot, but what do they know of beauty and meaning and choice — of the subtle richness of experience itself? Long before the coining of terms like "eliminative materialism," humanists have been afraid that the materialism inherent in science would eliminate all that they most value.

The Self at Risk

Today these fears seem more justified than ever. Science does seem to threaten many familiar concepts and values, in more ways than I can hope to review here. One danger, however, is particularly relevant to my argument. What will happen to the *self* in the coming neuroscientific age? To understand what is in danger, we must distinguish two complementary notions of the self that have been fundamental to Western thinking for centuries. One is that we are selves, the other is that we have selves. We *are* selves in the sense that each of us is an independently willing, acting, and responsible individual; we *have* selves in the sense of some inner Cartesian core that consitutes our real identity. These two ideas have not always been kept clearly distinct, but we can no longer afford to confuse them. While the fact that we *are* selves cannot be denied, the claim that we *have* selves seems increasingly dubious.

If there were an inner self, where might it be? Over the years, one frequently suggested answer has been "in the brain." The most recent news from neuroscience, however, makes that seem increasingly unlikely. So far as we can tell, there is no cerebral center, no key nucleus to which the afferent impulses go and from which decisions emerge, no critical locus that could play—even approximately—the role that Descartes originally assigned to the pineal gland. There is just an array of quasi-independent neural systems, operating roughly in parallel although they also affect one another in ways that are still largely unknown. Perhaps this discovery should not have surprised us (what other kind of brain could evolution have built?) but it is difficult to reconcile with the notion of an ultimately central and controlling self somewhere inside the head.

If the brain is the wrong place to look, how about the mind? Here too, there seems to be less self than one might have expected. David Hume was very clear about the negative outcome of his own introspective search:

> For my part, when I enter most intimately into what I call *myself*, I always stumble on some particular perception or other, of heat or cold, light or shade, love or hatred, pain or pleasure. I can never catch myself at any time without a perception, and can never observe anything but the perception. (Hume, 1888, p. 252)

The Buddhists, too, have long insisted that careful attention to inner experience never reveals anything like a Cartesian self (Varela, Thompson, & Rosch, 1991). And for what it's worth, many contemporary models of information processing are coming to the same conclusion. In a trend that seems likely to continue, multiple systems and parallel processing have replaced "Central Processing Units" as the most popular theoretical architectures in cognitive science. The fears of the humanists seem to be coming true, at least where the inner self is concerned. Neuroscience and cognitive science are apparently undermining it together.

The Gestalt Precedent

One does not need to be a philosopher or a post-modern critic to feel a bit nervous about the advances of science. The best way to deal with that anxiety is surely to be clear about it—to see which concepts are really at risk of "elimination" and which ones are not. Before trying to answer that question directly, I cannot resist a brief historical digression. As it happens, we are not the first generation to experience this particular anxiety. Some three-quarters of a century ago, a remarkable group of German experimental psychologists found themselves in a very similar situation. The founders of Gestalt psychology—Max Wertheimer, Wolfgang Köhler, and Kurt Koffka—were explicitly concerned with the potential conflict between science and value. Köhler devoted a whole book to it, which he called *The Place of Value in a World of Facts* (1938).

In the early twentieth century, the advance of science seemed at least as rapid and as inexorable as it does today. The dominant Newtonian conception of the world implied that all matter—including the matter of the brain—consisted of nothing but tiny particles blindly following mathematical laws. Beauty and meaning and choice were mere illusions, soon to be swept away by scientific progress. There seemed to be only one alternative to this dismal point of view: to exempt human nature from scientific study entirely by claiming that it was based on some obscure "vital essence." But this seemed even more repugnant. Caught between mechanism and vitalism, what was psychology to do?

The Gestalt response to this existential dilemma was not to yield but to fight. Instead of turning their backs on science, they tried to develop a whole new science with a place for value in it. Köhler did this in a particularly ingenious way, using assumptions about physics and physiology that were highly plausible in the nineteen-twenties. He started with the fact that simple physical forces often produce remarkably symmetrical patterns: for example, soap bubbles in air are always spherical. Given the electrical properties of neural tissue, there may be significant electromagnetic fields in the brain that also tend toward simplicity and elegance. This hypothetical tendency toward goodness of form in physical systems was called the *law of Pragnanz*. And since consciousness reflects the structure of brain activity, it too may be infused with the same tendency toward good form. Mental life thus would acquire meaning and value, not by opposing the laws of physics but by capitalizing on the elegant structures that those laws can produce.

It was a noble idea, but it didn't work. The concept of *Pragnanz* turned out to be a weak reed: essentially circular, it could not make testable predictions. Moreover, the field theory of cerebral activity turned out to be simply wrong. The brain is a set of specialized neural networks, not (in any important way) a volume conductor of electricity. These flaws were insuperable. Besides, there is a sense in which the whole enterprise may have been headed in the wrong direction. Despite their anti-vitalism Köhler and his associates were still phenomenologists, trying most of all to explain the special world of inner experience. From a Darwinian standpoint, that is the wrong place to begin. The study of consciousness should be only a subsidiary goal of psychology, not its primary focus. What we must explain first is the individual's ongoing commerce with the environment. Little is left of Gestalt psychology today except some familiar demonstrations of figure-ground and perceptual grouping—demonstrations that they hoped were manifestations of the law of *Pragnanz*. Köhler's attempt to find a place for value in a world of facts is all but forgotten. Perhaps the attempt was just ahead of its time, too early in the history of psychology to have had a chance of success. Now it's our turn.

Another way to think about this issue is in terms of Dilthey's dichotomy between the natural and the human sciences. The Gestalt psychologists tried to transcend that dichotomy and failed; today it seems sharper than ever. Envious and anxious by turns, the human sciences are in a sour and defensive mood. They are tempted to reject science entirely, fencing off little safe areas like "meaning" and "interpretation" where it is not allowed to enter. In the long run, this can't possibly work. The natural sciences are on a roll, and, if we turn our backs, they will roll right over *us*. What we must do instead is find some solid ground on which to stand, some domain of knowledge that is scientifically based and yet relevant to human life as we experience it. Perhaps the ecological approach to cognition can help to establish a place of that kind.

Some Things We Know

The beginnings of such an enterprise may already be under way. Before I describe those beginnings, it may be useful to review some of the empirical findings on which they rely. To do so will conveniently converge on the other aim of this chapter: to convince both neuroscientists and skeptics that there actually is an ecologically oriented cognitive science, and that it does make discoveries. Twelve such discoveries are listed in Table 17.1.

Perhaps for the sake of symmetry—an old Gestalt value—my list includes three items in each of four groups: 12 in all. This was hardly a necessary constraint. There is no magic in the number 12; six items or 15 might have done just as well. Moreover, I do not claim that these are the 12 most important findings ever made. Any such selection is a bit arbitrary, and everyone may have individual favorites. The list is basically just an existence proof. Cognitive science does find things out, and here are a few of them.

Visual kinesthesis: In the classical study of perception, the eye was regarded as located at a stationary point. But people move, and their eyes move with them. These movements create complex optical changes that not only produce experienced ego-motion but also control bodily posture (Lee & Aronson, 1974). We see where we are and how we are moving largely because such flows specify the position and movement of the self (Cutting, Springer, Braren, & Johnson, 1992; Warren, Mestre, Blackwell, & Morris, 1991). What we see in this way is not the murky inner Cartesian self but the real tangible ecological self, of whom I shall have more to say below.

Kinetic depth: The changing optic projections produced when an object rotates in the field of view (or when we walk around it) specify its real shape very precisely. That shape is what we perceive (Ullman, 1979; Wallach & O'Connell, 1953). The kinetic depth effect is responsible for much of the realistic depth that we see in movies and TV displays, which are in fact projected on flat screens.

Perception in infancy: From William James to Freud and Piaget, the infant's world was said to be a "blooming buzzing confusion." It was universally supposed that babies had no realistic visual perception; they could not even distinguish themselves from their mothers or from the world. Today we know that this assumption was deeply mistaken. Infants are ecological selves from a very early age, realistically aware of their own capabilities and of the three-dimensional environment around them (Baillargeon, Spelke, & Wasserman, 1985; Spelke, 1982).

Domain knowledge: Memory is only partly a matter of storing perceptual impressions. It is also a way of relating present information to past systematic regularities, namely, those established by extended experience in a given domain. Chess masters, having seen and played thousands of games, recall briefly presented chess

Table 17.1. Some Recent Discoveries at the Ecological Level

VISUAL PERCEPTION

1. *Visual kinesthesis*: Posture, balance, and perceived direction are largely dependent on the optic flows produced by locomotion.
2. *Kinetic depth*: The changing optic projections produced when an object moves are what enable us to perceive its real shape.
3. *Perception in infancy*: Babies see the world three-dimensionally and realistically from a very early age.

MEMORY

4. *Domain knowledge*: Background knowledge and expertise are major determinants of memory for material from a given domain.
5. *Confident errors*: Even vivid and confident recollections of personally experienced events may be entirely mistaken.
6. *Memory in infancy*: Very young children can recall events that happened months or years ago, although not in narrative form.

SPEECH AND LANGUAGE

7. *Infant phonemes*: Pre-linguistic infants can distinguish between phonemes that sound the same to their parents.
8. *Joint attention:* Social understanding (of what one's partner is attending to) is necessary for the acquisition of language.
9. *Sign language*: The manual languages of the deaf have a full, rich linguistic structure.

THINKING

10. *Basic-level categories*: Based on perceptual and action-related properties, these early categories have a prototype structure.
11. *Gains in IQ*: Scores on intelligence-related tests are rising all over the world, at a rate of about 3 "IQ points" per decade.
12. *Concept of consciousness*: Young children are slow to understand that they (and other people) have mental states—states that may not correspond to what is actually the case.

positions far better than novices do (Chase & Simon, 1973). Electrical engineers have good memories for wiring diagrams (Egan & Schwartz, 1979); for that matter, so do psychologists for cognitive experiments. Any complete theory of memory will have to include an understanding of environmental information structures, for it is those structures that shape and support what we actually remember.

Confident errors: Recent media coverage of incest accusations—often based on entirely false childhood memories—has drawn our attention to the ease with which such memories can be established (Loftus, 1993). Memory research, both in the field and in the lab, has confirmed that this is true (Hyman, Husband, & Billings, 1995; Neisser & Harsch, 1992). These findings mean, among other things, that the psychology of memory is no longer as comfortably encapsulated from social and motivational factors as we used to think.

Memory in infancy: Although two-year-olds never reminisce about old times, they can recall surprisingly much about past events when appropriately cued (Fivush, Gray, & Fromhoff, 1987). Even much younger children will imitate actions that they watched a week earlier (Meltzoff, 1988). Infants who have learned to produce an interesting visual effect will retain that knowledge for many days or weeks

(Rovee-Collier, 1989). Thus, memory does not appear suddenly at the offset of "childhood amnesia" in the third year (Usher & Neisser, 1993); it has a continuous course of development from the beginning.

Infant phonemes: It often happens that a difference between two speech sounds is "phonemic" in language A but not in language B. (A familiar example is the distinction between /r/ and /l/, which matters in English but not in Japanese.) Adult speakers of language B often have great difficulty in distinguishing these unfamiliar (to them) phonemes. Surprisingly, however, their own infant children do not share this difficulty. Studies of many languages and many phonemic contrasts show that infants under six months old, growing up with only language B spoken around them, can nevertheless distinguish the phonemes of A. This precocious ability does not last. As early as the end of the first year, infants have become just as insensitive to those distinctions as the B-speaking adults in their families (Werker & Tees, 1984). The implications of this surprising sensitivity and its equally surprising disappearance are still unclear.

Joint attention: Infants do not acquire language continuously, whenever adults happen to use it around them. They can learn the meanings of words only when they actually know what an adult is talking about. In early life these are the occasions when child and mother are both attending to the same object (or expecting/ thinking of the same event), and each of them knows that the other one is attending in this way (Bruner, 1983; Tomasello & Farrar, 1986). Thus, the acquisition of language depends on social as well as on ecological perception.

Sign language: The discovery that the manual languages of the deaf are as subtly and richly structured as any spoken language (Stokoe, 1960) seems enormously important, not least because of the increased respect we now accord the mental life of the deaf themselves (A. Neisser, 1983). It also has profound implications for the ongoing theoretical debate about the biological basis of language. "Certainly we would be the last to argue that speech does not constitute part of the biological foundations of language. But if speech is specially selected, if sound constitutes such a natural signal for language, then it is all the more striking how the human mind . . . seizes on, perfects, and systematizes an alternate form to enable the deeper linguistic faculties to give explicit expression to ideas" (Klima & Bellugi, 1979, p. 315).

Basic level categories: Objects (tools, furniture, and the like) can be categorized in many ways, but there is always a basic level where they are classified by their shapes and the actions they afford. ("Knife" and "hammer" are such categories; so are "chair" and "table.") Basic level categories have simple names, are learned early in childhood, are easily imaged, and have fairly characteristic prototypes (Rosch, Mervis, Gray, Johnson, & Boyes-Braem, 1976).

Gains in IQ: Mean test scores have been rising steadily, all around the world, at an average of about 3 points per decade. A person who scores 100 on an IQ test today would have scored about 115 on the tests that were current 50 years ago. Because it was James Flynn (1987) who first systematically explored this rise and recognized its significance, it is now often called the "Flynn Effect." One might think that such a rise would have been obvious, but the fact that most intelligence tests are periodically "re-normed" (to keep the mean at 100) has made it hard to see. Possible explanations of this increase, which is greatest on so-called "culture free" tests of abstract reasoning, include improved nutrition, better test-taking skills, and cultural change. I hold the third view. Throughout this century there has been a steady increase in schooling, in exposure to media, in overall awareness of what goes on outside one's own neighborhood. More complex environments may just produce more complex minds.

Development of the concept of consciousness: Young children do not understand that reality is one thing while how we experience reality is another. This failure shows up in two different experimental paradigms. With respect to their own awareness, children are confused by the "appearance/reality distinction"—they don't realize that an object may look like a rock but just be a painted sponge, or look pink (through tinted glass) and yet really be white (Flavell, Flavell, & Green, 1983). Where other people's consciousness is concerned, they have trouble understanding false beliefs: that someone may think X while Y is really the case (Astington, Harris, & Olson, 1988). It is often said that such children have not yet developed a "theory of mind" (though I am not entirely comfortable with that term). Note that we are talking about children who already have complete control of perception and action as well as an impressive grasp of language: it is minds they don't know about.

Ecological Interpretations

Perhaps the ecological aspects of these findings are self-evident, but it may be useful to review them nevertheless. There are no mental models on this list, nor are there any findings from neuroscience. All the entries can be regarded as facts about what people take from, do with, create in, or understand about their physical and social environments—that is, as basically ecological discoveries. This is perhaps most obvious for the examples from perception, which clearly reflect the influence of ecological theory. They show how easily and accurately people perceive the environmental layouts around them, as well as the possibilities for action that those environments afford. This is true of infants as well as adults, though infants have fewer action capabilities and notice less.

The existence of an ecological approach to memory is perhaps more surprising. Isn't memory—by definition—something stored away in the brain? Perhaps the plausibility of that definition helps to explain why the naturalistic study of memory is still controversial (Banaji & Crowder, 1989). In fact, however, much more than mere storage is involved. Remembering is always the act of a particular individual on a particular occasion—an individual who is more or less knowledgeable, more or less influenced by suggestions or preconceptions, more or less skilled in the use of memory itself. A great deal of ecologically sophisticated memory research will be needed if we are to understand how that knowledge develops, how those suggestions operate, how those skills develop.

Linguistics and psycholinguistics have not yet fully assimilated the discovery that infant phoneme-perception is universal. It suggests (at least to me) that speech perception begins as a form of general auditory event perception. The events in question are the movements of articulatory organs in the mouth of the speaker: movements that give rise to structured sound patterns and hence can be perceived. This general perceptual ability is then sharpened by processes of learning that apparently have a negative as well as a positive side.

Joint attention is important because it highlights the importance of social perception for the acquisition of language. Indeed, first-language learning is the paradigmatic example of genuinely interpersonal cognition. Children do not blindly learn "associations" between vocal sounds and objects, nor (in my view) is their learning guided by innate "constraints." It is because they know Mommy is trying to communicate with *them*, about something to which *she* is attending and to which they attend as well, that they can learn the new words she utters.

The linguistic structure of American Sign Language was first established by William Stokoe in 1960 and has been confirmed by many other investigators. It has little to do with ecological psychology; I mention it mostly to show how full of surprises the cognitive world can be. Note that this discovery could not have been made by neuroscience: it is a fact about the world, not about the brain. It has, however, inspired studies of the brain localization of sign language.

Eleanor Rosch's studies of basic-level categories were examples of ecological research at its best. These mental categories, fundamental to our understanding of the world, are anything but arbitrary. They are based on perceiving what J.J. Gibson (1979) called the *affordances* of objects: what we do with them, how we use them. They are acquired very early—in what Mervis (1987) calls their "child-basic" form—because those characteristics are especially salient for children. By now, of course, there are competing information-processing models of the representations that underlie this form of categorization. The models are endlessly disputable, but the existence of basic level categories is not.

The steady rise of IQ scores around the world is a truly remarkable demonstration of environmental influence on mental process. In today's intellectual climate, where neuroscience and genetics have such a strong hold on public imagination, this demonstration seems especially important. The psychometric difference between generations only half a century apart amounts to a full standard deviation—at least as much as the heavily publicized difference between black and white Americans today.

I rounded out the list with young children's concept of consciousness largely because it is relevant to my own theory of the self (to be considered below). The findings show that—at an age when they are keenly "self-conscious" in the ordinary sense of that term—young children can still be confused about the relations of mind and world. The exquisitely private self of inner experience, so important for many philosophical and cultural accounts of selfhood, is a fairly late intellectual development. Self-awareness does not begin with introspection.

Ecological Theories

These phenomena were selected primarily to illustrate the possibility of significant findings at the ecological level of analysis. But even granting that such findings exist and that they are interesting, they are still just facts, observations. Facts alone do not establish a science; we need theories too. But what constitutes a theory? In modern cognitive science that term is typically used for block-box mental models—so much so that the term "ecological theory" sounds almost like an oxymoron to some of my friends. Nevertheless, not all scientific theories are models of covert processes. (Newton's, Einstein's, and Darwin's theories would all be counter-examples to such a claim.) Theories at the ecological level can present new causal explanations, enriched accounts of development, insightful redescriptions of actual situations, and unexpected integrations across what seem to be distinct domains. Several such theories already exist.

The most important of these, at least from my perspective, is James J. Gibson's (1979) ecological theory of perception itself. He insisted that the fundamental task for perceptionists is not to model mental processes but to study the objectively existing information structures on which perceiving depends. For vision, these are the structures in the optic array. Gibson made many contributions to that study during his lifetime, and his successors are making many more. There is also Eleanor Gibson's (1969) long-established theory of perceptual development, which has largely shaped recent studies of infant perception and action and profoundly influenced our conceptions of learning itself (E. J. Gibson, 1982; 1988). In a related vein are various new approaches to movement and motor control, especially the dynamic systems theory of Esther Thelen (1993). There is the increasingly influential Vygotskian approach to development, with its premise that all cognitive activity is social and public at first and "privatized" only later (Vygotsky, 1978). Related to that approach is the social-cognitive account of language acquisition, first proposed by Bruner (1983) and substantially extended by Tomasello and his collaborators (1993). None of these formulations proposes mental models, but all of them are theories in the best sense of the word: they propose hypotheses and can be tested against data. Whatever their future may be, they cannot be outdated by new developments in neuroscience.

I have a theory too. Although in many ways less developed than those mentioned above, it may be more relevant to the topic at hand. Mine is a theory of the *self*; more precisely, of self-knowledge. Briefly, I suggest that people have access to five different kinds of knowledge about themselves (Neisser, 1988; 1991; 1993). Two of these are founded on perception, the other three on reflection and thought. These modes of knowing are very different—so different that we can usefully think of them as establishing different "selves." On this list:

The *ecological self* is the individual considered as situated in and acting on the immediate physical environment. Ecologically speaking, I am the person who is sitting here at my desk, punching keys on my computer, very much aware of where I am. Infants are ecological selves from a very early age.

The *interpersonal self* is the individual engaged in social interaction with another person. The behavior directed toward me by others confirms my existence and my personal effectiveness very directly: I can easily see that they are responding to *me*. The interpersonal self, like the ecological self, is given in perception and available from early infancy.

The *conceptual self* or self-concept is a person's mental representation of his own (more or less permanent) traits and characteristics. I am a man, a husband, a professor, a bit slower now than I used to be but probably no less contentious. Such representations vary from one culture to another as well as from one person to the next. The self-concept is heavily dependent on verbal formulations, and probably develops in the second year of life.

The *temporally extended self* is the individual's own life-story as he or she knows it, remembers it, tells it, projects it into the future. I am the person who grew up in such-and-such a house in such-and-such a town, who has had these successes and experienced those disasters, who—now engaged in writing this very chapter— has some fairly specific plans about what he will do next. In contemporary American culture, this "narrative" self seems to appear about the fourth year.

The *private self* is given in introspection. No one else can know my thoughts, my dreams, or even my headaches just as I do. These inner experiences become relevant to self-definition only when the child comes to appreciate the privacy of conscious life, when the child understands that no one else has access to his or her own thoughts and dreams. This must happen rather late in childhood.

This is not a list of selves we have, but of selves we are and selves we know. They vary in their degree of congruence with reality. The ecological and interpersonal selves, given in perception, correspond closely to the real state of affairs. Ecologically, I can see (and hear and feel) exactly where I am and what I am doing; interpersonally, I can see (and hear and feel) that the person I am with is attentive and responsive to me. The conceptual self, in contrast, is much less tightly constrained. I may think of myself as generous when in fact I am stingy, or feel fat when in fact I am thin. In today's epidemic of "multiple personality disorder," I may even hold the curious view that several different persons inhabit my body at once. Self-narratives, too, are subject to serious distortion: under some conditions, people confidently "remember" events that never happened at all. The private self, of course, may be entirely disconnected from one's actual situation. Sitting here in Atlanta, I can easily imagine myself in Moscow or on the moon. Veridical or not, all these forms of self-awareness are certainly *me, myself,* and *I.*

Having several sources of self-knowledge does not make individuals fragmented or disjoint. These aspects of self-awareness are typically coherent, because there is exactly one real perceivable person—in exactly one environment—to whom they all refer. What holds individuals together is not some ultimate neural center in the brains but their commerce with their environments: what they know and what they do. The self may be disappearing from neuroscience, but it is alive and well in ecological psychology.

It would be a mistake to dwell too long on this theory of self-knowledge. It is only one of the many lines of ecological thinking that flourish today; others are far better established and may deserve more careful attention. The important point is that the study of individuals in their environments continues to make progress at many levels and in many domains. We need not fall prey to post-modern cynicism, and we are in no danger of being eliminated by our neuroscientific friends. The ecological alternative clearly has the three main assets that all scientists need: theories and data and hope. That is fortunate, for in the twenty-first century it may be the only psychology we have.

References

Astington, J. W., Harris, P. L., and Olson, D. R. 1988. *Developing Theories of Mind*. New York: Cambridge University Press.

Baillargeon, R., Spelke, E. S., and Wasserman, S. 1985. "Object Permanence in Five-Month-Old Infants." *Cognition*, 20, pp. 91–208.

Banaji, M. R., and Crowder, R. G. 1989. "The Bankruptcy of Everyday Memory." *American Psychologist*, 44, pp. 1185–1193.

Bruner, J. 1983. *Child's Talk*. New York: Norton.

Chase, W. G., and Simon, H. A. 1973. "The Mind's Eye in Chess," in W. G. Chase (ed.), *Visual Information Processing*, pp. 215–81. New York: Academic Press.

Churchland, P. M. 1988. *Matter and Consciousness*, rev. ed. Cambridge: MIT Press.

Cutting, J. E., Springer, K., Braren, P. A., and Johnson, S. H. 1992. "Wayfinding on Foot from Information in Retinal, Not Optical, Flow." *Journal of Experimental Psychology: General*, 121, pp. 41–72.

Egan, D. E., and Schwartz, B. J. 1979. "Chunking in Recall of Symbolic Drawings." *Memory and Cognition*, 7, pp. 149–158.

Fivush, R., Gray, J. T., and Fromhoff, F. A. 1987. "Two-Year-Olds Talk About the Past." *Cognitive Development* 2, pp. 393–409.

Flavell, J. H., Flavell, E. R., and Green, F. L. 1983. "Development of the Appearance-Reality Distinction." *Cognitive Psychology* 15, pp. 95–120.

Flynn, J. R. 1987. "Massive IQ Gains in 14 Nations: What IQ Tests Really Measure." *Psychological Bulletin*, 101, pp. 171–191.

Gibson, E. J. 1969. *Principles of Perceptual Learning and Development*. New York: Appleton-Century-Crofts.

Gibson, E. J. 1982. "The Concept of Affordances in Development: The Renascence of Functionalism," in W. A. Collins (ed.), *The Concept of Development: Minnesota Symposium on Child Psychology*, Vol. 15, pp. 55–81. Hillsdale, NJ: Erlbaum.

Gibson, E. J. 1988. "Exploratory Behavior in the Development of Perceiving, Acting, and the Acquiring of Knowledge." *Annual Review of Psychology*, 39, pp. 1–41.

Gibson, J. J. 1979. *The Ecological Approach to Visual Perception*. Boston: Houghton Mifflin.

Hume, D. 1888. *Treatise on Human Nature*, L.A. Selby-Bigge (ed.). Oxford: Oxford University Press.

Hyman, I. E., Husband, T. H., and Billings, F. J. 1995. "False Memories of Childhood Experiences." *Applied Cognitive Psychology* 9, pp. 181–198.

Klima, E. S., and Bellugi, U. 1979. *The Signs of Language*. Cambridge: Harvard University Press.

Köhler, W. 1938. *The Place of Value in a World of Facts*. New York: Liveright.

Lee, D. N., and Aronson, E. 1974. "Visual Proprioceptive Control of Standing in Human Infants." *Perception and Psychophysics*, 15, pp. 529–532.

Loftus, E. F. 1993. "The Reality of Repressed Memories." *American Psychologist*, 48, pp. 518–537.

Meltzoff, A. N. 1988. "Infant Imitation After a 1-week Delay: Long-term Memory for Novel Acts and Multiple Stimuli." *Developmental Psychology* 24, pp. 470–76.

Mervis, C. B. 1987. "Child-Basic Categories and Early Lexical Development," in U. Neisser (ed.), *Concepts and Conceptual Development: Ecological and Intellectual Factors in Categorization*. New York: Cambridge University Press.

Neisser, A. 1983. *The Other Side of Silence*. New York: Knopf.

Neisser, U. 1988. "Five Kinds of Self-Knowledge." *Philosophical Psychology*, 1, pp. 35–59.

Neisser, U. 1991. "Two Perceptually Given Aspects of the Self and Their Development." *Developmental Review*, 11, pp. 197–209.

Neisser, U. (ed.). 1993. *The Perceived Self: Ecological and Interpersonal Sources of Self-Knowledge*. New York: Cambridge University Press.

Neisser, U. and Harsch, N. 1992. "Phantom Flashbulbs: False Recollections of Hearing the News about Challenger," in E. Winograd and U. Neisser (eds.), *Affect and Accuracy in Recall: Studies of "Flashbulb" Memories*, pp. 9–31. New York: Cambridge University Press.

Rosch, E., Mervis, C. B., Gray, W. D., Johnson, D. M., and Boyes-Braem, P. 1976. "Basic Objects in Natural Categories." *Cognitive Psychology* 8, pp. 382–439.

Rovee-Collier, C. 1989. "The Joy of Kicking: Memories, Motives, and Mobiles," in P. R. Solomon, G. R. Goethals, C. M. Kelley, and B. R. Stephens (eds.), *Memory: Interdisciplinary Approaches*, pp. 151–80. New York: Springer-Verlag.

Spelke, E. S. 1982. "Perceptual Knowledge of Objects in Infancy," in J. Mehler, M. Garrett, and E. Walker (eds.), *Perspectives on Mental Representation*. Hillsdale N.J.: Erlbaum.

Stokoe, W. C. 1960. "Sign Language Structure." *Studies in Linguistics: Occasional Papers* 8.

Thelen, E., Corbetta, D., Kamm, K., Spencer, J. P., Schneider, K., and Zernicke, R. F. 1993. "The Transition to Reaching: Mapping Intention and Intrinsic Dynamics." *Child Development* 64, pp. 1058–98.

Tomasello, M., and Farrar, J. 1986. "Joint Attention and Early Language." *Child Development*, 57, pp. 1454–63.

Tomasello, M., Kruger, A.C., and Ratner, H.H. 1993. "Cultural Learning." *Behavioral and Brain Sciences*, 16, pp. 495–552.

Trevarthen, C. 1993. "The Self Born in Intersubjectivity: The Psychology of an Infant Communicating," in U. Neisser (ed.), *The Perceived Self*, pp. 121–75. New York: Cambridge University Press.

Ullman, S. 1979. "The Interpretation of Structure from Motion." *Proceedings of the Royal Society of London*, B 203, pp. 405–25.

Usher, J. A., and Neisser, U. 1993. "Childhood Amnesia and the Beginnings of Memory for Four Early Life Events." *Journal of Experimental Psychology: General*, 122, pp. 155–65.

Varela, F. J., Thompson, E., and Rosch, E. 1991. *The Embodied Mind: Cognitive Science and Human Experience*. Cambridge: MIT Press.

Vygotsky, L. S. 1978. *Mind in Society*. Cambridge: Harvard University Press.

Wallach, H., and O'Connell, D. N. 1953. "The Kinetic Depth Effect." *Journal of Experimental Psychology*, 45, pp. 205–17.

Warren, W. H. J., Mestre, D. R., Blackwell, A. W., and Morris, M. W. 1991. "Perception of Circular Heading from Optical Flow." *Journal of Experimental Psychology: Human Perception and Performance*, 17, pp. 28–43.

Werker, J. F., and Tees, R. C. 1984. "Cross-Language Speech Perception: Evidence for Perceptual Reorganization During the First Year of Life." *Infant Behavior and Development*, 7, pp. 49–63.

18

Edward Reed

The Cognitive Revolution from an Ecological Point of View

The Failure of the Information-Processing Approach to Psychology

Within psychology, the so-called cognitive revolution was spearheaded by the *information-processing approach*. This approach to psychology emerged in the wake of World War II developments in the engineering theory of information and control. Human beings were conceived of as special elements in communication and control systems (Broadbent, 1958; Leahey, 1992), elements that receive, process, and transmit information, sometimes acting to change the control parameters of the system within which they are embedded.

In my opinion, this version of cognitivism is now dead. Not a single one of its scientific pretensions has proved itself. Information theory does not characterize and quantify human perceptual processes (contra Attneave, 1959). No one nowadays even tries to connect Shannon's information theory with human cognition, and the use of information theoretic quantification in the psychology laboratory is a thing of the past. The attempt to explain behavioral activities as resulting from information-based programs or plans (Fitts & Posner, 1967; Miller, Galanter, & Pribram, 1960) has also failed. Even the most cognitivist students of human motor action realize that symbolic programs alone do not and cannot explain the complex dynamical movements of bodies, and there is a veritable renaissance of biomechanical and "applied physical modeling" work in this area (e.g., Thelen & Smith, 1994).

Within psychology itself these failures are not yet readily and openly acknowledged. Nevertheless, the deficiencies of information-processing cognitivism are so many and the problems so deep that a number of critiques and counter currents have begun to achieve considerable prominence in the past decade. These critiques of information processing cognitivism are coming from many quarters, and especially from:

- Connectionists (e.g., Rumelhart & McLelland, 1986)
- Situated activity theorists (e.g., Johnson, 1987; Varela, Thompson, & Rosch, 1991)
- Activity theorists in the tradition of Vygotsky and Leont'ev (e.g., Wozniak & Fischer, 1993)
- "Theory of mind" theorists (e.g., Astington, Harris, & Olson, 1988)

To some degree, the ecological critique of information processing offered here has elements in common with all these other critiques. However, I also believe that all these other critiques have missed a fundamental point, which will therefore be emphasized in this chapter. *At least some of the goals of the information-processing revolution were the right goals for moving psychology forward. If information processing has failed (as I believe it has), then so will we fail unless we do a better job of meeting these important goals.*

I will go further than this, however, and argue that the revolution attempted by information-processing psychology and by cognitive science has in fact been made by ecological psychology.

And I will take one step further to say that the revolution accomplished by ecological psychology does even more than envisaged by the cognitive revolution, so that, properly developed, the ecological revolution will encompass not only what was good in the cognitive revolution, but will go beyond this in certain important ways, thus incorporating some of the important points raised by critics of cognitivism.

The following chapter is thus divided into two parts: a first part assessing both the goals and failures of the information-processing revolution in psychology, and a second part showing how the ecological approach to psychology accomplishes much of what was best among the intentions of the cognitivists, and then some.

Goals of the Cognitive Revolution

The primary goal of the information-processing revolution was to bring the study of mental processes (back) into psychology. A secondary aspect of this goal was to rethink the concept of mental processes and, in particular, to broaden existing notions of mental processes.

It is a fallacy to think that behaviorism did not deal with mental processes. After all, Watson (1914) explicitly argued that thought was a kind of subvocalization, and Skinner (1957) had his tacts and mands. (And these are only the two most famous examples.) But these behavioristic notions of mental processes are unsatisfactory from the point of view of trying to do research. One might demonstrate the existence of subvocalizing activity in the larynx during thought—but then what? How could such a line of research hope to inform us about the nature, organization, and scope of thought?[1]

The important and proper goal of the post World War II "cognitive revolution" in psychology was to try to broaden the concept of mental processes in such a way that research programs could be designed for studying those processes in the laboratory. The processes involved in thinking, problem solving, prejudice, stereotyping, and mental imagery were among the phenomena emphasized in the early 1950s as being unsusceptible to experimental analysis via the tools of behavioristic learning theory, but which, it was claimed, would be brought into the lab by the cognitivists.

It was Noam Chomsky—a linguist, not a psychologist—who realized more clearly than anyone else that the key to this successful broadening of the concept of mental process was to reject the associationist analysis of all mental processes, and emphasize their intrinsic generativity (see Brewer, 1974). I would add that for such a rejection to succeed psychologists would have to abandon the entire S-R framework as well (see Asch, 1952: Part 2 for a cogent—but unheeded—contemporary analysis along these lines). As I will show, it is the inability of the cognitive revolutionaries to live up to this goal of transcending associationism that has resulted in the ultimate failure of the cognitive revolution.

Why is overcoming associationism so central? Because associationist analyses of either mental states ("ideas" or "representations") or of behavior (stimuli, responses, internal mediators of stimuli and responses) are so powerful that *any* psychological process can be described *as if* it were accomplished by an associative network. But if associationist principles (e.g., contiguity, similarity, strengthening of activation and inhibition through repetition) are sufficient to describe a process, then no mental activity need be involved at all. This kind of approach to psychology is currently being actively promoted by neuroreductionists (e.g., P.S. Churchland, 1991), but its classical form is found in those American behaviorists who extended Pavlov's conditioning model of associationism into all areas of psychology—most especially, Clark Hull (1943) and his students.

A cognitivism that describes the phenomena of behavior and mental processes in such a way as to allow for such an associationist account is asking for trouble. Any putative mental activity can, under these circumstances, be reduced to non-mental processes of association. Mental states might therefore be allowed to exist, but only as epiphenomena of non-mental processes. And the goal of cognitive psychology would be to find *non-psychological* processes (whether computational, neurophysiological, or something else) to account for resultant mental states or behavior.

There are many who find such a cognitive psychology unproblematic or even desirable. But even those who like to explain away mental processes in this way cannot argue that this sort of cognitivism succeeds in bringing mental processes back into psychology. On the contrary, to the extent that this kind of cognitivism succeeds, it succeeds because it is *eliminating* mental processes from psychology! This serious internal inconsistency has plagued information-processing psychology from its outset. Ironically, then, the cognitive revolution brought the seeds of its own destruction into psychology.

It is striking that even close followers of Chomsky did not see the destructive genie they were bringing into their field under the rubric of "psychological explanation." For example, one of the first serious attempts to bring Chomsky's ideas into psychology is found in Miller, Galanter, and Pribram's (1960) *Plans and the Structure of Behavior* (especially Ch. 11). In this widely influential theory, Miller and his colleagues proposed that all skilled action is based on a process they dub "a plan." A plan is instantiated as a set of rules for ordering and reordering stimulus-response relations. Under Chomsky's influence, these theorists suggested that plans for speaking (and, *inter alia*, plans for thinking) involved hierarchies of plans: plans for ordering and reordering plans. Like Chomsky, they used arguments based on a Markov chain analysis to show that associative mechanisms are inadequate here. Without rules and plans to guide the concatenations, the associations of words will be so unconstrained as to be quite unlike the complex patterning of syntax found in natural language. They go to great lengths to reject trial-and-error learning theories, but do nothing to reject framing their own theory in terms of stimulus-response relations amplified by sets of rules constraining those relations. They end the book calling themselves "subjective behaviorists" (a term that was passed over in favor of "cognitivists" for obvious PR reasons). For Miller, Galanter, and Pribram cognitive processes should not be reduced to mere behavioral inputs and outputs. Instead, cognitive processes should be reduced to behavioral inputs and outputs as constrained by specific rules.

Similarly, one of the most influential early philosophical accounts of the allegedly revolutionary theorizing—Fodor's (1968) *Psychological Explanation*— offered what amounted to a subjective behaviorist account of psychological explanation. For Fodor, the whole purpose of cognitivism is to explain the "disparity between input and percept" or output (p. 85). Fodor writes that "what happens when one hears a sentence in a language one knows is apparently that the acoustic signal is integrated by reference to some set of rules that the hearer has internalized during the course of learning the language" (p. 84). Such rules explain how and why a particular interpretation is given to the "stimulus." Again, the target of Fodor's attack is random trial and error S-R connections, and his goal is to enunciate the rules intervening between input and output. Fodor is right to emphasize that the operation of such rules need not implicate any overt behavior, and therefore that explanations that make an appeal to such operations cannot be called (objective) behavioristic. But whatever one calls this theory, subjective behaviorism, or cognitivism, or something else, it has not proved very helpful in analyzing mental processes. This is because most of our mental processes have nothing at all to do with mediating between stimuli and responses.

The rule-based associationism that lies at the heart of cognitive psychology is so widely assumed to be *the* way to analyze cognitive processes that my criticism will undoubtedly sound surprising. It sounds scientific to analyze mental states in terms of inputs and outputs (stimuli and responses) and unscientific to reject these terms. But sophisticated behaviorists already had extended S-R theory to include "goal-directed" (= "rule driven") internal associations (see Hull, 1943); and the difference between information-processing cognitivism and this kind of behaviorism has proved to be largely verbal (Hintzman, 1993). In fact, it is really not difficult to make the case that the kind of information-processing psychology that embraces input-output associationism cannot, in principle, be very informative about mental processes. The argument can be sketched as follows.

Assuming the truth of something like Miller, et al.'s "plans for organizing plans" as the basis of cognition, how would one study a mental process? One would start by finding a particular pattern of input-output relations. In this part of the work, there is little difference between cognitivists and behaviorists. The experimentalist creates some "stimuli" and puts them to the subject along with a task, which results in a "response." One difference is that cognitivists are in principle more interested in responses that are essentially just "interpretations" of the stimuli (e.g., sentence meanings). However, in actual practice, such interpretative responses have been downplayed in favor of responses for which "harder," more quantitative dependent measures can be found, such as reaction times and patterns in errors.

Having found this set of S-R relations, one then would attempt to derive a rule that operates on the total possible set of Ss to produce just those Rs observed in the experiments. An obvious problem is that it is almost always possible to produce more than one such rule. New experiments can be designed to test between the two or more competing rules, only if the following conditions are met.

(1) If the two (or more) rules make demonstrably different predictions about how S-R relationships would emerge from a novel set of Ss or a novel task; and (2) if the two (or more) rules make demonstrably different predictions about measurable properties of the responses. In practice, the first case has boiled down to the use of catch trials, error analysis, or analysis of transfer effects; the second case has boiled down to chronometric analysis, trying to track predicted differences in response time or total time taken. Thus, the much heralded cognitive revolution was certainly not a revolution in the psychology laboratory. On the contrary, the "brass instrument" psychology of Wundt and his cohort simply was updated to use computer technology (Leahey, 1992).

The direct study of mental processes through introspection or phenomenological analysis has not played a significant role in cognitive psychology, because it was ruled out from the very beginning. These mental processes, being neither stimuli nor responses are not objects of study. Indeed, they are widely considered to be unconscious, with only their outputs accessible to consciousness (Kihlstrom et al., 1992).

In other words, within cognitive psychology, when two theories make conflicting claims about the nature of mental processes, the ways of assessing which claim is correct are very limited. The claim cannot in principle be assessed through introspection. And if the mental processes do not in some way produce differences in behavior, either in a verbal interpretation, or through time taken, or through the pattern of correct and incorrect performance, then the conflicting claims cannot be evaluated. Information-processing psychology appears to be a subdivision of behaviorism.

As Fodor (1968, p. 79) rightly noted, behaviorism as a general research strategy can be ruled out only "if it is possible to demonstrate the occurrence of psychological phenomena for which the simplest available explanation requires us to hypothesize the occurrence of mental events that do not exhibit behavioral correlates." Ignoring Fodor's caveat about "simplest possible explanations" (for the good reason that no two psychologists or philosophers are likely to agree on the reference of that phrase) we can see that mainstream information-processing psychology does *not* suffice to rule out behaviorist explanations. "What if the neobehaviorists had called the unobservable events they postulated 'representations,' 'units,' or (even more vaguely), 'processes'?" asks Hintzman (1993, p. 382). Then the continuity between behavioristic S-R association theory—the real kind, with its host of intervening variables and internal S-S connections—and information-processing psychology would be obvious. We should not be fooled by a few labels into missing the strong relationship between behaviorism and cognitivism as practiced by information-processing psychologists.

It might be argued that the key difference between information-processing psychology and any form of behaviorism—even Miller, Galanter, and Pribram's "subjective behaviorism"—is precisely the emphasis on "information" as opposed to stimuli. Yet this turns out to be an empty boast. Modern writers either confess that they have no account of what "information" is (e.g., Palmer & Kimchi, 1986) or insist that, whatever it is, it cannot be information in the cognitive sense of referring to things in the world (Jackendoff, 1992). (This does not apply to Gibson's concept of information that we will discuss later.)

It also might be argued that the concept of mental representation serves to move cognitivism beyond the framework of S-R psychology. But this concept has remained essentially undefined for so long that the only useful response to such a claim is to ask for a detailed account of "representation" that both fits what cognitive psychologists do and offers more than a gloss on the phrase "representations are the contents of cognitive states." Until such an account is forthcoming, this argument for saving cognitivism is nothing more than a promissory note—a dismal state for a field to be in after 40 years.

For three decades now information processing psychologists have offered us a "cognitive psychology" that reduces the phenomena of human mental life to differences in reaction time, or to different patterns of erroneous judgments. When this "achievement" is compared to the desiderata of the early cognitive theorists (Postman, 1951; Asch, 1952), the disparity is enormous and unsettling. The call for a cognitive theory came from scientists who wanted to understand human thinking, planning, and deciding as they function in the real world, solving problems that arise as people try to make their way through the world, dealing with each other and with all the various aspects of daily life. The input-output framework that has so dominated experimentation and theorizing about cognition has simply stymied almost all research that might have explored the intricacies of the phenomena of human thought and judgment in everyday life. Can anyone honestly claim that we have made significant progress on the kinds of questions and concerns that motivated the development of cognitivism in the first place?

The present analysis of cognitivism as a set of epicycles on S-R theory—to put it crudely—is not original with me. Much of the present argument appeared in D. O. Hebb's (1960) presidential address to the American Psychological Association, "The American Revolution." Hebb insisted on the continuity between behaviorism and the emerging cognitivism of those times. "It may even be thought," he wrote "that the stimulus-response idea was a mistake in the first place, and is no part of a good psychology now (a good cognitive psychology, that is). This is absurd: the whole meaning of the term 'cognitive' depends on it, though cognitive psychologists seem unaware of the fact" (p. 737). Hebb went on to argue that "cognitive" has come to mean anything that deviates from simple S-R relations. The "cognitive" process is just whatever set of internal rules and relationships "must be" hypothesized to account for those deviations.

Hebb held that the S-R ideal was foundational even for cognitivists because it provided the descriptive framework within which experimental tests of competing ideas could be produced and assessed: "Much of the progress [in psychology] that has been made in this century is evident in the codification of ideas and terminology. The extent to which there is a common behavioral language today is remarkable, in contrast to the pet terms of the various schools 40 years ago. This partly reflects and partly creates an increased agreement concerning the facts, what the essential problems are, and what would constitute crucial evidence. . . . In this address I have tried to show that the clarification originated with behaviorism broadly speaking: on the one hand from the devoted effort to reduce all to the S-R formula, and on the other to search for the unambiguous experiment by which to refute that effort and on which to base the postulate of an ideational process" (Hebb, 1960, p. 744).

I agree with much in Hebb's astute analysis, except that I disagree with his belief that S-R psychology is good science, and therefore also disagree with the implications he claimed for his ideas. (An easy thing to do: hindsight being far more acute than foresight.) Cognitivism as it has been practiced has been little more than the "flip side" of behaviorism, trying to establish "mental processes" as anything that is left over after one tries to stuff all psychological phenomena into the S-R box. But Hebb's optimism about common language and goals was obviously delusional. Instead of the pet terms of different theoretical "schools," psychologists now have pet terms for every model-building enterprise; and there is literally no hope of integrative translation. Where it used to be that the three major schools of behaviorists had difficulty communicating, now each group of researchers working on a single cognitive process (and there are hundreds of such processes) uses their own hermetically sealed vocabulary. Instead of agreement on terms and concepts with unambiguous reference, no one can even agree on what processes cannot be reduced to the S-R formula! Our experimental paradigms, with their very limited sets of dependent measures (chronometry and error analysis being the best ones) are just not up to the task envisioned by Hebb. When Anderson (1978) showed that the supposedly monumental difference between analog and propositional coding could not be made to render clear-cut measurable differences, this should have been read as a warning that something was very wrong. But the warning went unheeded. Few cognitivists have shown a serious interest in reviewing their first principles.

What Ulric Neisser (1979) once wrote in commenting on some mental imagery research might well be generalized: "Why does the theory suggested . . . strike the reader as clever rather than insightful, as cute model making rather than serious psychology? I think it is because the thinking [of cognitivists] is completely detached from everything we know about human nature or about perception, thinking, and the nervous system. Like much contemporary work in 'information processing' it attempts to 'account for' a sharply restricted body of experimental results (usually reaction latencies) by relating it to an equally restricted class of models (usually computer programs or something similar). The effect is often as if the baby had been discarded and only the bathwater remained" (p. 561).

But if Hebb and I are right, then information-processing cognitivism has had no choice. If to study mental processes is to put people into experimental paradigms where they are stimulated and must respond within a certain time-frame, if reaction times and errors are the only reliable dependent measures, and if "rules" for operating on stimuli to produce response patterns are what counts as cognitivist explanations, then, dirty bathwater or not, this is the only cognitive "game in town," as Jerry Fodor might say. But whether or not this is the only cognitive game in town (obviously I don't think so), let us at least not perpetuate the patently false claim that this kind of psychology studies mental processes. The only people who would consider the behaviors and mental states studied in these experiments "mental processes" are the professional information-processing psychologists who perform these studies.[2] The cognitive revolution has failed at its most important goal. It does not offer us viable ways of studying human mental life. Whatever information-processing psychology does, it cannot make that claim in good faith. Therefore the time is ripe to consider how one might go about constructing a more viable psychology of cognition.

The Ecological Approach to Mental Processes

If the above analysis is correct, the root of the malaise in modern psychology is really our inability to transcend our associationist heritage. George Miller (1986) calls this an "analytic pathology" and has argued that psychological theory periodically destroys itself in this pathological way.

The heart of the pathology lies in the belief that one "must" describe behavior in certain ways—ways that always turn out to be reducible to input-output analysis. (It makes no difference how big a central representation unit one adds to these input-output descriptions, as I hope the above argument has made clear.)

The one major psychological theorist who saw this pathology most clearly, and who spent the last two decades of his life working to overcome these problems, was James Gibson. While others were developing a "subjective behaviorism" by other names, and building elaborate processing models that concealed slight variations on associationistic themes, Gibson was making a truly radical break with behaviorism, associationism, and S-R analysis. I have analyzed the steps in this break in detail elsewhere (Reed, 1986, 1988). In what follows, and given space constraints, I am essentially doing little more than offering an outline of a very rich set of concepts and empirical results.

The rejection of the concept of the stimulus. In the same year that Hebb lauded S-R psychology in his lecture to the American Psychological Association, Gibson (1960/1982) attacked the concept of the stimulus in a presidential lecture to the Eastern Psychological Association. Gibson showed that psychologists worked with at least a half dozen different definitions of the word stimulus, and that some of these definitions prove to be circular when scrutinized. He attempted to salvage something by redefining "molar stimulus," emphasizing that many molar stimuli were also "potential stimuli" and asking the important question "What is the informative value of a stimulus?" Within a few years, however, Gibson (1963/1982) had entirely given up on keeping "stimulus" as a central concept in psychology. (The reasons for this are described at length in Reed, 1986.) Physiologists can impose stimuli on the passive animal preparations they use in their laboratories, but perceptual psychologists cannot and should not hope to follow suit. A perceiving organism is and should be an active, motivated observer, one that is hunting for stimulation, not passively receiving stimuli. Stimuli may exist for receptors, but they do not exist for perceivers.

Ambient information is what exists for perceivers. In perceiving, the hands or eyes of an observer explore the available stimulation in an environment, running over the complexly structured energy fields. These fields of ambient structured energy are an environmental resource to be exploited by motivated, active observers. Hence, the methodology of imposed stimulation, which psychologists borrowed from physiologists, is inappropriate to the task of investigating real psychological processes. These processes begin with the obtaining of stimulation by the organism, not with the imposition of stimulation by an environment (or an experimenter).

Thus, the distinction between stimulus and response disappears within psychology, to be replaced by a distinction between two modes of activity: Exploratory *versus* Performatory. Exploratory activity is the motivated obtaining of stimulation. Gibson hypothesized that it is regulated by the function of information pickup. By this he meant that the organization of the activities involved in obtaining stimulation are determined by two things: the information for which the system is looking and the information that in fact is available in that environment. Performatory activity is the actual behavior deployed to achieve a goal, to create a change in the environment. Such activity requires information for its regulation, not in the sense of merely detecting the information, but in the sense of using whatever information is available about the relationship between the observer and its surroundings. I have hypothesized that there are a number of distinctly different *action systems*, or ways of regulating performatory behavior, each of which has evolved under different selection pressures (Reed, 1982; 1996a).

On the basis of the above concepts and a series of experiments on both vision and active touch, Gibson (1966, 1979) offered a new theory of perception that was explicitly presented as a theory of the basic cognitive process. Again, I have not the space here to discuss the empirical motivation for the theory (for which see especially Chapters 12–16 of Reed, 1988), so instead I will simply outline the novel conceptual steps in Gibson's work.

A new definition of cognition. Cognitivists, as we have seen, have tended to define cognition by exclusion—cognition is whatever behaviorism cannot explain, or so it has seemed. This is a remarkable piece of foolishness, which forever ties the success of cognitivism to the fortunes of behaviorism, a theory cognitivists claim to disdain. Alternatively, one finds casual, "folk theoretic" definitions of the "cognitive processes" used in both the philosophical and psychological literature. This also is foolish, for it simply lumps undigested concepts like "thought" and "emotion" as if they formed some meaningful category. Worse, the important distinction between "cognitive" in the strict sense and "non-cognitive" is lost. In the strict sense, a "cognitive" state is one in which an observer correctly knows or understands something about the world. When I see you walking down the street, that is a cognitive state. When, because of the distance between us, I merely guess that it is you, and in fact guess wrong, that is a non-cognitive state. For ecological psychology, the basic cognitive process is perceiving, because perceiving serves to keep us in touch with our surroundings. (This does not mean that perceiving is never subject to illusion, but only that it often is not illusory, and therefore is often cognitive, in the strict sense.)

Cognition, in this ecological sense, has many and varied uses. The regulation of thought and action by information—the ability to be in touch with our surroundings—turns out to be quite useful in such matters as feeling as well as thinking. My feelings about you depend to a considerable degree on my cognition of you. And even my fantasizing about you has at least some roots in my perception of what you have done and are doing.

If all "higher thought" is based on basic cognition (perceiving), then there can be no absolute separation between action and thought (because exploratory activity never can be eliminated without eliminating cognition). The active pickup of information—exploratory activity—thus becomes a kind of bedrock for all keeping in touch with the world, from the simplest case (e.g., standing) to even the most sophisticated kinds of thought, cognitive or non-cognitive (such as my planning to interact with you or fantasizing about it). Note that because cognition is conceived of as the ability to use ecologically specific information in controlling action—not as filling in some sort of "gap" between stimulus and response—all action involves some amount of awareness, as well as vice versa.

Humans have the capacity to select, transform, and display information to one another. Picture-making, gesture, and language are among the many universal or near-universal human habits that go beyond mere exploratory activity. In all these cases, one of the primary functions is to create some sort of "display" of information for someone else to explore—in short, to make another person aware of something. Thus, cognition becomes socialized, something I have especially emphasized in my recent work (Reed, 1993; 1996a, b). The developing human infant is thus not merely exposed to an inanimate environment in which he or she makes her way. On the contrary, the infant is not "exposed" to any environment. Instead, he or she develops within a populated environment; caregivers within that populated environment not only act to structure the infant's surroundings, but selectively to make certain kinds of information available. This complex of activities and processes I have dubbed "The Field of Promoted Action" and it is within this field that both cooperation and competition emerge. Hence, the activities of planning and thinking are typically forms of socialized cognition, not only involving the exploitation of ecological information, but also the selection and transformation of that information, and the complex interactive processes involved.

Some Tentative Conclusions

Although the cognitive revolution has failed, we should not lose sight of some of its important successes. Most important, we should not lose sight of the important goals it set for itself. Since its inception, experimental psychology has had little constructive to say about mental processes; and the attempt by cognitivists to rectify this failure is important, and to be lauded.

A corollary of the cognitivists' attempt to bring mental processes into psychology is the kind of interdisciplinary focus that has emerged occasionally throughout cognitive science's short history. From its early days in the 1950s to more recent times, there have been small groups of cognitive scientists who have tried to keep alive serious intellectual contact among a variety of disciplines, from neuroscientists and computer scientists, to linguists, anthropologists, and even historians. This is a truly admirable goal, even if in most cases the results of these interdisciplinary contacts have been disappointing. If the above analysis is correct, this is not to be wondered at—how could a cultural anthropologist interested in, say, varieties of visual thinking use even the most elegant of psychological experiments on imagery, such as those of Roger Shepard? The S-R methods we have insisted on using in our laboratories as well as in our theories simply obliterate much of the phenomena we wish to study.

The ecological approach to cognition points to a way to break out of the straitjacket imposed by input-output thinking. The study of perceiving as the active obtaining of stimulus information by an organism motivated to achieve some end places a new definition of cognition at the center of psychological study. On this new definition, cognition is a keeping-in-touch with the world around us, a keeping-in-touch that always includes both behavior and awareness, and which requires the utilization of ambient information. Mental processes that do not serve to keep us in touch with our surroundings are not denied, although they probably ought not to be called "cognitive" as they are all too often in existing theories.

The ecological approach puts mental life back into the environment as part and parcel of the ordinary vicissitudes of living. The information that regulates action and the information that gives rise to mental content are both parts of the same rich ambient information. Researchers thus can actually test hypotheses about both action and cognition directly. Different kinds of activities and different kinds of information produce a number of various cognitive functions (as described above), although all of them have their basis in perceptually guided encounters with the environment. Researchers can modify ambient information (precisely because it is ambient, unlike the hypothetical internal representations of traditional cognitivism); and they can modify task demands as well when they attempt to study cognition. Because behavior itself is considered part of the cognitive process, it should be possible to look at organizational and functional aspects of behavior (and not merely at durations) as evidence for and against hypotheses about the cognitive aspects of those behaviors.

Ecological psychology thus has begun the difficult task of accomplishing the goals set forth by the cognitive revolutionaries of the early 1950s. Mental life, and especially perceptual awareness, should be brought into psychology, not as a kind of spectator sport, but as an aspect of everyday psychological processes that deserves empirical study. The hypothesis of ecological information, and the theory of exploratory activity might together prove adequate for the making of a truly experimental phenomenology. Albert Michotte (1991) and James Gibson both made significant efforts toward this goal, but it is unclear whether these gains will be consolidated. An information-based, non-associative theory of cognition also is possible using these ecological ideas, as well as the concept of selecting and presenting information to others, as outlined above (see Reed, 1991). Finally, both observational and experimental methods can and should be used to study how humans organize their environments into fields of promoted action that encourage their offspring to develop particular skills of thought, action, and interaction (Reed, 1993; 1996a). Whether all these ecological trends taken together will effect a revolution in psychology—which has always been an exceedingly conservative science—is to be hoped for, but remains to be seen.

Notes

1. This essentially phrenological concept of psychological explanation is with us still, promoted by unabashed cognitivists (e.g., Fodor, 1983). For example, at present many researchers seem very excited about using modern imaging technology to "locate" the neural substrate of thought, emotion, or whatever. The only sense in which such "discoveries" are explanatory is in that they reveal a correlation between one kind of explicit behavior (whatever the patient thinks he or she is doing) and observable neural activity. Without controlled experimentation based on testable hypotheses derived from theories of how "thought" varies in different circumstances, such correlations are uninterpretable.

2. The obvious—and important—exception to this critique is the "protocol analysis" studies that emerged from the work of Newell, Simon, and their students.

References

Anderson, J. 1978. "Arguments Concerning Representations for Mental Imagery." *Psychological Review*, 85, 249–277.

Asch, S. 1952. *Social Psychology*. New York: Oxford University Press.

Astington, J., Harris, P. and Olson, D. 1988. *Developing Theories of Mind*. New York: Cambridge University Press.

Attneave, F. 1959. *Applications of Information Theory to Psychology*. New York: Holt.

Brewer, W. 1974. "There Is No Convincing Evidence for Operant or Classical Conditioning in Adult Humans," in W. Weimer and D. Palermo (eds.), *Cognition and the Symbolic Processes*. Hillsdale, NJ: Erlbaum.

Broadbent, D. 1958. *Perception and Communication*. London: Pergamon.

Churchland, P.S. 1991. *Neurophilosophy*. Cambridge: MIT Press.

Fitts, P. and Posner, M. 1967. *Human Performance*. Belmont, CA: Brooks-Cole.

Fodor, J. 1968. *Psychological Explanation*. Englewood Cliffs, NJ: Prentice-Hall.

Fodor, J. 1983. *The Modularity of Mind*. Cambridge: MIT Press.

Gibson, J. J. 1960. "The Concept of Stimulus in Psychology." Reprinted in Reed and Jones (1982).

Gibson, J. J. 1963. "The Useful Dimensions of Sensitivity." Reprinted in Reed and Jones (1982).

Gibson, J. J. 1966. *The Senses Considered as Perceptual Systems*. Boston: Houghton Mifflin.

Gibson, J. J. 1979. *The Ecological Approach to Visual Perception*. Boston: Houghton Mifflin.

Hebb, D. O. 1960. "The American Revolution." *American Psychologist*, 15, 735–45.

Hintzman, D. 1993. "Twenty-Five Years of Learning and Memory: Was the Cognitive Revolution a Mistake?," in D. Meyer and S. Kornbluh (eds.), *Attention and Performance* XIV. Cambridge: MIT Press.

Hull, C. 1943. *Principles of Behavior*. New York: Appleton-Century.

Jackendoff, R. 1992. *Languages of the Mind: Essays on Mental Representations*. Cambridge: MIT Press.

Johnson, M. 1987. *The Body in the Mind*. Chicago: University of Chicago Press.

Kihlstrom, J., Barnhardt, T., and Tataryn, D. 1992. "The Psychological Unconscious: Found, Lost and Regained." *American Psychologist*, 47, 788–801.

Leahey, T. H. 1992. "The Mythical Revolutions of American Psychology." *American Psychologist*, 47, 308–18.

Michotte, A. 1991. *Michotte's Experimental Phenomenology of Perception*, G. Thines, G. Butterworth and A. Costall (trans. and eds.). Hillsdale, NJ: Erlbaum.

Miller, G.A. 1986. "Dismembering Cognition," in S.H. Hulse and B.F. Green, Jr. (eds.), *One Hundred Years of Psychological Research in America; G. Stanley Hall and the Hopkins Tradition*, pp. 277–98. Baltimore: Johns Hopkins University Press.

Miller, G.A., Galanter, E., and Pribram, K. 1960. *Plans and the Structure of Behavior*. New York: Holt, Rinehart-Winston.

Neisser, U. 1979. "Images, Models, and Human Nature." *Brain and Behavioral Sciences*, 2, 561.

Palmer, S. and Kimchi, R. 1986. "The Information Processing Approach to Psychology," in T. J. Knapp and L. C. Robertson (eds.), *Approaches to Cognition: Contrasts and Controversies*. Hillsdale, NJ: Erlbaum.

Postman, L. 1951. "Toward a General Theory of Cognition," in J. Rohrer and M. Sherif (eds.), *Social Psychology at the Crossroads*. New York: Harper.

Reed, E. S. 1982. "An Outline of a Theory of Action Systems." *Journal of Motor Behavior*, 14, 98–134.

Reed, E. S. 1986. "James J. Gibson's Revolution in Psychology: A Case Study in the Transformation of Scientific Ideas." *Studies in the History & Philosophy of Science*, 17, 65–99.

Reed, E. S. 1988. *James J. Gibson and the Psychology of Perception*. New Haven: Yale University Press.

Reed, E. S. 1991. "James Gibson's Ecological Approach to Cognition," in A. Still and A. Costall (eds.), *Against Cognitivism*. New York: Harvester Wheatsheaf.

Reed, E. S. 1993. "The Intention to Use a Specific Affordance as a Framework for Psychology," in Wozniak and Fischer (1993).

Reed, E. S. 1996a. *Encountering the World: Toward an Ecological Psychology*. New York: Oxford University Press.

Reed, E. S. 1996b. *The Necessity of Experience*. New Haven: Yale University Press.

Reed, E. S. and Jones, R. K. (eds.). 1982. *Reasons for Realism: Selected Essays of James J. Gibson*. Hillsdale, NJ: Erlbaum.

Rumelhart, D. E. and McLelland, J. L. (eds.). 1986. *Parallel Distributed Processing*. 2 vols. Cambridge: MIT Press.

Skinner, B. F. 1957. *Verbal Behavior*. New York: Appleton-Century-Crofts.

Thelen, E. and Smith, L. 1994. *A Dynamic Systems Approach to the Development of Cognition and Action*. Cambridge: MIT Press.

Varela, F., Thompson, E., and Rosch, E. 1991. *The Embodied Mind*. Cambridge: MIT Press.

Watson, J. B. 1914. *Psychology from the Standpoint of a Behaviorist*. New York: Lippincott.

Wozniak, R. and Fischer, K. (eds.). 1993. *Development in Context*. Hillsdale, NJ: Erlbaum.

PART FIVE

Christina Erneling

CHALLENGES TO COGNITIVE SCIENCE

The Cultural Approach

All the chapters in this part recognize the large impact of the shift in research strategies in psychology that occurred in the 1950s. Yet all of them also call for a re-evaluation and, in some cases, an out-right rejection of the research program of cognitive science, arguing that many fundamental assumptions of this research program are flawed.

Although they do not say so explicitly, the main criticism they make goes beyond cognitive science. It also is a criticism of the dominant tradition of psychology, which, for example, takes meaning, understanding, and language acquisition to be processes or mechanisms internal to the individual, and denies that they are social achievements. Psychology in general and cognitive science in particular fit in with the dominant trend in the Western philosophical tradition, according to which the basic epistemological problem is one that exists for the individual mind in relation to the external world. To be more precise, this tradition supposes that interaction with other people is not a condition for knowledge; and therefore the individual is not someone who fundamentally shares life, knowledge, and language with other human beings. In the early modern period Descartes sanctified the "I" and "I think" in his search for ultimate certainty. That is, he claimed that knowledge begins with the awareness of one's own existence as a thinking being, and that knowledge of the world, other minds, and language is founded on this. Hume's associationism and Kant's account of knowledge in terms of the synthesizing unity of the individual understanding, although different from each other in many ways, also are similar in this respect.

This kind of individualism and cognitivism is prevalent in cognitive science and, in Shotter's phrase, is the center of its ideology. For example, Chomsky (in his two chapters in this volume) speaks of I-language—that is, an internal, intentional, and innate language competence—as the only proper object of the scientific study of language. Instead of studying I-language as Chomsky recommended, Shotter believes E-language (E for "external") is the proper study of cognitive science, thus standing Chomsky and the whole traditional cognitive science on its head. What Chomsky takes to be mysterious and impossible to study scientifically is precisely what one should study, according to the chapters in this part.

Jerome Bruner in his book *Acts of Meaning* (1990), and in his chapter here, shows in some detail what an alternative approach would be like. For him, the crucial failure of the cognitive revolution of the 1950s was its inability to account for meaning-making and symbol use in different societies and contexts, which is essential to all mentation. The fundamental question is how human beings create or infuse meaning—that is, how we make sense of what our perceptual organs provide in the form of meaningless, physicalistic input. Aristotle's common sense or *sensus communis*, Hume's laws of association, and Kant's categories are all examples of structures internal to the mind postulated to account for how humans achieve meaning. But, according to Bruner, these attempts as well as contemporary cognitive science have all missed the diversity and, in particular, the cultural aspects of the human mind. His chapter specifies four different patterns or modes of making and construing meaning, or processes of interpretation that account for the ability of the human mind to go beyond information given by the senses. The first mode, which also is ontogenetically primary, is the cognitive activity of instituting, organizing, and maintaining intersubjectivity. It involves the recognition that other people have minds like one's own. This particular ability is in place very early in infancy, and is manifested in such activities as following the caregiver's gaze and adjusting one's activity to the intonation of the caregiver's voice. Later it develops into an implicit theory of folk psychology about feelings, beliefs, etc. guiding everyday interactions. The other modes of meaning-making Bruner discusses have to do with how humans make sense of and structure experiences and behavior into actions, as well as how normative and narrative structures, and later also logical concepts, have come to play a role both ontogenetically and phylogenetically.

Several of the authors in this part take the later writings of Wittgenstein (1953) as their starting point, arguing that cognitive science and the general tradition out of which it grew are confused. For example, Coulter argues that remembering, recognizing, perceiving, and so on are not inner mental processes. Rather, it only is possible to understand these phenomena by putting them in a wider context. That is, we should not think of them as something an individual mind does by itself, but as parts of an interpersonal context that "defines" them and refers to projects in which the whole person is engaged along with other humans, and which are not just (or limited to) activities of his or her brain.

Shotter, for example, points to another dimension one needs to take into account—namely, the normative character of human mentality. To describe someone as perceiving or remembering is not just a description, but also involves a judgment with normative force. That is, the truth of a claim that the person is actually seeing something that is there, and not only thinks he or she is seeing something, cannot determined by looking into his or her brain or at an abstract mathematical description of his or her alleged mental processes, but only by judging the person's performance against a socially negotiated standard.

Stenlund argues that the strong commitment of cognitive science to inner mental processes is a result of lingering behaviorism. (For the role of behaviorism in cognitive science see Shanker's chapter in Part One.) According to Stenlund, cognitive science and behaviorism have a shared view of how to account for the observable—namely, in terms of physical descriptions. According to this shared assumption, when I see a robin, for example, I do not see a bird but something like a patch of reddish and grayish colors; and this input has to be interpreted to yield the perception of a robin. In the same way, when I hear speech, I hear various sounds, not sentences; and on basis of these I construct or interpret the sounds as meaningful, grammatical sentences. The additional mental process thus seems to be a necessity. How otherwise could I recognize the reddish moving patch as a bird, which I clearly do, or make sense of the sounds I hear? But, Stenlund claims, if we reject this view of the observable, then the seeming necessity to postulate inner mental processes disappears. Such processes are idle wheels that do not contribute to the individual's achievements. In this context, it is interesting to note that Stenlund thinks Bruner in his analysis of meaning-making is still captured by the traditional approach to experience, implying that Bruner has not gone far enough in his rejection of cognitive science.

Harré, like the other authors in this section, takes mental life to be thoroughly social or discursive; and this, he argues, calls for a rejection of the Cartesian (and Newtonian) ontology that cognitive scientists have taken for granted. For Harré, the paradigmatic example of a mental phenomenon is conversation, where two or more people interact discursively, by utilizing their individual skills (innate and learned), and the socio-linguistic and cultural structure for certain kinds of conversations that their particular culture provides. For instance, in order for an utterance (which presupposes linguistic skills like the ability to utter sounds in an ordered way) to be taken as an apology, it has to be part of an ongoing discourse in which one assigns it the role of an apology. And this in turn presupposes the social rules or conventions that give meaning to the social practice of apologizing. Furthermore, speech-acts or other interpersonal acts are not casually related to one another, but semantically related. Therefore, even if all cognitive processes require bodily and brain activity, the latter are not causes of the cognitive processes. Harré argues that the mistake of traditional cognitive science is of the same kind as that which Berkeley attributed to Locke—namely, that it makes no scientific sense to postulate a meaningful, discursive, and unobservable process as an explanation of an observable discursive process. The unobservable discursive process is as much in need of explanation as the observable. Instead, according to Harré, we should develop an ontology that is capable of capturing the meaningful relations between discursive skills occurring in patterns of mutual activity. The task of psychology is to construct formal models of interpersonal relations—that is, structures and skills that constitute and go into producing mental phenomena, both public ones like conversations, and private ones like thinking. (For an accessible introduction to this approach, see Harré and Gillett, 1994.)

To conclude: the chapters in this part focus on meaning and meaning-making as the central aspect of psychological phenomena. They call for quite a radical revision of the assumptions that formed the basis of the cognitive revolution of the 1950s—namely, a rejection of the traditional focus on individual, inner mental processes as the proper subject matter of cognitive science.

References

Bruner, J. 1990. *Acts of Meaning*. Cambridge: Harvard University Press.
Harré, R. and G. Gillett. 1994. *The Discursive Mind*. Sage: London.
Wittgenstein, L. 1953. *Philosophical Investigations* (trans. G. E. M. Anscombe). Oxford: Basil Blackwell.

19

Jerome Bruner

Will Cognitive Revolutions Ever Stop?

I do not write this chapter as an historian of the cognitive revolution. I'm not very good at that sort of thing, and besides, we're all still too much on the barricades to be historians. Modern historiography teaches one big truth that should sober us. The Past (with a capital letter) is a construction: how one constructs it depends on your perspective toward the past, and equally on the kind of future you are trying to legitimize (see, e.g., Mink, 1978).

My own perspective is that a cognitive revolution is always incipiently in progress and always has been so throughout the entire history of psychology—necessarily so in the very nature of psychology's subject matter and, possibly, in the very nature of human experience itself. This circumstance inheres, I think, in the bifurcation between what we take to be private to our own consciousness and what we take to be "outside us" and therefore public, reproducible, and transmissible (Popper, 1972). By whatever means we officially try to banish the former, the private, from "real" psychology (and banishment proceedings have always been at the top of our agenda), some of us, perhaps all of us, remain residually itchy about these exclusionary tactics. Even the most puritanically "scientific" of our behaviorist forbears suffered covert revolutionary pangs expressed in such oxymoronic concepts as "implicit responses,"[1] or "pure stimulus acts"[2] and the like. The contemporary counterpart is, in some ways, even more dismissive of the private sphere. It relegates privacy to the purdah of "folk psychology" which then can be either dismissed as "irrelevant and immaterial" or treated as a phantom epiphenomenon to be explained away later as an inconsequential squeak in the computational cogs (Stitch, 1983; see also Christensen and Turner, 1993).

I guess it was Aristotle's *De Sensu et Sensibili* that made me aware long ago of the perpetuity of cognitive revolutions. The Old Fox begins this intriguing book by posing the question how we know that it is Cleon's son descending the steps of the Parthenon, given only the testimony of our special senses, each limited to its own distinctive quale—blips of light, sound, touch, etc. How do we make a world of that multisensory jumble? Aristotle first demolishes the old *eidolon* theory, that stimuli are copies of things in the real world, though he never doubts that there is a real world even after proving succinctly that we can never know it directly. To cover up the gap of ignorance into which his reasoning has led him, he forthwith creates something new and wondrous: a Common Sense or *sensus communis* to do the work the special senses could not do, to make a world that *means* something, Cleon's son, say, rather than flashes and beeps and skin sensations.

How does the sensus communis bring this off? Well, to begin with, it partakes of NOUS, the soul or mind, one of whose powers is "association." The laws of association presumably guarantee that our private experience *reflects* the "real" world that is only transmitted piecemeal by the special senses. Aristotle was too shrewd to fall into that epistemological trap of pure correspondence. To avoid it, he equips NOUS with Reason, a means whereby the *sensus communis* is assured a knowledge of *necessary* truths rather than only the contingent ones afforded by the association of sensations. Now, behold, we can both sense and reason our way to Cleon's son on the Parthenon's steps. We have, *mirabile dictu*, achieved meaning while avoiding the scent of private experience.

But obviously, there had to be more to it than that, or else Aristotle would not have had to write the *Rhetoric* to explain, for example, why some accounts or views of the world were more compelling or believable than others even when equally "true," or why some representations of it were more "mimetic" than others, which was his preoccupation in the *Poetics*. Indeed, the *Poetics* contains the first real discussion on record of *narrative necessity* (as in "tragedy"). What could *that* mean? And how could dramatic necessity be so real as to effect *catharsis* in people so firmly held in the real world by reason and the laws of association?[3]

Never mind: he was making a cognitive revolution. After all, he rejected Plato's cranky notion that true knowledge could never be gained through the senses, but only through reflection on a world of ideal forms whose truths were necessary truths like those of the revered geometers of his day.[4] Now, on Aristotle's account, rational mind lived not just on reflection but on the testimony of the senses: input was in, and stayed in forevermore. Many centuries later, Leibniz (1898) was still echoing that first cognitive revolution in his famous "Nothing gets into the mind save through the senses—except mind itself." And though John Locke (principally for rather regicidal political reasons) tried to minimize its intervention into the world of sense, Mind *per se* has managed to persist as the neatly concealed staging area for all future cognitive revolutions.[5]

But the British empiricists had a pretty good inning. Their revolution, to use Charles Taylor's term (1985), was "against enchantment," against the concoctions of mind left to itself. The scientific revolution had begun and sheer speculation about the world was to be brought to empirical heel. For Locke, even complex ideas could be reduced to composite primary sensations that then could be subjected to direct verification, a game all could play, commoner or King, naturalist or metaphysician (Locke, 1959). Further north in Edinburgh, the new doctrine took a more skeptical turn, as it were, particularly in David's Hume's hands. Not everything could be reduced to primary sensations, and what couldn't just had to be chucked (Hume, 1888).

But, as always, one cognitive revolution breeds another. Hume awoke Immanuel Kant from his dogmatic slumbers. If empiricist analysis of sensation could not yield such concepts as time, space, causality, and normative obligation (as Hume argued, advising that they be forthwith "cast unto the flames as naught but sophistry and illusion"), then their reality must be intuited by mind itself. These were properties the mind imposed on nature to give it cognitive order (Kant, 1934). This particular style of backing-and-forthing is, of course, quite familiar to us as children of a later revolution.

It does no good to say on the basis of all this that the whole lot of us are hung up in Cartesian "substance dualism," a dualism that we are perpetually trying to winkle out of by postulating such odd devices as the *sensus communis* or the pineal gland or the brain-mind interface or the comparable interface between a computer's hardware and its program. We are still faced with the puzzling problem of how to reconcile *explanations* of our world in public, measurable, reproducible terms with *interpretations* of it couched in the rich language of meanings that yield the subjective landscape in which human beings live, for which they fight and die (often privately and without flags), and around which they construct cultures for shaping and regulating their "public privacies." In our post-modern age we want both of them (and probably always have, even in our evolutionary history, a point I will come to at the end). Only lunatic fringes reject the reality of one approach as a price for embracing the other: the contest between explanation and interpretation, we go on believing, is not a zero-sum game. And so we go on fomenting cognitive revolutions.

Our recent ancestors—our fathers, uncles, and elder brothers—illustrated the revolutionary discontent in many fruitful ways, but the issue never could get settled. Freud (1949) scored by making experience itself—psychic reality as he sometimes called it—a process of construction, a compromise between conflicting constituting processes. Bartlett (1957) scored by introducing the idea of schematization (rather than association) as a principle of memory (mental?) organization. Even pawkish old Pavlov scored by introducing the Second Signal System, the input of the special senses shaped in the cookie cutter of language (1949). And his fellow countryman, Vygotsky (1962), and ours, G.H. Mead (1934), scored with the notion of the internalization of dialogue as providing a format for the operation of thought—of mind itself. Tolman's big bang was getting rid of what he called a "telephone switchboard" theory of mind and substituting a "map room" model governed by the instrumentalism of "means-end-readiness," a kind of heuristic teleology (1948).

And, finally, computational models, about whose virtues and shortcomings the chapters of this book are largely concerned. I am not going to say much about them. I am a well-wisher, all right, but not much of a participant in the game. The reason is simple enough: computational models are not, on the whole, much concerned with my principal interest, my version of the perpetual cognitive revolution. My version focuses on how human beings *achieve* meanings and do so in a fashion that makes human culture possible and effective. Computationalism, in the very nature of things, takes meanings for granted as given, inscribes them in the system at an address, to be retrieved and processed as demanded by a program in force. Computationalism does not and cannot, also in the nature of things moreover, deal with the messy, ill-formed interpretive procedures involved in constructing contexts and construing meanings in terms of them; not just word meanings, but the meanings carried by syntactic forms as with case and cognitive grammars, or the meanings carried by the tropes and figures of poetry and narrative. But please do not take all this as a complaint about computationalism. It has plenty to do, even though it helps me only on the margins. Perhaps the creation of a coherent *cultural* psychology, with its emphasis on meaning making, can help the computationalist become more helpful to the likes of me. So let me dive right in.

II

Let me start by proposing that there are three primitive modes of making meanings, three distinctive patterns of placing events, utterances, particulars of all kinds into contexts that make possible their construal as "meaning something." Each leads to a form of understanding. I think of them as comprising three distinctive forms of human cognitive activity required for living under cultural conditions. Each is sustained by a body of folk-theoretic beliefs and—not that it matters all that much for our purposes—each is made possible by preadapted cognitive dispositions that reflect the evolution of man in the primate order. It would be astonishing if man's major modes of meaning making had no assist at all from the human genome.

The first mode of meaning making is directed to the establishment, shaping, and maintenance of *intersubjectivity*. It grows from an initial though primitive human capacity to "read" other minds, a virtual compulsion to do so. From this obligatory simple presumption of consciousness or intentional states in others grows a highly acculturated and complex folk theory of other minds (Astington, 1994). Without such a folk theory there could be no presupposition about the intentions of others, including their communicative intentions, or about their beliefs, feelings, or whatever. Without such intersubjectivity, we could not develop the conventions and felicity conditions for managing illocutionary speech acts (Austin, 1969; Searle, 1969), nor could we grasp the distinction between what is said and what is meant, nor master the rich lexicon of psychological verbs, nor construe Gricean implicatures (Grice, 1989; esp. chap. 2). The coherence of our intersubjective meaning making inheres in a more or less internally congruent theory of other minds, about which we are presently learning a great deal. But what also gives it coherence are culturally transmitted strategies for using our presumptions about other minds in discourse—like Sperber and Wilson's "presumption of relevance" (1986) that leads us to presume that whatever an interlocutor says is relevant to the context of the encounter, with the first burden of proof resting on the listener to determine how our interlocutor is expressing his or her notion of relevance. Intersubjective meaning making, in a word, is an elaborated expression of our recognition of the mental processes of our conspecifics. I can tell you autobiographically that among my own research studies of the last decade, the one that most knocked me for a loop was finding that young infants followed an adult's gaze direction in search of an object of attention, and when they failed to find one, looked back at the adult to recheck on gaze direction—a study with Scaife (1975). Then Anat Ninio and I (1978) discovered that infants grasped the distinction between given and new for labels, signaled by the mother's use of rising intonation for new and unsettled matters ("What's *TIup* that?"), versus falling intonation for already negotiated labels ("What's *TIdown* that?").

Let me briefly note in passing that intersubjective meaning-making is the more interesting for its lack of precise verifiability. It depends enormously on contextual interpretation and negotiation. This may explain why Anglo-American philosophical theories of meaning, so reliant on verificationist notions of reference and sense, have paid so little attention to it.

A second form of meaning making is concerned with relating events, utterances, acts, or whatever to the so-called arguments of action: who is the agent of what act toward what goal by what instrumentality in what setting with what time constraints, etc. I refer to it as the *actional* mode (Fillmore, 1968). Again, it is a form of meaning attribution that begins very early and appears initially in a surprisingly accomplished form. As I tried to say in my last book, *Acts of Meaning* (1990), it is as if the case structure of language were designed to reflect a "natural" grasp of how action is organized, as if the young child's theory of action were a prelinguistic prerequisite for language mastery.

The third mode of meaning making construes particulars in normative contexts, a matter about which we know altogether too little. Here we are dealing with meanings relative to obligations, general standards, conformities, and deviations. This is the *normative* mode. The adult linguistic vehicle for normative meaning making is the deontic mode, concerned with the nature and limits of obligation, the territory beyond the optative. It has to do with *requiredness*, a matter to which we will return later. Again, we know from Judy Dunn's work and other studies of "social understanding" that the young child early develops a sense of the canonical status of various ways of doing, of feeling, even of seeming to feel as in pretend play. The child soon learns what is required of her (Dunn, 1988). Normative expectations develop initially in the informal settings of the infant's or young child's intimate world, but they are soon reconstituted and elaborated through encounters with institutionalized forms—as in the law, in religious doctrine and practice, in custom and precedent, even in the poet's tropes ("I could not love thee, dear, so much/Loved I not honor more").

The normative mode is not detached: it expresses itself by imposing constraints on the first two modes. Both intersubjective and actional meanings are shaped by canonical expectations: what is a fit, an appropriate, or a required state of mind, intention, line of action, etc. Again, language serves powerfully both to transmit and constitute normative standards through its deontic mode, its distinctions between the obligatory and the optative. The same can be said of the symbolic forms that underlie cultural institutions for controlling the exchange of respect and deference, goods and services, and so on. Normative meaning making delineates a culture's standards of fitness or appropriateness—whether in setting discourse requirements, felicity conditions, or limits on so-called self-interest.

Intersubjective, actional, and normative meaning making are all highly tolerant with respect to verifiability, truth conditions, or logical justification. While meanings relating to intentional states, to human action and its vicissitudes, and to cultural normativity can, within limits, be translated into the propositional forms of a logical calculus (to which I turn shortly), they risk degradation in the process. For while propositional translation always works toward the decontextualization of meanings, intersubjective, actional, and normative meaning making remain stubbornly context dependent. Construing the meaning of the condolence "I know how awful it feels to lose a close friend" is no mere exercise in propositional calculus. It requires pure "psychologism." Its "meaning" lies in its appropriateness. And its appropriateness is context bound. Its context, moreover, is the story into which it can be fitted. And this leads me to an interesting matter that requires immediate attention: the role of story or narrative in meaning making.

Narrative or story is one of the second-order forms (see Chapter 1) for achieving coherence in the three primitive forms of meaning making. A story in its very nature involves action by an agent in a setting where normative expectations have been breached or otherwise brought into question. A story, moreover, is played out on a dual landscape, the landscape of "reality" as set forth by a narrator or as canonically presumed, and a subjective landscape in which the story's protagonists live. Stories are the vehicles *par excellence* for entrenching the first three modes of meaning making into a more structured whole, a whole that widens the interpretive horizon against which the construal of particulars is achieved. "What's going on here?" is a bid to get the particulars into the diachronic meanings of a narrative.

But stories accomplish something else as well: they provide the means for transcending particularity. Stories are not isolated recountings: they are also instantiations of broader genres. The events and characters of any particular story are, as Vladimir Propp put it (1968), "functions" of a genre of which the particular story is an instance. Nobody has a notion of how many fundamental genres there are, even in a particular culture: all that we know is there are not many and that they both die out and keep getting invented and reinvented by the rare likes of Herodotus, Augustine, Cervantes, Shakespeare, Sterne, Flaubert, and Joyce. Narrative genres like, say, Northrop Frye's tragedy, comedy, romance, and irony (1957) typically represent canonical human plights, but they also trigger ways of thinking and talking about those plights (and what they "stand for"). Genres are, as it were, in the text and in the reader's head. You can "read" an intended comedy as a tragedy or irony. Genres, in a word, are modes of thought (Iser, 1978; see also Feldman, 1994). Whether in the head or on the page, stories have a certain psychological or cultural necessity about them: romantic heroes "deserve" their reward; tragic ones fall of an excess of their own virtues, and so on. Obviously such "necessities" are not of the same order as causal necessities or logical entailments. Yet they have an enormous power in providing coherence and generality to meanings constructed under their contextual control. Such power does not rest on either empirical demonstrability or on necessary logical truth, but only on "verisimilitude." And verisimilitude cannot be reduced either to tests of inference or to operations of a logical calculus.

Meaning making, given the narrative condition, is *interpretation*. So when the recipient of the condolence mentioned earlier interprets it as "a hypocritical offering motivated by toadyism," her interpretation involves not only notions about states of mind, ways of striving in the world, and normative expectations, but also certain *necessities* inherent in the culture's ways of dealing with life and death, necessities that derive from the genres on offer in the culture: *narrative* necessities.

This brings us directly to the fourth mode of meaning making, the *propositional* one. In this mode, meaning making is dominated by the formal necessities imposed by the rules of the symbolic, syntactic, and conceptual systems that we use in achieving decontextualized meanings. These include an abundance of rules not only of contingent or causal inference and of logical justification, but also simpler, almost "invisible" ones. These include such rule-bound mundane distinctions as object-attribute, identity-otherness, whole-part, and so on. That something can be treated as an attribute of something else, or as a part of it, or as an instance of a category, or as an opposite of something else, these are "simple" rules that compel their own forms of logical necessity (Beckwith, Fellbaum, Gross, & Miller, 1991). Yet we identify, place, and define things steadily and easily by hyponymy, meronymy, antinomy, and the like. And we know intuitively that an arm is "part of a body" in a different, more strictly logical sense than, say, an insult is "part" of a plot in a picaresque revenge narrative whose constituents are determined not by strict "logic" but by cultural convention. Indeed, "insult" and "revenge" are themselves constituted by violation of cultural norms and not by the decontextualized necessity that requires a part to be subsumed by a whole. When we subsume such things under a "part-whole" structure, we leave something crucial behind.

Even granting that formal "logic" may derive from "natural" features of cognition (the Piagetian as well as the Kantian view of the matter), its rules are said to be autonomous and context free. Their application leads to a unique solution, not to alternative interpretations. What justifies a claim of logical "necessity" is an appeal, say, to a syllogism or to *modus ponens* or to a test for contradiction. And in the case of causal "necessity," the appeal is either directly to the rules of inference or indirectly to the logical rules imported into a causal theory from a mathematical model.

It takes boldness and cheek to propose to a logician that logical necessities may grow out of efforts to refine, decontextualize, and universalize the intersubjective, the actional, and the normative modes. Such a proposal would provoke the thundering accusation of "psychologism." But let me speculate on this point, with gratitude to my bold assistant David Kalmar. It may well be that the three procedures of logic—deduction, induction, and abduction—are directed to the "taming" and decontextualizing of the modes earlier described. That is, deduction involves an effort to impose an abstract norm on a set of particulars. "All men are mortal"; therefore it is required that, since Socrates is a man, he too be mortal. Suppose Socrates goes on living longer than Methuselah. The natural language response, of course, is that he *should* be dying—as if there were a hidden deontic in the universal syllogism, as if the syllogism were normative in root. Induction involves, one might speculate, something of the actional. Formally, it includes "all cases of nondemonstrative argument in which the truth of the premises, while not entailing the truth of the conclusion, purports to be a good reason for belief in it" [Enycl. Phil., IV, p. 169]. Does this not involve the placement of an array of particulars in a putatively common action context—what a set of things can do, how they can be situated, what agent affected them, in what aspectual point in a sequence they occurred, and so on? As for abduction—the term Charles Sanders Peirce used for hypothesis making—it is a way of externalizing and decontextualizing one's notions about what others (or "other minds") will find believable if tests support them. Peirce once wrote a famous paper about the "if-then" query as the essential vehicle for "making our ideas clear," to quote the title of his famous paper.[6] It is how to make the "intersubjective" public, testable, and beyond prejudice. Please take all of this in the speculative sense in which it was offered to me by David Kalmar! The upshot of the proposal is that propositional thinking did not grow out of the blue or spring from the "organ of language," but emerged as a way of going beyond the particularities of the intersubjective, the actional, and the normative modes.

Let me step back for a moment and ask the functional question about decontextualized, propositional meaning making. What function is served by "going Pythagorean" (to use John Bruer's happy expression for it)? Perhaps we save ourselves the trouble of having to learn over and over again, as I once suggested (1957). Or perhaps, to echo an old Basil Bernstein (1973) point, to get wider social travel from universals? But does propositional understanding always provide a better guide to the world? Well, the proper answer is "It depends." It depends, for example, on whether you are conducting a love affair or writing a paper for the *Philosophical Review*.

A close look at propositional efforts to understand (leaving aside the scientist and the logician) tells us that they are mostly used to tame context effects and to obviate negotiation—with all the risks entailed. Our own work as cognitive psychologists suggests just this. Consider what we have learned these last decades about such seemingly formal matters as category rules. As Douglas Medin notes (1989), concept attainment studies have steadily retreated from formalistic conclusions since the Bruner-Goodnow-Austin book (1956) of 40 years ago. Back then, it was proposed that categories be envisaged as governed by formal rules for combining the defining attributes of class exemplars.[7] Making sense or meaning had to do with placing events in categories according to such rules. Then Rosch and her colleagues (1978) noted that natural-kind categories were not constructed that way, but were organized around the similarity of instances to a good or conventional prototype exemplar (a sparrow is a better bird than an awk). Situating things to determine what they were often seemed less a matter of rule following than similarity matching. But similarities are notoriously protean. Smith and Medin (1981) then demonstrated that categories were not "defined" just by a single basic level prototype, but by prototypical exemplars that typified different contexts of encounter—prey birds in the wild, garden bird in the back yard, sea birds on the water, etc. Finally, both Keil (1979) and Carey (1985) showed that what held a category or a category system together was neither attribute rules, base-level prototypes, nor context-sensitive exemplars. What mattered was a *theory*: if a paramecium were an animal, for example, it had to have ways of sensing its environment, of eating, of taking up oxygen, of eliminating waste, and so on. It is a theory of "aliveness" that dictates a category's properties, not a formal conjunction of attributes or some exemplifying prototypes. A categorial placement, then, is the terminal step in constructing and applying a theory.

But how, typically, do most theories come into being? It is often by taming and attempting to formalize a narrative, or so we learn from Misha Landau's work (1991) on the origins of evolutionary theory or Horward Gruber's (1981) on Darwin's way of thinking. "Survival of the fit" is a stock, almost mythic narrative from which Darwin began. Granted there are branches of mathematics, and of the physical and biological sciences, that are so formally or propositionally entrenched as to permit derivations without the heuristic support of folk tales. But even that is a little dubi-ous. Niels Bohr once told me that the "inspiration" for the Principle of Complemen-tarity grew by analogy with his recognizing that he could not simultaneously under-stand a petty theft to which his small son confessed both in the light of love and in the light of justice. While this "fact" may not "explain" why you cannot include terms both for a particle's position and its velocity in the same equation, it goes a way toward explicating the interplay of different modes of meaning making in our construction of theories.

The deep issue for the student of cognition is how the meaning seeker proceeds in getting to a final formulation. We get trapped by the ideals of Science when we insist on an exclusive role for well-formed computation, verifiability, and truth con-ditionality. It was a triumph in the propositional mode when Newell and Simon devised a program for proving a Whitehead-Russell theorem from the *Principia* (1972). But how about the more interpretive insight of a myth maker like Homer, honing a story to its intersubjective, actional, and normative meaning essentials?[8] And what of the interpretive activities of a Vladimir Propp (1968) in delineating the morphology of the millennia-old folk tales in the Helsinki corpus? Or of James Joyce in recognizing that the epiphanies that move us are not the marvels of life but its very ordinariness?

I am not going literary or anti-science. These are the things we see when we observe how people construct their meanings in the world. These are what compel my excursion around the well-formed and propositional. The three primitive modes we have discussed—the intersubjective, the actional, the normative—probably all have biological roots in the genome. They certainly have elaborated support systems in the cultures that humanize us. They yield meanings as forms of intimacy, as requirements of action, as norms of appropriateness, and as knowledge *eo ipso*. Their incorporation into the narrative mode and propositional thinking is no less remarkable an evolutionary achievement. I want to end with a brief account (as promised) of how this set of meaning-making processes might have grown out of human evolution.

III

Let me begin with a brief comment on what might be called the cultural idealization of meaning making. Two have already engaged us: "literature" and its forms of story and drama; and "science" with its procedures and proofs. The first "personalizes" meaning by anchoring it in what people do, feel, believe, hope for, and so on. It defines what is expectable and canonical and assures cultural solidarity through myths, legends, genres, and the like. These nourish folk meanings. The propositional idealization of meanings is radically more impersonal. In the guise of truth, it seeks to transcend both the private individuality of the listener and the nature of the occa-sion on which such meanings are told. Truth is through proof: Proof yields unique, aboriginal truth. The meaning of "hypotenuse" in plane geometry is given by oper-ations performed on an idealized "right triangle," operations indifferent to who car-ries them out under what circumstances—a king or a commoner, a Hottentot or a Harvard mathematician, whether hypotenuses are ritually holy or repellent. Truths reveal themselves because they are there and/or formally necessary, not because they are sponsored by believable stories or compelling narrators. At least so goes the standard version.

The conflict between the two ideals has sometimes been incendiary—particularly in the nineteenth century, but let us not forget the sixteenth. However, the contemporary stance has softened. A modern poet could even proclaim that "Euclid alone has looked on beauty bare." Is this all the swing of a cultural pendulum or can we trace evolutionary roots in this struggle?

Let me take a final moment to sketch a possible evolutionary context. Merlin Donald's recent book (1991) reminded us of some relevant matters. Recall there were two explosions in hominid brain size: one coincident with the emergence of *homo Erectus* about a million and a half years ago; the other with *homo Sapiens* three-quarters of a million years later. Both involved disproportionate growth not only of the cerebral cortex but of the cerebellum and hippocampus. The swollen cerebellum ensured bipedal agility. The cortex was the instrument of more abstract intelligence. The best guess about the enlarged hippocampus is that it provided the basis for vastly increased human affectivity. Professor Donald is of the view that the evolutionary step to *homo Erectus* witnessed the emergence of self-triggered motor routines, skilled routines called up on command rather than in response to appropriate triggering situations. This "rehearsal loop" made retrieval possible in the interest of practice and play, or ritual. Self-mimesis can also serve as the basis for the joint performance of group rituals organized around communally significant skilled actions. What of the enlarged hippocampus in all this? Professor Donald speculates that there was probably a good deal of affect invested in these ritualized skills—throwing rituals, spearing rituals, whatever. And what we know about higher primate immaturity would also suggest that the young of the emerging species performed these motoric skilled routines in play, and immaturity grew longer. In all of this, we have the beginning of enactive representation (Bruner, Olver, & Greenfield, 1966) not only by the individual but by the group. It is a step toward "externalizing" memory and knowledge, to quote Merlin Donald.

We know nothing about early morpho-phonemic or lexicogrammatic language in *homo Sapiens* three-quarters of a million years later, but what we can reasonably suppose is that it was initially used as an adjunct to and a way of further representing skilled action but, more notably, ritualized skilled action sequences. Professor Donald takes speech communication as a major step in the *externalization* of memory—he uses the expression "exogram" to refer to such externally carried memory, in contrast to "engram." One of the chief forms of "externalizing" memory is by story, by shared stories built around performance rituals. This is the classic transition "from ritual to theater" that Victor Turner (1982) has so brilliantly described: the emergence of a designated story teller as a vicar of group ritual. On this account, oral culture's chief "external memory device" becomes the story or narrative. And the study of aboriginal narrative—as in Carol Feldman's (1994) paper on genres—reveals that narratives beget genres, or perhaps vice versa.

Perhaps the greatest next step forward in memory externalization is crude literacy. For an inscribed record of the past provides the vehicle par excellence (Olson, 1994) for reflection or metacognition—although Cole and Scribner (1974) remind us that it is not always used in that way. In any case, on Merlin Donald's account, evolving *homo Sapiens* now has three ways of representing the past in an externalized way: through ritualized mimesis of accumulated skill, through oral narrative, and through external literate representation. Cultural institutions emerge to practice and exploit each of these: those who *do* with artful artisanly skill, those who tell narratives of traditional legitimacy, and those who *reckon* by manipulating the externalized record. The three domains remain in chronic overlap—like Leonardo's notebooks.

Contemporary classical historians, like Vernant in France and Geoffrey Lloyd in England, give vivid accounts of this overlap—as when fourth-century Greek thinkers tried to enrich Homeric narrative conceptions of virtue with geometrical concepts for deducing the nature of goodness as forms of harmony and symmetry (Vernant & Vidal-Naquet, 1988).[9] But it was not zero-sum: Euclid did not preclude Homer. That was to come later. The Greeks were still tolerant toward all the ways of meaning making that came naturally.

The post-modern view—represented by the likes, say, of Richard Rorty (1979), Paul Ricoeur (1984), Thomas Kuhn (1962), or Nelson Goodman (1978)—is rather more akin to the Greeks, though hardly as innocent. As noted, they take the view that meanings are always relative to the perspective from which they are derived. Their radical anti-reductionism and anti-positivism seems to live more easily with our evolutionary past. It does not urge that what cannot be proved logically or demonstrated empirically should, in Hume's harsh words, be "cast unto the flames."

Which leads me to end this evolutionary diversion with a proposal and plea. Given human evolution and human history, we cognitive scientists err in insisting on one model of cognition, one model of mind. *Any* one model. And by the same token, we do well to avoid theories of meaning making tied exclusively to the needs and perspectives of science and analytic philosophy. The current cognitive revolution began well: to explain how people came to *understand* something rather than simply how they responded. It is time we now turned more vigorously to the *different* ways of understanding, to different forms of meaning-making. I've suggested several of them. Cognitive science should become the repository of our knowledge about *possible* uses of mind. If it seems at times to reflect literary theory, at times historiography, at times anthropology and linguistics, perhaps that is how it may have to be.When George Miller and I went to our Dean at Harvard, McGeorge Bundy, to discuss founding a Center for Cognitive Studies and told him something like that, he replied merrily, "But how does that differ from what Harvard as a whole is supposed to be doing?"A fortnight ago I attended the Stubbs Lecture at the University of Toronto given by the great Cambridge classicist Geoffrey Lloyd. Its title was "The Rise of Science in Classic Greece and China: Problems in Cognitive Representation." The most daring law review article on jurisprudence of the last two decades was by the late Robert Cover of Columbia (1983), centering on the issue of how communities convert their norms into presuppositions for guiding legal interpretations, an issue that had occupied a distinguished anthropologist, Clifford Geertz, a few years earlier in his Storrs Lectures at Yale (1983). It is a big world out there, and a varied one. No reductionist theory of mind in the old psychological or the new computational style will do it proper justice.

Notes

I thank the Spencer Foundation for support of this work through its grant, "Studies in Cultural Psychology."

1. See, for example, the definition of "implicit behavior" (including "internal speech") in English and English (1958). *A Comprehensive Dictionary of Psychological and Psychoanalytical Terms*. London: Longmans, Green.

2. This is Clark Hull's term (1952) for an act that sets up the proprioceptive stimuli in preparation for an operant response.

3. The best source for Aristotle's works is W. D. Ross (ed.) (1908/1952), *The Works of Aristotle Translated into English*. Oxford: Oxford University Press.

4. See particularly Plato's *The Phaedo*, in Irwin Edman (1928).

5. For a discussion of the regicidal intent of Locke's position, see Brinton (1965).

6. The general reader should not be expected to find his or her way around in Peirce's *Collected Papers*. A brilliant exposition of Pierce's pragmatism is to be found in Gallie (1966).

7. In fact, we proposed three types of categories—formal, functional, and affective—only the first of which was so constituted.

8. For a particularly interesting perspective on this problem, see Auerbach (1953).

9. See also Geoffrey Lloyd's Stubbs Lectures at the University of Toronto in 1993: "Modes of thought in classical antiquity," in preparation for publication.

References

Astington, J. 1994. *The Child's Discovery of the Mind*. Cambridge: Harvard University Press.

Auerbach, E. 1953. *Mimesis*. Princeton, N.J.: Princeton University Press.

Austin, J.L. 1969. *How to Do Things with Words*. Oxford: Oxford University Press.

Bartlett, F.C. 1957. *Thinking: An Experimental and Social Study*. Cambridge University Press.

Beckwith, R., Fellbaum, C., Gross, D., and Miller, G. 1991. "WordNet: A Lexical Database Organized on Psycholinguistic Principle," in U. Zernick (ed.), *Using On-Line Resources to Build a Lexicon*. Hillsdale, N.J.: Erlbaum.

Bernstein, B., and Henderson, D. 1973. "Social Class Differences in the Relevance of Language to Socialization," in B. Bernstein (ed.), *Class, Codes, and Control, Vol. II: Applied Studies toward a Sociology of Language*. London: Routledge.

Brinton, C. 1965. *The Anatomy of Revolution*. New York: Random House.

Bruner, J. 1990. *Acts of Meaning*. Cambridge: Harvard University Press.

Bruner, J.S. 1957. "Going beyond the information given," in H. Gruber et al. (eds.), *Contemporary Approaches to Cognition*. Cambridge: Harvard University Press.

Bruner, J.S., Goodnow, J.J., and Austin, G.A. 1956. *A Study of Thinking*. New York: Wiley.

Bruner, J., Olver, R., and Greenfield, P.M. 1966. *Studies in Cognitive Growth*. New York: Wiley.

Carey, S. 1985. *Conceptual Change in Childhood*. 1985. Cambridge: MIT Press.

Christensen, S.M., and D.R. Turner. 1993. *Folk Psychology and the Philosophy of Mind*. Hillsdale, N.J.: Erlbaum.

Cole, M., and Scribner, S. 1974. *Culture and Thought: A Psychological Introduction*. New York: Wiley.

Cover, R. 1983. "Nomos and Narrative: The Supreme Court 1982 Term." *Harvard Law Review* 97.

Donald, M. 1991. *Origins of the Modern Mind*. Cambridge: Harvard University Press.

Dunn, J. 1988. *The Beginnings of Social Understanding*. Cambridge: Harvard University Press.

Edman, Irwin. 1928. *The Works of Plato.* New York: Modern Library (Random House).

English, H.B., and English, A.C. 1958. *A Comprehensive Dictionary of Psychological and Psychoanalytical Terms.* London: Longmans, Green.

Feldman, C. 1994. "Genres as Mental Model," in M. Ammaniti and D. Stern (eds.), *Psychoanalysis and Development: Representations and Narratives.* New York: New York University Press.

Fillmore, C.W. 1968. "The Case for Case," in E. Bach and R. Harms (eds.), *Universals in Linguistic Theory.* New York: Rinehart and Winston.

Freud, S. 1949. *An Outline of Psychoanalysis.* New York: W.W. Norton.

Frye, N. 1957. *Anatomy of Criticism.* Princeton, N.J.: Princeton University Press.

Gallie, W.B. 1966. *Peirce and Pragmatism.* New York: Dover.

Geertz, C. 1983. *Local Knowledge.* New York: Basic Books.

Goodman, N. 1978. *Ways of Worldmaking.* Hassocks, Sussex: Harvester.

Grice, P. 1989. *Studies in the Way of Words.* Cambridge: Harvard University Press.

Gruber, H.E. 1981. *Darwin on Man: A Psychological Study of Scientific Creativity,* 2nd ed. Chicago: University of Chicago Press.

Hull, Clark. 1952. *A Behavior System.* New Haven: Yale University Press.

Hume, D. 1888. *A Treatise of Human Nature.* Oxford: Clarendon.

Iser, W. 1978. *The Act of Reading.* Baltimore, Md.: Johns Hopkins University Press.

Kant, I. 1934. *Critique of Pure Reason.* London: Dent.

Keil, F.C. 1979. *Semantic and Conceptual Development: An Ontological Perspective.* Cambridge: Harvard University Press.

Kuhn, T. 1962. *The Structure of Scientific Revolutions.* Chicago: University of Chicago Press.

Landau, M. 1991. *Narratives of Human Evolution.* New Haven, Conn.: Yale University Press.

Leibniz, G.W. 1898. *The Monadology and Other Philosophical Works.* Oxford: Oxford University Press.

Lloyd, G. 1993. *Modes of Thought in Classical Antiquity.* Stubbs Lecture at the University of Toronto.

Locke, J. 1959. *An Essay Concerning Human Understanding.* New York: Dover.

Mead, G.H. 1934. *Mind, Self, and Society.* Chicago: University of Chicago Press.

Medin, D.L. 1989. "Concepts and conceptual structure," *American Psychologist* 44(12), pp. 1469–81.

Mink, L.O. 1978. "Narrative Form as a Cognitive Instrument," in R.H. Canary and H. Kozicki (eds.), *The Writing of History: Literary Form and Historical Understanding.* Madison: University of Wisconsin Press.

Newell, A., and Simon, H.A. 1972. *Human Problem Solving.* Englewood Cliffs, N.J.: Prentice Hall.

Ninio, A., and Bruner, J.S. 1978. "The Achievement and Antecedents of Labelling." *Journal of Child Language* 1, pp. 1–15.

Olson, D. 1994. *The World in Print.* Chicago: University of Chicago Press.

Pavlov, I.P. 1949. *Complete Collected Works,* vol. III. Moscow: SSSR.

Popper, K. 1972. *Objective Knowledge.* New York: Oxford University Press.

Propp, V. 1968. *Morphology of the Folktale,* 2nd ed. Austin: University of Texas Press.

Ricoeur, P. *Time and Narrative,* vol. 1. Chicago: University of Chicago Press.

Rorty, R. 1979. *Philosophy and the Mirror of Nature.* Princeton, N.J.: Princeton University Press.

Rosch, E. 1978. "Principles of Categorization," in E. Rosch and B. Lloyd (eds.), *Cognition and Categorization.* Hillsdale, N.J.: Erlbaum.

Ross, D.W. (ed.). 1908/1952. *The Works of Aristotle Translated into English.* Oxford: Oxford University Press.

Scaife, M., and Bruner, J.S. 1975. "The Capacity for Joint Visual Attention in the Infant." *Nature* 253(5489), pp. 265–266.

Searle, J. 1969. *Speech Acts.* Cambridge: Cambridge University Press.

Smith, E., and Medin, D.L. 1981. *Categories and Concepts.* Cambridge: Harvard University Press.

Sperber, D., and Wilson, D. 1986. *Relevance: Communication and Cognition.* Oxford: Blackwell.

Stich, S.P. 1983. *From Folk Psychology to Cognitive Science: The Case Against Belief.* Cambridge: MIT Press.

Taylor, C. 1985. *Philosophy and the Human Sciences.* Cambridge: Cambridge University Press.

Tolman, E.C. 1948. "Cognitive Maps in Rats and Men," *Psychology Review* 55, pp. 189–208.

Turner, V. 1982. *From Ritual to Theater: The Human Seriousness of Play.* New York: Performing Arts Journal Publication.

Vernant, J.P., and Vidal-Naquet, P. *1988. Myth and Tragedy in Ancient Greece.* New York: Zone Books.

Vygotsky, L. 1962. *Thought and Language.* Cambridge: MIT Press.

20

Jeff Coulter

Neural Cartesianism

Comments on the Epistemology of the Cognitive Sciences

Introduction

A wide range of logical problems beset the heterogeneous field we call cognitive science. In this brief review, I will characterise several recurrent and consequential forms of logical error. The problems and fallacies on which I focus are spread across the intellectual landscape of contemporary cognitivism, whether the theoretical structure is of the "rules-and-representations" variety or the new connectionism. In isolating and exposing these logical anomalies, I will try to show that their removal facilitates a clearer picture of the nature of the phenomena in question, phenomena conceptualized in often misleading and confusing ways within existing frameworks of theory.

What sort of recurrent logical anomalies are to be found? Consider, for example, the following remark, which appeared in a recent technical overview of connectionist research:

> the brain is capable of recognizing visual scenes or understanding natural language within around 500 milli-seconds—a performance inconceivable for today's computers. (Pfeifer, 1989, p. xiii)

Aside from the (probably unintended) claim that something called "natural language" (as distinct from a natural-language expression) can be "understood" in 500 milliseconds, there are other errors here that can be found repeated throughout the research and theoretical literature. These are, first, the fallacy of treating "recognizing" and "understanding" as predicable of someone's brain, when they are person-level predicates. (Kenny, 1985, has discussed this at length under the title of the new "homunculus fallacy"). Second, there is the fallacy of treating "recognizing" and "understanding" as "performance" verbs when they are, in Ryle's (1973) terms, "achievement" verbs.

In this chapter, I will lay out schematically some of the major logical anomalies besetting cognitive science, especially as these pertain to treatments of perception, memory and mental images.

Problem One: The Global Specification of the "Explananda" in Cognitive Science

A principal source of confusion is the tendency to postulate a globally defined *explanandum* for the purpose of modeling or theory construction. Typically, one finds references to a theory of cognition, or of memory or perception or language, as if these were relatively straightforward generic terms yielding discrete referents or phenomenal domains. Because these terms are used as "master-categories," there is an effort to subsume *under* them a motley of diverse phenomena with little heed paid to the consequential nature of the diversity. We are not dealing here with genuine natural kind terms. Indeed, I will show that by mistaking a global specification of an explanandum for the specification of a "phenomenon," presumably neural in form, cognitivists routinely reify and homogenize the properties of mental and experiential predicates.

The chief category deployed within the field is the hybrid, the mind-brain. Its workings are then said to be the target for explanatory work. Commenting on the logical anomalies engendered by conjoining the diverse concepts of mind and brain, Hacker (1990a, pp. 135–36) observes:

> What may grow in the brain, e.g., a tumor, cannot grow in the mind, and what may grow in the mind, e.g., suspicion, cannot grow in the brain. One may have an idea in the back of one's mind, e.g., that perhaps one ought to go to London next week . . . But *that perhaps one ought to go to London* . . . cannot be in one's brain. . . . When one speaks of someone as having a devious mind one is talking of his character, namely that he is cunning, not of his brain. When one says that a person has an agile or quick mind, one is speaking of his intelligence, his capacity to grasp and solve problems, not about his cerebral cortex.

When we assert, quite reasonably, that someone has something on his mind, that something has slipped his mind, that he bears things in mind, etc., within these language-games, where the concept of "mind" has its normal part to play, we are not claiming *anything* about neurophysiological states, processes, or mechanisms. We are saying that someone is preoccupied, that he has forgotten something or that he is taking specific things into consideration.[1] It is only via the "neuralization" of the Cartesian concept of the mind as a *res cogitans* (thinking entity) that we might entertain the notion of linking the ordinary concept of mind (with its ramifying uses) to the biological concept of the brain. Neural Cartesianism is, however, no less of a myth than its mentalistic precursor.

Problem Two: Reifications of Complex Predicates into State and Process Concepts

Naturally, denying Neural Cartesianism is not to deny the fact that neural states and processes of various kinds (electrochemical, metabolic, etc.) *facilitate* our conduct, such that the mental predicates apply to us *as persons* in virtue of our sayings and doings in everyday-life contexts. What is being rejected is the logical fallacy of thinking that the applicability of mental predicates (understanding, thinking, remembering, forgetting, intending, believing, etc.) is an exclusive function of the existence of neural phenomena, which these predicates somehow "indirectly" identify. Searle speaks of a principle of "neurophysiological sufficiency" in this regard (1987), and I will discuss his views on this further on.

Reification of complex predicates into "entity" or "state" concepts is a common error in much recent philosophy of mind. To "have an intention" is, *inter alia*, to be committed to some course of action, and to "have a delusion" is falsely to believe something. But neither of these is to possess some hitherto undetermined (neural) phenomenon, nor to be in some identifiably correlatable neural "state." Similarly, we also can subject the whole array of mental predicates to analysis, and reveal their contextual conditions of ascription/avowal/detection. In undertaking such an inquiry, we find that many existing cognitivist stipulations about the "mental" are exactly that—*stipulations*—and not *elucidations* of the domain under investigation. In their well-justified efforts to escape the restrictions and distortions of behaviorist forms of reductionism and reification, the cognitivists entertain their own varieties of the same logical fallacies.

Refusing to treat all human actions as "habits" or "conditioned responses," and arguing correctly that the concepts of rational human actions and capacities will not yield to any such restrictive characterizations, the cognitivists nonetheless exempt themselves from the constraints of logical grammar when they take up the question of the nature and conceptualization of the "mental." Consequently, they conceive of "perceptions" as "neural representations" arrived at via "computations on sensory input" (Gregory, Marr), "memories" as neurally encoded traces, "engrams" or inner "representations" of experiences (Booth, Deutsch), "understanding" as a neural-computational "process" (Fodor, Chomsky), "imagining (something rotating)" as "mentally rotating a neurally-realized image" (Shephard), and so on.

These are all complicated matters, and, in the space allotted me, I can summarize only schematically some results of relevant Wittgensteinian-ethnomethodological studies.[2] It will appear that cognitivist misconceptions of the mental largely turn on ignoring the dimensions of social praxis essential to their nature. Consider, first, perception.

Perception as a Misleading Master Category

The concept of perception as deployed in much cognitive-science theorizing actually functions as a misleading gloss for a wide range of diversely rule-governed visual (and other) activities and achievements. Not all of these are interchangeably predicable of persons; but all are applicable only in contextually circumscribable, socially organized ways. For example, we cannot transpose the object-complements of (verbs of) visual action and achievement generically, without loss of sense or validity (Thus, the sorts of things one may ogle may not be noticed; the things looked at may not be (have been) seen; the things seen occasionally not recognized; the things seen as something not thereby seen *simpliciter*; and so on) (see Coulter & Parsons,

1990). In light of this, even if perception were amenable to the kind of empirical investigations to which it is commonly thought to yield, explanations targeting this concept could not logically encompass the many other concepts of visual orientation. Efforts to reduce all of the perceptual verbs to a homogeneous core—perceiving or seeing, reducible analytically to some singular, subserving, physiological process (e.g., photon-photoreceptor interaction and subsequent transduction)—cannot work.

The further cognitivist proposal to assign representations of any perceptible phenomenon within the cortex as interior end-products of neural processing is also flawed. It obliterates the distinction between, for example, seeing the depiction of something and seeing the thing depicted; between the image of something observed and the actual thing observed; or between a replica and the original—distinctions we need to respect, on pain of losing or arbitrarily reconceiving the phenomena to be explained. Much of the talk about mental representation as a necessary (omnipresent) intermediary between perceivers and things perceived begs the question. For the category of representation to be usable, there must be criteria for distinguishing between correct representation and misrepresentation, which presuppose direct access to phenomena in order to apply such criteria. Thus, the thesis that reality is available exclusively through inner representations of it is self-subverting. Of course, by direct access, I do not mean to deny the mediating, facilitating role of neural events and processes, only the incoherent interpretation of these biological occurrences as comprising an interior scenic depiction of what is being, for example, directly looked at.

Many perceptual verbs are verbs of rule-governed action (e.g., watching, scanning, looking for/around/up(at)/into, watching, scrutinizing, staring, glancing, gazing, etc.). But some of the most common among them are not. Indeed, studies focusing generically on perception, which is not an activity but an achievement, cannot logically illuminate the vast variety of perceptual activities. Take the example of seeing.

To say that "I see or can see *X*" is to make a perceptual claim that is susceptible to intersubjective ratification or defeat. There is no subjective sovereignty in relation to perceptual claims. Saying that one sees (saw), being wholly convinced that one sees, having not the slightest doubt that one sees, whatever the object-complement, is not *eo ipso* to see it. All are logically compatible with seeming to see, thinking that one sees, hallucinating, and so on, for any given occasion. "I (can) see *X*" is not performative in the sense in which "I apologize" or "I promise" can be. The fact that, for the most part, normally sighted people's claims are not questioned or defeated by others does not affect the point that claiming does not itself constitute the achievement of the thing claimed. The fact that, ordinarily, our claims are (usually tacitly) ratified does not mean that perception (seeing) is an activity we engage in individually. It is, rather, a clear achievement on our part, whose ratifiability is contingent on the satisfaction of intersubjective criteria.

A Memory Is Not an Entity of Any Kind

Next, consider memory. Remembering is not detachable from its complementizers and object-complements; and, as such, it encompasses very diverse and disjunctively characterizable ranges of phenomena. We have remembering-how-to (usually an ability, and hence not susceptible to storage theorising, since one does not store *abilities* anywhere); remembering to (do *X* or *Y*), which also is not a candidate for any sort of storage model); and the canonical case of remembering that (*X* or *Y*), which is the one we will consider in this brief discussion.

A memory, then, in the sense of "a memory of . . . ," is usually a correct account of something in the past, personally witnessed. Cognitivists, however, tend to think of a memory as something distinct from, lying behind or in back of any such account. The account itself is thought of as merely portraying (in words or images)[3] the real memory that is elsewhere (e.g., in the brain). This is mistaken. If my suggested account is correct, satisfying the intersubjective criteria for genuine recollection, *then it is* the memory. What could a discursive recollection of something (memory of *X*, that *X* occurred, etc.) consist of, other than a ratified-as-correct account of some antecedent state of affairs personally witnessed? If one proposes that the memory lies *behind* the (correct) account, thus allowing the intensely problematic issue of retrieval from storage to arise, in what could such a prior memory phenomenon consist?

Whatever theoretical models one constructs to accommodate this pseudo-problem, it is easy to see that they all need to individuate specific recollection accounts in a decontextualized way, prior to their "translation" into a favored form of hypothetical encoding (e.g., in protein molecules, RNA, neural nets, glial cells, etc.). This move reifies components of the contextually variable elements of an audienced "telling" of some past state of affairs into a privileged version. But it is one whose criteria of adequacy are indeterminable, because the criteria in question are severed from the social purposes that alone could give actual accounts their intelligibility, relevance, and truth-value, along with the further criteria for their assessment along those dimensions. As I have argued elsewhere,

> Rather than construe memories as *themselves* neurally-encoded phenomena, we should think instead of neural structures, states or events as (co-)enabling, facilitating, *the situated production of memory-claims* (to oneself or others) in all their variety. (Coulter, 1991, p. 188)

Imagining Is Not Seeing Neural Images

Finally, but more extensively, let us consider mental images and imagining. This is a complex topic but, in my view, it is possible to establish certain points to counter the conventional cognitivist account.

According to most versions entertained by cognitivists, having a mental image is analyzable as having a mental picture; that can, in turn, be construed as akin to generating an analog representation in the cortex. I will not examine all of the problems involved in this reductionist move. Instead, I will focus on a preliminary step, which appears quite reasonable at first—namely, treating the having of a mental image as the conjuring up of an interior picture.

The identification of images with pictures is problematic. "Picturing something to oneself" is an idiomatic expression that causes theorists a great deal of trouble when they fail to attend to what it would entail, if taken literally. Hacker, again, is helpful here:

> Why should we say that when we visually imagine a robin, for example, we give our- selves a *picture* of a robin, rather than that we give ourselves a robin? For one can imagine (have a visual image of) a picture of a robin no less than one can imagine a robin, and these are by no means the same. If one objects that when one imagines a robin, one does not *really* "give oneself" a robin, one may reply that that is obvious, but that, equally, one does not *really* "give oneself" a picture of a robin either! To picture something to oneself is not to give oneself a kind of picture (viz., a "mental" one). (Hacker, 1990b, pp. 413–414)

That I imagine X or Y, or X as Y, etc., is not established by anything that goes on inside me but by the account I (can) give. To "have" a mental image is not to "have" anything at all in the sense of possession (whether inner or outer), just as to "have" obligations is not to possess queer sorts of intangible objects. Similarly, to see something in the mind's eye is not to see something simpliciter in the cranium. Where is the mental image you are seeing located—is it three inches behind your nose or two inches under your left eye? The question has no empirical answer; it does not make sense as a serious inquiry at all. One should not conclude from this, however, as Ryle did in his *Concept of Mind*, that all visualizing amounts only to "seeming to see" something, for while a person may be asked (and grant the request) to "visualize" something, one cannot be commanded, requested, or asked to "seem to see" something and proceed to do so. Further, one may have seemed to see some- thing in a surrounding environment, when it turns out that it wasn't there; but this does not mean that one visualized it.

There is still a propensity to hear these arguments as parts of a case for a reac- tionary revival of behavioristic theorizing. Nothing of the sort is intended here. Behaviorists have tended to deny visualization itself in (correctly) denying the reifi- cation of the mental image concept. To deny a mistaken conception of a phenome- non, however, does not put an end to inquiry, especially if there are alternative ways of elucidating it. The problem has been that, in resisting the behaviorists' efforts to throw the baby out with the bathwater (i.e., in this case, to reduce visual imagination to a myth or, at best, to a strange *façon de parler*), a space is opened up for creating neural Cartesian myths of an exactly opposite kind—namely, the myth of construi- ing visual imagination as the interior production of *neural* representations. This cog- nitivist gambit fares no better. It is similarly predicated on construing mental images as types of pictures with genuine spatial extension, and then trying to reason from such a preconception to a (hypothetical) *physiological* characterisation.

I do not wish to demean the many interesting efforts theorists (e.g., Block, Kosslyn, or Shephard) have made to render such a notion—of a neurally generated picture—intelligible. But I suggest that we are not dealing here with a phenomenon, but a chimera. Just as (discursive) memories are not themselves neural phenomena lying behind correct accounts of personally witnessed past states of affairs, neither are mental images analyzeable into neural phenomena lying behind the characteri- zations we give of what we visualize. For when I describe what I visually imagine to you, I am not reading off a description from a perception of a neural entity.

I can imagine a closed box, and I can draw a picture of a closed box, but a picture of a closed box could also represent several other things. For this configuration of lines on paper could represent a wire structure or three adjacent quadrilaterals, etc., hence it might need an interpretation. *But a mental image does not, in that sense, consist of a configuration of lines*; it is a . . . property of my visual image of a closed box that it is of a closed box. A mental image is, as it were, all message and no medium—like a thought. (Hacker, 1990b, p. 413; emphasis added)

If we try, as cognitivists do, to decompose visualizing into perceiving (inner phenomena) and subsequently describing them, then we have opened up a space for possibilities that are logically excluded from genuine instances of visualization. For to perceive is *not* to *mis*perceive: each allows for the logical possibility of its contrast. However, in the case of visualizing, I cannot misperceive my mental image. That is, I cannot mistakenly judge my mental image of, say (to use a Wittgensteinian example) a man walking up a hill to be one of a man backing down a hill, even though in pictorial form both might be represented the same way.

Consider: I can claim to have seen something, be shown to have been wrong, and then subsequently claim only to have thought that I saw it and not actually to have done so. But there is no analogue in the case of imagining. I cannot claim merely to have thought that I imagined/visualized something and not to have done so at all. I cannot "misimagine" or "misvisualize" as I can misperceive. To think that such-and-such is an *X* (often) is to imagine/visualise *X* or something *as X*. This is shown in and through my situated, communicative, and other forms of conduct.

Concluding Remarks: Why Searle's Principle Will Not Do

In these comments on aspects of the psychological—namely, perceiving, remembering/recollecting and imagining/visualizing—the following point emerges. Contrary to John Searle (1987), the principle of "neurophysiological sufficiency" in their elucidation, analysis, and explanation is fallacious. For Searle, to claim that "social relationships are necessary for, or constitutive of, mental life" is a mistake "because social relationships are relevant to the causal production of intentionality only insofar as they impact on the brains of human agents; and the actual mental states, beliefs, desires, hopes, fears and the rest of it have causally sufficient conditions that are entirely internal to the nervous system" (Searle, 1987). Here, Searle's conception of social relationships is entirely too concrete: he lacks a concept of intersubjective criteria as the grounds of sense. Moreover, he treats conceptual investigations as exclusively concerned with the conditions of possibility for our having a certain vocabulary, as if this were separable from grasping the conditions of possibility for its application! Along with many other philosophers, especially those who seek to raise the biological and physical sciences to panchrestonic (omni-explanatory) status, he misunderstands the whole point of Wittgensteinian logical grammar and its connection to the phenomenological exploration of the constitution of phenomena.

Grammar—the rules of concept-formation and conceptual deployment—provides for us the (logical) possibilities of *phenomena*, not the other way around. Without denying a causal role for neural states and processes, the issue before us all is to give a clear characterization of what is thereby to be explained—the mental. It often appears, as above, that neuropsychological accounts presuppose incorrect or incoherent depictions of the explananda, and that their scope is far less wide-ranging than might be thought, when we settle for simple reifications (and other logico-grammatically anomalous treatments) of the mental predicates.

Notes

1. Hacker adds: "In these cases, one is no more 'speaking at some level of abstraction of yet unknown physical mechanisms of the brain' than when one orders eggs and bacon for breakfast one is speaking at some level of abstraction of the chemical structure of proteins" (1990a, p. 136).

2. More detailed treatments can be found in my *Social Construction of Mind* (Rowman & Littlefield, 1979); *Rethinking Cognitive Theory* (N.Y.: St. Martin's Press, 1983); *Mind in Action* (Atlantic Highlands, N.J.: Humanities Press, 1989): " 'Recognition' in Wittgenstein and Contemporary Thought," in M. Chapman and R. Dixon (eds.), *Meaning and the Growth of Understanding: Wittgenstein's Significance for Developmental Psychology* (N.Y.: Springer-Verlag, 1987); "Two Concepts of the Mental," in K. J. Gergen and K. E. Davis (eds.), *The Social Construction of the Person* (N.Y.: Springer-Verlag, 1985); "On Comprehension and 'Mental Representation' " in G. N. Gilbert and C. Heath (eds.), *Social Action and Artificial Intelligence* (Nottingham: Gower Press, 1985); "Materialist Conceptions of Mind: A Reappraisal." *Social Research*, Vol. 60, No. 1, Spring 1993; and "Cognition in an Ethnomethodological Mode," in G. Button (ed.), *Ethnomethodology and the Human Sciences* (Cambridge: Cambridge University Press, 1991).

3. I will not elaborate on the arguments specifically developed to handle the problem of memories expressed in images or in imagistic form. Suffice to note here that no mental image, in and of itself, independent of purpose and context, depicts what it is an image *of*. And no mental image in itself constitutes a memory-image. For that to occur, it should depict correctly some past state of affairs, personally witnessed by the one entertaining the image. Thus, some discursive characterization of any such image must be producible and intersubjectively ratifiable; otherwise we cannot draw a distinction between someone's image expressing an apparent recollection (as in, e.g., "I had this vivid image of him standing there, but he was elsewhere at that time so I couldn't have remembered him there: I only seemed to remember him") and someone's image expressing a genuine memory.

References

Coulter, J. 1993. "Materialist Conceptions of Mind: A Reappraisal." *Social Research*. 60, 1.

Coulter, J. 1991. " Cognition in an Ethnomethodological Mode," in G. Button (ed.), *Ethnomethodology and the Human Sciences*. Cambridge: Cambridge University Press.

Coulter, J. 1989. *Mind in Action*. Atlantic Highlands, NJ: Humanities Press.

Coulter, J. 1987. " 'Recognition' in Wittgenstein and Contemporary Thought," in M. Chapman and R. Dixon (eds.), *Meaning and the Growth of Understanding: Wittgenstein's Significance for Developmental Psychology*. New York: Springer-Verlag.

Coulter, J. 1985. "On Comprehension and 'Mental Representation,' " in G.N. Gilbert and C. Heath (eds.), *Social Action and Artificial Intelligence*. Nottingham: Gower Press.

Coulter, J. 1985. "Two Concepts of the Mental," in K.J. Gergen, and K.E. Davis (eds.), *The Social Construction of the Person*. New York: Springer-Verlag.

Coulter, J. 1983. *Rethinking Cognitive Theory*. New York: St. Martin's Press.

Coulter, J. 1979. *Social Construction of Mind*. Totowa, NJ: Rowman and Littlefield.

Coulter, J. and E.D. Parsons. 1990. "The Praxiology of Perception: Visual Orientations and Practical Action," *Inquiry*, 33, 3.

Hacker, P.M.S. 1990a. "Chomsky's Problem." *Language and Communications*, 10, 2.

Hacker, P.M.S. 1990b. "Images and the Imagination," in P.M.S. Hacker, *Wittgenstein: Meaning and Mind*. Oxford: Basil Blackwell.

Kenny, A. 1971/1985. "The Homunculus Fallacy," in A. Kenny (ed.) *The Legacy of Wittgenstein*. Oxford: Basil Blackwell.

Pfeifer, R., Schrefer, Z. Fogelman-Soulié, F., and Steels, L. 1989. "Putting Connectionism into Perspective," in R. Pfeifer et al. (eds.), *Connectionism in Perspective*. New York: Elsevier.

Ryle, G. 1973. *The Concept of Mind*. Harmondsworth: Penguin.

Searle, J. 1987. "Minds and Brains Without Programs," in C. Blakemore and S. Greenfield, Mindwaves: *Thoughts on Intelligence, Identity and Consciousness*. Oxford: Basil Blackwell.

21

Sören Stenlund

Language, Action, and Mind

Introduction

The philosophical claims of cognitive science are the subject matter of this chapter. I would like to discuss a few of the philosophical presuppositions that seem to be among the common assumptions underlying cognitive science, and which are still alive, despite recent attempts to reorientate cognitive science away from computationalism and naturalism. Some of these assumptions are explicitly stated by cognitive scientists, while others manifest themselves indirectly in certain characteristic features of their discipline's established vocabulary and technical notions. They also manifest themselves in certain typical attitudes toward alternative views, and toward the scientific tradition in which cognitive scientists place themselves. My main concern will be with some of these tacit presuppositions that I find problematic.

To question these philosophical presuppositions is not to question the whole enterprise of cognitive science as an existing field of scientific research. My criticism affects the work of cognitive scientists only insofar as it really depends on these presuppositions. And I believe that there is a lot of technical and scientific work done in the field that is quite independent of the philosophical views and assumptions I will be discussing. This is particularly true of a lot of empirical studies and experimental work in psychology, linguistics, and computer science—perhaps including the best work in these fields. However, the independence is not always obvious. It is concealed by many statements of the significance of results, by those who do their best to present the results as being significant for the original philosophical aims and claims of cognitive science. What I hope to do in this chapter is provide some grounds for reassessing such interpretations.

What makes it difficult, from a philosophical point of view, to form an opinion about some of the explicit claims and general assumptions made by cognitive scientists is that the assumptions are usually articulated and defended within a vocabulary that is already informed by and committed to these same assumptions. The difference between what is put forward as empirical hypotheses and what should be taken as conceptual claims is often not clear. In the literature of cognitive science, one and the same assumption, the assumption of the mental reality of rules and representations, for instance, often has a double role. It is sometimes treated as though it expressed a necessary truth; but at the same time it is discussed as a contingent truth amenable to empirical justification. The doctrine of the existence of mental representations is sometimes put forward as a conceptual principle with *a priori* validity expressing an essential feature of the cognitivist approach to the problems of explaining behavior (which, I believe, is the correct account). But, in the same context, it may be treated as a hypothesis to be established by scientific proof or by future empirical research, or even as an empirical claim for which we already possess sufficient experimental evidence. Thus, for instance, some of the notions and assumptions of Chomsky's theory of generative grammar, like the notion of a syntactic transformation, had the role of defining the purely theoretical framework created by Chomsky for the study of language. These notions, one might say, had the role of grammatical rules for the technical vocabulary of generative grammar; and, as such, they could be followed and applied like the rules of an abstract calculus, wherein assumptions about the psychological reality of the rules were neither used nor needed. This is a common way of working with generative grammar in so-called "natural language processing." But it was part of Chomsky's doctrine of generative grammar to claim to have *demonstrated* the existence of "mental counterparts" to these notions and rules, or to have given evidence for "the necessity of postulating the mental reality of the rules." Ken Wexler expresses this as follows:

> Chomsky's early demonstration of the need for syntactic transformations was the demonstration of the existence of a particular kind of mental linguistic entity, one not definable in terms of stimuli and responses. (Wexler, 1991, p. 253)

What about the claim to *necessity* that is implied by this talk of demonstration? How can we understand the "must" with which cognitivists often have expressed the need for postulating mental representations and cognitive processes? This claim of necessity appears to me to be one of the most characteristic features of the movement that started with what has been called "the cognitive revolution." It is a feature that distinguishes tendencies toward cognitivistic ways of thinking within behaviorism (in Tolman, for instance) from the cognitivism and mentalism that began what is now called cognitive science. What is the significance of this claim of necessity? It does not seem to be just a kind of empirical issue, a conviction based on experimental evidence that certain notions and methods will turn out to be more fruitful than alternative ones. One claims to know that "an adequate theory" *must* postulate mental operations.

In discussing such operations designed to explain sentence perception, Jerry Fodor remarks:

> We posit such operations simply because they are required for the construction of an adequate theory of speech perception. (Fodor, 1968, p. 84)

And Fodor adds that

there appears to be no alternative to invoking such operations if we are to explain the disparity between input and percept. (p. 85)

This last remark gives a clue to the significance of the claim that mental operations *must* be postulated, it seems to me. The postulating is felt to be needed in order to handle problems arising out of the behaviorist project of giving a general causal theory of behavior, *a project that cognitivists wish to pursue.*

On the Notion of the Observable

Thomas H. Leahey (1987) introduces the term *behavioralism* as an umbrella term denoting the trend toward defining psychology as the science of behavior in contrast to the conception of psychology as the science of consciousness, which we find, for example, in Wilhelm Wundt's and William James' writings. According to Leahey, this trend emerges around World War I and comes to dominate the science of psychology in this century. This development was to a great extent a result of psychologists' desire to be socially useful, their "desire to be practical and succeed in business, industry, government, and among other professional providers of services" (p. 287). Different forms of behaviorism were only parts of this larger trend toward behavioralism, which also includes cognitivism. Leahey explains:

> The behavioralist seeks to predict, control, explain, or model behavior, and to do so he may or may not refer to conscious or unconscious mental processes. Behavioralism is aimed at behavior; consciousness—the mind—is not the object of study, although it may be called upon to explain behavior. (p. 260)

The mainstream of cognitive science is a continuation of this trend, because the cognitive scientist is willing to postulate the existence and functioning of unconscious processes, which are seen as internal, causal agencies that produce behavior. The true behaviorist, on the other hand, is unwilling to admit any nonobservable influence on behavior; but any influences observed may be used in predicting, controlling, and explaining behavior. Leahey's notion seems to me to be well motivated, since it softens the significance of the idea that the cognitive revolution was a conceptual change, and puts the opposition between behaviorism and cognitivism in a more accurate historical perspective.

Following Leahey, I want to suggest that it is more rewarding to look for *common* features of cognitivism and behaviorism rather than differences, if we are interested in the conceptual presuppositions and philosophical claims of cognitive science. I will argue that both these approaches make important common assumptions concerning the nature of language, mind, and human action. Advocates of cognitive science tend to stress the differences in a way that brings technical matters into focus, and that conceals the basic assumptions they share with behaviorists.

In the quotation from Wexler we find a reference, similar to Fodor's, to the behaviorist conceptual framework. He says that Chomsky's "demonstration" proves the existence of a certain mental linguistic entity, by showing that it is *not definable in terms of stimuli and responses*. For the demonstration to prove what it is supposed to prove, it appears to be important that our linguistic practices are conceived within a perspective that is as much materialistic and mechanistic as the behaviorist one. (Cf. Chomsky's notion of "the data of linguistic performance.") What Chomsky opposes is rather the one-sided empiricist and positivistic approach of behaviorism. It constitutes an obstacle to the formal notions and methods from mathematical logic that he wants to introduce, with strong realist claims, into linguistics. It is clear from his criticism of Skinner's *Verbal Behavior* that he accepts and respects Skinner's problem of giving a causal or functional account of "the actual behavior of speaker, listener and learner" (in Chomsky's own words, 1964). What he objects to is Skinner's way of dealing with the problem. Instead of renouncing the project of giving a causal account of verbal behavior, Chomsky's point is that this urgent task cannot be accomplished unless we take into account the speaker's, listener's, and learner's "knowledge of grammar"; this knowledge, he maintains, does not manifest itself completely in the observable verbal behavior of speakers, listeners, and learners. What needs to be taken into account, according to Chomsky, is thus other kinds of causes, which are said to be unobservable and mental.

Several of Chomsky's assertions in his review of Skinner show that he does not really reject the physicalistic "data-language" of behaviorism as an altogether inappropriate and misplaced perspective in the recording of facts about human practices of language use. The following assertion, from the end of the review, is a good example:

> The grammar must be regarded as a component in the behavior of the speaker and listener which can only be inferred . . . from the resulting physical acts.

And in *Aspects of the Theory of Syntax* he contrasts the mentalist and the behaviorist attitudes to the study of language in following question:

> [D]o the data of [linguistic] performance exhaust the domain of interest to the linguist, or is he also concerned with other facts, in particular those pertaining to the deeper systems that underlie behavior? (1965, p. 193)

What Chomsky is attacking is not the *concept* of observation and observability that goes with the behavioral data-language rooted in scientific empiricism. His point is that the one-sided emphasis on observable features of verbal behavior, which is a signpost of the behaviorist approach, is too restrictive. Chomsky, like other cognitivists, manifests an essential agreement with behaviorism about what constitutes observable properties and features of behavior and about what observability means. I think that this is a conceptually important fact because it means that the concept of *un*observable or *abstract* phenomena—among which are the cognitive processes and mental representations of cognitivism—are conditioned by the physicalistic notion of observation and observability of behaviorism.

The "poverty of stimulus" described in the Chomskyan argument for mentalism (which has been called "the argument from the poverty of stimulus") is, it seems to me, very much a "poverty," but one that is imposed through the behaviorist notion of what it is to observe and describe facts about human action and human language use. As far as I can see, this notion of observation and observability was not questioned in the cognitive revolution, but rather was imported into the cognitivist conceptual framework. This is evident, for instance, in the frequent use of the notion of *behavior* itself as a technical term in the cognitivist literature. The very notion of behavior (as opposed to action) as well as notions related to it, such as "stimulus," "response," "organism," "environment," "environmental factors," etc., were each informed by this notion of what constitutes the observable features of the external world. This notion had its roots in scientific empiricism; so it seems to me that the conceptual framework of scientific empiricism, as defining the notion of the observable, was not questioned by those who wanted to introduce unobservable, mental causes of behavior. They were rather *extending* the empiricist conceptual framework.[1]

The Chomskyan "argument from the poverty of stimulus" was not an argument against the general scientific materialism and mechanism of behaviorism. It was rather a proposal designed to show how to pursue these philosophical views in the face of apparent limitations in the behaviorist ways of developing them. In a certain sense, one might say that Chomsky's program, as well as other forms of information-processing cognitivism, was a way of developing and *supplementing* the behaviorist approach in order to save some of its essential features.

I think it is important to notice that there is a general agreement between behaviorism and cognitivism about the use of the dichotomy between the concrete or observable and the abstract or unobservable. The abstract or unobservable is what behaviorists traditionally have rejected, as much as cognitivists have argued for its necessity. But a common ground for this disagreement is a (general) agreement about what constitutes the concrete or the observable, and how the phenomena of language and human action are given in perception. It is important to notice this general agreement because, as I already have suggested, it seems to be a principal source of the problems that motivated the cognitive turn. The "disparity between input and percept in sentence perception," which Fodor puts forward as a main reason for the necessity of postulating "mental operations," is a disparity that depends very much on the fact that the sentence is conceived as an input (or stimulus) of a certain perceptual mechanism rather than as an expression whose essential features are determined by the use it has in some normal practice of linguistic communication. The disparity will depend on how we conceive the input, or how we describe "the information given" in a process of sentence perception that is defined in abstraction from the normal, actual use of the sentence. We certainly will have such a disparity if we conceive of a sentence as being basically a sequence of sounds with a certain grammatical structure—that is, as an object of study of empirical linguistics—because the disparity is basically a disparity between, on one hand, the sentence as an expression with a certain use in language and, on the other hand, the sentence as an object of study described in the "data-language" of linguistics. Fodor's mental operations are introduced as a corrective for this disparity, so the claim about the necessity of postulating the mental operations reflects the privileged or normative role given to the framework of empirical linguistics that is used in describing "the data of linguistic performance."

Experimental results of George Miller and others have been reported as showing that words masked by noise are more easily recognized when they are in a meaningful context than when they are presented in isolation. (see, e.g., Bever, 1988). This is put forward by cognitivists as a striking and almost surprising fact, which requires an explanation in terms of some mental process. It is a conceptually significant fact that the cognitivist explanations one is looking for tend to leave unquestioned the linguistic notion of a word as being basically an isolated acoustic event. It obviously was against the background of this linguistic notion of a word that the result was experienced as striking and in need of an explanation. A reaction to this result, different from the cognitivist one, would have been to say that we need to question our basic conception of what is essential about a word as an expression of a language. A word is not primarily a physical object. The kind of features that are essential to it as an expression of a language are determined in the context of its meaningful use.[2]

The dichotomy between the observable data and the unobservable or abstract entities that "underlie" and explain them occurs frequently in the literature, not least of all in statements of the significance of experimental results. In cognitivist literature we often find reports of experimental results that are supposed to show the necessity of appealing to unobservable or abstract entities in order to account for the results, and in order to explain how "it becomes possible for people to go beyond the information given," as Jerome Bruner expressed it in a paper from 1957 (1973). As already indicated, I think that this use of the dichotomy between the concrete and the abstract, or between the observable and the unobservable, is highly questionable. The notions of observation and the observable, on which the dichotomy is based, are very special and technical. They result from giving an unjustified logical priority to the attitude and vocabulary used in reports of observation and experiment in empirical science. This vocabulary is taken to have a privileged position over all other practices of language use, as though these other, everyday practices all started out from given phenomena and facts about the situation as conceived in this observational vocabulary. When an "organism's" response deviates from the "input" or "stimulus" so described—for example, when the response appears to involve something more or something else—one accounts for the disparity in terms of some unobservable internal process or mechanism that "organizes the information given," or "interprets it," or "makes inferences from it." And when one supposes that the cognitive processes must be postulated, *this "must" reflects the cognitivist attitude to the observational vocabulary*. It is the adherence to the causal and physicalist nature of the notions of the observational vocabulary that is felt to necessitate the postulating of mental processes.

Our Ordinary Language of Mind and Action

These notions of what constitutes the given, observable features of someone's behavior, and of the situation in which he or she acts, make it appear as though inference and interpretation are *always* involved in observing and judging what the person is doing, believing, desiring, or feeling. What I would like to point out is that we are all familiar with ordinary notions of observing (or seeing or hearing) that someone is saying something, or doing something, or that someone believes or desires something; and the observation is immediate in the sense that there is no inference or interpretation involved. My judgment or reaction to the person's behavior is not the result of an inference or interpretation based on the external or physical properties of his or her behavior. And such "immediate" hearing of what someone is saying, or seeing what someone is doing, is not an exceptional case; it is what we are normally engaged in, in our everyday linguistic practices and interaction with other people.

I am sure that most parents would agree that there are numerous common situations in which they can see that their child desires something, or that the child is afraid or is surprised or is in pain; and there is no room for doubt about it. Doubt would not make sense in the situation. The idea that a parent does not immediately see that his or her child is afraid, but merely infers it, is related to claims made earlier in this century by sense-data theorists. They too had a restrictive and technical notion of observation—namely, a language for describing the form and location of color patches in a two-dimensional plane. Everything, which is not merely a color patch in a two-dimensional plane, such as tables or trees, children, and, of course, frightened children, had to be inferred—or to be "logically constructed." When I oppose such claims by saying that we normally see that someone is afraid "immediately," I am not saying that this act of seeing is some sort of intuition, or insight. I am just saying that the word "afraid" has a use in our language, just as the words "table" and "color patch" have. In certain situations we say that someone "is afraid," and in doing so we do not take a roundabout way through a physicalist description or observation of the situation.

This is not to deny that there are many situations where doubt occurs in our interaction with others, where we do indeed make inferences or interpretations about what other persons are up to, and about their mental life—for example, about their beliefs, desires, and feelings—on the basis of their ways of acting and reacting. This is particularly true of people who are strangers to us, or in situations where someone we know well is behaving in an unexpected or surprising way. But the considerations that eventually lead us to the inference or interpretation that the person believes thus and so, or that he or she is afraid of this or that, presuppose that there are normal situations where such judgments are *not* the results of inferences or interpretations. It presupposes that there are certain *logical* criteria for our use of words such as "believes," "is afraid," "is in pain," etc. That the criteria are logical does not mean that they are formal or that they could be formally expressed. On the contrary, they are bound up with human circumstances and situations where our ordinary psychological terms have their proper use. The criteria are as complex and unsurveyable as are the circumstances of human life. In the cognitivist conception, however, where the agent is conceived as a mechanism or as a functionally defined system, such criteria would be understood as symptoms or evidence supporting a hypothesis about something partly unobservable that goes on inside the agent.

It is of course true that we sometimes ask someone whose behavior puzzles us "What is on your mind?" meaning "What are you thinking about?" or "How do you feel?" when we want him or her to make his or her actions or reactions intelligible. But we should not forget the enormous number of cases in which we do not have to ask these questions, and where it even would be insulting or absurd to ask them. In such cases, it might be said that we see immediately what he or she is thinking or feeling; we *react* to the other person. On the classical cognitivist conception of mental causes, however, it appears as though there were no such situations. It is as though our judgments about the mental life of others were always based on hypotheses of what "goes on inside them," hypotheses that could always be doubted.

What prevents us from recognizing the familiar, ordinary senses of observing, seeing, or hearing is thus very much the notions belonging to the causal or observational vocabulary rooted in the behaviorist tradition. These notions are used in an attitude that abstracts away from human circumstances (like the relation between parent and child) to which the notion of seeing the mental states of another person is internally related. The causal notions make it seem as though we would have to observe something that is partly unobservable in order to see that someone is doing something familiar with a certain familiar intention (such as putting on one's shoes in order to go to work) or that one is in a certain mental state (such as being afraid). In the cognitivist conception, the intention of the action and the beliefs or desires of the person appear as abstract causes of behavior in a wider causal framework. But there is a familiar and ordinary sense in which my judgment that someone wishes to say such and so, or intends or believes or desires something, is not a judgment about the (hidden) *causes* of his or her behavior.

We have a rich and very complex language for giving reasons for what we are doing, for describing human action and human reality from within, so to speak. The content of the expressions of this language, and its significance for us, derive from our familiarity with various human practices and situations, rather than from subsumption or arrangement of particular cases under generalities (as in *theories* of behavior or of mind). The notion of understanding what someone is doing or what someone intends, believes, desires, or feels, which belongs to this everyday "language of action" or "language of mind," is not the one that corresponds to causal explanation. The mental terms of this language of ours are not the theoretical terms of a "folk science." In this language, when we say that someone is doing something because she believes so and so, or because she desires this or that, our attitude to the person is the attitude that we normally have toward a *person*, an individual, who is acting as she does for her own reasons. It is not the attitude we take toward what we conceive of as a mechanism, or an organism, or some functionally defined physical system (as might be the attitude a doctor temporarily adopts toward a patient in examining him, for example).

It is part of our attitude toward an agent as a *conscious individual* that he or she should know the reason for his or her action. This is part of what we often mean by "reason," and then it is something that distinguishes the notion of reason for action from the notion of cause (see Johnston, 1989). Our interest in the reason for someone's action is our interest in the individual's own account of that action. We are interested in that reason when we want to make the action intelligible for ourselves as the action of a conscious individual, as the action of a human being. If we misconstrue the reason as a cause, it appears as though the reason were a report about something that occurred inside the person and caused the person's behavior, something about which the person could, in principle, always be mistaken.

Our ordinary psychological notions have the sense and significance they have for us only against the background of our experience of and familiarity with various practices, circumstances, and facts of human life. The internal dependence of our psychological notions on these experiences is cut off in the causal point of view. It is cut off by the adoption of a formal attitude toward our ordinary psychological language, and the treatment of it as a calculus of words and sentences having their (semantic) content in themselves as mere linguistic entities. It is in this formal attitude that we find the treatment of beliefs as "propositional attitudes."

Yet it might seem as if some cognitive scientists acknowledge this dependence of psychological notions on our experience of various human practices and on the world in which we live, when they assert the necessity of the level of explanation called "the knowledge level" (or "intentional level" or "semantic level"), where it is explained how mental representations "are about" what they are claimed to be about (Pylyshyn, 1984).[3] That conception of the dependence is doubtful, however, because these explanations construe the dependence as an external dependence seen from a spectator's perspective of the situations. Experience is usually construed as *knowledge*, with scientific or theoretical knowledge as the paradigm for this concept. But the experience and familiarity to which I am referring here is not knowledge in that sense. The experience of human realities that is crucial for the sense of psychological terms is not knowledge of causal relationships or of general laws for predicting and explaining behavior. It is rather familiarity with and experience of *what it is* to be in certain situations, *what it is* to do certain things or to be engaged in certain activities—that is, *what it is* when we conceive of these situations and activities as we normally do, and not according to some imposed observation language.

Cognitive scientists often contrast their point of view with behaviorism by opposing the latter's stubborn rejection of "intentional notions." But by introducing the knowledge level (or intentional level) of explanation, one has not really admitted and taken into account the "intentionality of human action," because this knowledge level is conceived within a conceptual framework where intentions, reasons, and mental states are construed as causes of behavior. As I have already suggested, it would be more correct to say that one has *extended* the behaviorist framework for causal explanation by giving up the radical empiricism of behaviorism and by adding a richer supply of possible causes of behavior.

The Idea of Folk Psychology

Part of what is involved in treating others as human agents or conscious individuals is expecting their actions to manifest a certain coherence and intelligibility, and presuming they will be able to offer an account in terms of reasons, intentions, desires, attitudes, etc. that will make their behavior intelligible to us in those (exceptional) cases where their behavior, as a matter of fact, appears unintelligible. But it has become part of the established practice in psychology and philosophy to treat such accounts as though they were disguised causal claims. The idea of "folk psychology" or "commonsense psychology," as cognitive scientists and philosophers of mind are currently using these terms, seems to me more or less the result of extending this causal view to our ordinary psychological language. For instance, the conceptual feature of an individual's account of his action, which may be described as his knowing what he is about to do, is misrepresented as the general empirical hypothesis or psychological law that "people generally do what they say they will do" (Fodor, 1987, p. 8).

The idea of folk psychology is the result of rendering our ordinary language about action and mind in causal terms, by focusing on the forms of words and sentences and ignoring the circumstances of their normal use. A formal procedure for paraphrasing and reading our everyday descriptions of people's actions and mental states is imposed, on the basis of a dichotomy between what people do and what *makes* people do what they do. In this reading, descriptions employing words such as "intention," "belief," "desire," and words for feelings and attitudes, seem to be nothing but causal explanations suggesting certain hidden mental causes of people's behavior. This way of thinking is another manifestation of what Leahey calls behavioralism. Our ordinary psychological language is conceived as though it were an unsystematic conceptual framework or an implicit, incompletely articulated theory for causal explanation of behavior.

This is not always as explicit or extreme as in the passage from Fodor referred to above, where he suggests that a piece from Shakespeare's *A Midsummer Night's Dream* is "a convincing (though informal) piece of implicit, nondemonstrative, theoretical inference" (p. 1). Fodor's object is to defend "good old commonsense belief/desire psychology" (as he calls it) that we use, he claims, "to predict one another's behavior." He wants to defend it against those philosophers of mind who think folk psychology is a "sterile theory." As an example of commonsense psychology's "predictive power," he says it tells us

> how to infer people's intentions from the sounds they make ("I'll be at your airport on the 3 p.m. flight," then, ceteris paribus, he intends to be at your airport on the 3 p.m. flight) and how to infer people's behavior from their intentions.

If someone intends to do something, our commonsense psychology allows us to infer, says Fodor, that the person will "produce behavior of a sort" that realizes the intention. The causal point of view, and the physicalist notions I have been questioning, are obvious in Fodor's use of expressions like "sounds they make" and "produce behavior." He adopts them without any doubt about whether it makes sense to use them in the non-scientific situations he claims to describe although, in my opinion, his account begins to look like a parody of philosophical explanation.

Folk Psychology as Cultural Psychology

We find an apparently different conception of folk psychology in Jerome Bruner's book *Acts of Meaning* (1990). There it has been detached from the computational conception of mind, and is conceived of as a basic ingredient of cultural psychology. The central concept of folk psychology so conceived is construction of meaning rather than processing of information. A folk psychology is a system of interpretive and organizing principles, on the basis of which the people in a culture interpret and organize their experiences and conduct their interactions with other people in social life. Bruner explains:

> All cultures have as one of their most powerful constitutive instruments a folk psychology, a set of more or less connected, more or less normative descriptions about how human beings "tick," what our own and other minds are like, what one can expect situated action to be like, what are possible modes of life, how one commits oneself to them, and so on. (1990, p. 35)

Obviously this is a different notion of folk psychology from the one or ones we find in mainstream cognitive science; but my question is, How different is it? Does it is represent a break with behavioralism? Does it respect the conceptual difference between reasons for action and causes of action? This is not at all clear, it seems to me.[4] As in classical cognitive science, there is still the idea that an element of organization and interpretation is *always* involved in our actions, and in what people see, hear, or feel. Obviously, a folk psychology is still conceived of as a cognitive system, a kind of mental representation, on the basis of which we can explain why the people of a culture behave as they do, even if this notion of explanation is not mechanist or naturalist. Bruner explicitly dissociates himself from the view that the causes of human behavior lie in biologically determined human nature, and asserts against this idea "that culture and the quest for meaning within culture are the proper causes of human action" (1990, p. 20).

This notion of a folk psychology, as much as the computationalist notion, is based on the idea that dichotomies such as

- What people do/what makes people do what they do
- Human action/the principles underlying human action
- Human action/the beliefs that determine human action

are universally applicable to human actions and practices, as much as the notion of folk psychology itself is claimed to apply universally to human action. This is true even if the words "make," "underlie," and "determine" are used in a sense that involves elements of "meaning-making" (e.g., interpretation and evaluation). But the idea that it is possible to explain or account for human action in general on the basis of these, or similar, dichotomies is a philosophical claim I want to question.

These dichotomies are not applicable to the practices of language use, which constitute, so to speak, the bedrock of our language. It makes no sense to try to explain, psychologically or in some other way, what *any* kind of explanation must presuppose, whether it be explanation that takes the form of formal theories or the form of narratives of a cultural psychology. It makes no sense to explain the many patterns of acting and behaving that constitute the base of our forms of language use. The sense of our ordinary notions and expressions is ultimately determined in these practices as *ways of acting*, and not as ways of organizing or interpreting experience. There is no problem of interpreting and understanding these actions, or making them intelligible, if that means just knowing how to do something, knowing what it is to be engaged in certain practices as an agent. We learn to call people by their names; we learn to describe the colors of objects around us; we learn the language game of reporting the time by means of a watch; we learn to count the number of objects in a box; and we learn to calculate according to a mathematical formula. Later on, such techniques will be parts of more complicated social practices and cultural contexts, where principles of interpretation, evaluation, and organization may be involved; but there is an ocean of patterns of acting and behaving that we have acquired as a part of learning language and learning to behave like normal human beings. Questions concerning the psychological causes or interpretive principles underlying these ways of acting are misleading, because these are the very bedrock of language. Language would collapse without them.

The Influence of Hermeneutics

The idea of folk psychology as cultural psychology, whose organizing principle is narrative rather than computational or theoretical, is inspired by the hermeneutic idea of understanding human life and action, as opposed to explaining the phenomena of nature. Current interest in this hermeneutic approach is connected with increasing awareness of the influence that natural science and economic and technological outlooks have had in our century. This influence has resulted in a tendency to favor forms and patterns of explanation connected with the above mentioned outlooks; this is true of psychology as well as of the human sciences. The dominance of computationalism in cognitive science is obviously one manifestation of this tendency. The hermeneutic tradition was, to a great extent, a revolt against the one-sided influence of the forms of explanation of natural science. And, in recent decades, this revolt has continued by the introduction of alternative approaches to the study of human life and action—approaches that are meant to do justice to neglected aspects, by taking historical, cultural, and literary points of view into consideration.

Bruner's notion of a cultural psychology is another manifestation of this revolt. But, as with other philosophical revolutions that involve the political motive of trying to change a certain scientific practice, one risks accepting too much of what one opposes—in particular if the struggle is successful. What tends to be unquestioned in these new hermeneutic approaches is *science as truth-teller*, science as a substitute for traditional metaphysics, science as the modern institution that has taken over the truth claims of traditional philosophy.

I want to question hermeneutic approaches to human science, insofar as they still aim at a *general* notion of what it is to understand human life and action—a notion that would manifest itself in the methods and principles of a cultural psychology, one task of which would be "the search for human universals."[5] The hermeneutic distinction between explanation and understanding is somewhat misleading from the point of view of the generality claims made, since it is a distinction between different kinds or forms of explanation corresponding to the differences between the natural and the human *sciences*. It is explanation in the sense that it imposes and applies a pattern or model for giving accounts of people's actions and ways of life, a model based on "human universals" that are supposed to apply to people of different cultures and societies, in the past as well as in the present.

The hermeneutic distinction is meant to be enlightening for those who are concerned about human action and mental life *as objects of scientific study*. The same thing is true, it seems to me, of Bruner's related distinction between "two modes of thought," the narrative mode and the paradigmatic mode (of which the latter is said to operate in theoretical argument and logical reasoning). These are "two modes of cognitive functioning . . . each providing distinctive ways of ordering experience, of constructing reality. The two (though complementary) are irreducible to one another" (Bruner, 1986, p. 11). What Bruner is giving us is a rough description of the difference between two intellectual traditions (manifest in our education systems and academic organization) that, according to him, also have psychological reality as basic "cognitive functions." I would say rather that these two modes of thought and their separateness are something that we acquire through the training and education we get in our Western culture, and not something that is rooted in a certain kind of psychological necessity. Bruner's distinction, as well as the hermeneutic one, it seems to me, rest on a political point. Bruner is concerned to widen the sense of scientific study to include literary and hermeneutic aspects, and to change the scientific practices of psychology accordingly. Against the psychology that is too much influenced by natural science, and that even wanted to *be* natural science, he launches a new form of scientific study in psychology. But, as in other attempts to invoke hermeneutic ideas in human science, the endeavor is governed by an interest in human action and mental life as objects of scientific explanation—though in a new and widened sense of "science." The interest in the human mind and human action is still a scientific interest with its characteristic cravings for general principles and notions that apply uniformly.

Current philosophical vocabularies, which employ notions such as "the conceptual scheme or system of beliefs of a culture," "the moral views of a society," "the narrative of a society that contains its self-conception" and the like, are forms or patterns of explanation as much as the forms of explanation they were intended to replace or supplement. Cultures, societies, narratives, and folk psychologies are spoken of as systems, as delimited objects of study, in about the same schematic way as philosophers of science talk about scientific theories and conceptual systems. The notions are not introduced in relation to a problem of accounting for some *particular* piece of human behavior of some particular culture or society, but are meant to apply uniformly to any culture, any society, and any human practice on the basis of the dichotomies previously mentioned. These systems are supposed to contain the underlying principles that *make* people do what they do. The ordinary language we use to talk about action and mind, in which we speak as agents and participants in activities and communities, and give reasons for our actions, does not work that way. It does not rest on the notion of generality that belongs to the intellectual or scientific enterprise of explaining what it is to understand human action and mental life in general.

A cultural psychology in Bruner's sense wants to determine a general notion of "how human beings 'tick,' what our own and other minds are like, what one can expect situated action to be like, and what are possible modes of life" (Bruner, 1986, p. 35). Bruner expresses himself as though human life and action were governed by an underlying cultural or hermeneutical (if not causal) mechanism. In trying to account for the way this mechanism works, for the ways of "meaning-making," we would be forced to conceive of human life *in general* as a closed and delimited phenomenon, as though it were the life of a people of a foreign culture presented in an historical account, or as the life of persons described in a finished story. But the human life that includes our own lives, the lives we live, is not a closed and delimited phenomenon in that sense; and the same is true of course of the culture in which we live. Psychological thinking that emanates from *this* fact, like the psychological insights expressed in the novels and literary works of Dostoyevsky and Kafka, for instance, is something quite different. Like other authors, they were writing for individuals who could experience the situations expressed *very concretely* on the basis of the readers' experiences in their own personal lives. It seems to me that we should be very cautious with the hermeneutic idea that novels and literary works contain something in addition to what they say to their intended readers, that they also communicate general, impersonal psychological knowledge or general knowledge of human life, in the academic sense of "knowledge." Most literary works were not written with the subsidiary motive of satisfying an interest in human action and mental life as objects of scientific study. Many of them, rather, were rather written *against* too strong an influence of that interest, and against the bad consequences of it.

The notion of a folk psychology as a "folk social science" is one such questionable consequence. What is needed for the future development of cognitive science is a conceptual clarification of the nature of our actual, everyday language of human action and mental life. The temptation to exaggerate the intellectual perspectives that Bruner calls "two modes of thought," and to conceive of our everyday psychological language either as an informal scientific language or as an imperfect literary language, can only be resisted by taking a closer look at the facts of our ordinary use of psychological terms.

Notes

1. Other ideas in the philosophy of language that depend on a technical notion of observation rooted in scientific empiricism are Quine's so-called principle of indeterminacy of translation (1960) and Kripke's ideas about rule-following (1982). Quine's argument is carried out within a strict causal and empiricist framework; and Kripke's sceptical conclusions obviously depend on his conceiving rule following in terms of an agent's input/output behavior.

2. Similar problems rooted in the empiricist and naturalist notions of linguistic expressions are discussed in Stenlund (1990).

3. Pylyshyn attributes the term "knowledge level" to Allen Newell.

4. Further reasons for this doubt can be found in Shanker (1992) and (1993).

5. Bruner (1990, p. 20. We are urged to accept the general principle that "[All] cultures have as one of their most powerful constitutive instruments a folk psychology," and Bruner has a lot to say about the general structure of such a folk psychology. We are told, for instance, that its organizing principle is narrative, and Bruner refers to Paul Ricoeur, who likewise has a lot to say about the structure and function of narrative in general (Ricoeur, 1981).

References

Bever, T.G. 1988. "The Psychological Reality of Grammar: A Student's-eye View of Cognitive Science," in W. Hirst (ed.), *The Making of Cognitive Science. Essays in Honour of George A. Miller.* New York: Cambridge University Press.

Bruner, J. 1990. *Acts of Meaning.* Cambridge: MIT Press.

Bruner, J. 1986. *Actual Minds, Possible Worlds.* Cambridge: Harvard University Press.

Bruner, J. 1973. *Beyond Information Given. Studies in the Psychology of Knowing.* London: Allan & Unwin.

Chomsky, N. 1964. "A Review of B.F. Skinner's "Verbal Behavior,'" in J. Fodor and J. J. Katz, *The Structure of Language. Readings in the Philosophy of Language.* Englewood Cliffs, NJ: Prentice-Hall.

Chomsky, N. 1965. *Aspects of the Theory of Syntax.* Cambridge: MIT Press.

Fodor, J. 1987. *Psychosemantics. The Problem of Meaning in the Philosophy of Mind.* Cambridge: MIT Press.

Fodor, J. 1968. *Psychological Explanation: An Introduction to the Philosophy of Psychology.* New York: Random House.

Johnston, P. 1989. *Wittgenstein and Moral Philosophy.* London: Routledge.

Kripke, 1982. *Wittgenstein on Rules and Private Language.* Oxford: Basil Blackwell.

Leahey, T.H. 1987. *A History of Psychology. Main Currents in Psychological Thought,* 2nd ed. Englewood Cliffs, NJ: Prentice-Hall.

Pylyshyn, Z. W. 1984. *Computation and Cognition. Toward a Foundation of Cognitive Science.* Cambridge: MIT Press.

Quine, W.V. 1960. *Word and Object.* Cambridge: MIT Press.

Ricoeur, P. 1981. "The Narrative Function," in P. Ricoeur, *Hermeneutics and the Human Sciences.* (trans. and ed., J.B. Thompson) Cambridge: Cambridge University Press.

Shanker, S. 1992. "In Search of Bruner." *Language & Communication,* 12, 1.

Shanker, S. 1993. "Locating Bruner." *Language & Communication,* 13, 4.

Stenlund, S. 1990. *Language and Philosophical Problems.* London and New York: Routledge.

Wexler, K. 1991. "On the Argument from Poverty of Stimulus," in A. Kasher (ed.), *The Chomskyan Turn.* Oxford: Basil Blackwell.

22

John Shotter

Cognition as a Social Practice

From Computer Power to Word Power

. . . in philosophy we often compare the use of words with games and calculi which have fixed rules . . . as if it took the logician to show people at last what a proper sentence looked like. . . . All this, however, can only appear in the right light when one has attained greater clarity about the concepts of understanding, meaning, and thinking. For it will also become clear what can lead us (and did lead me) to think that if anyone utters a sentence and means or understands it he is operating a calculus according to definite rules. (Wittgenstein, 1953, no.81)

The real foundations of his enquiry do not strike a man at all. Unless that fact has at some time struck him.—And this means: we fail to be struck by what, once seen, is most striking and most powerful. (Wittgenstein, 1953, no.129)

Over the last 25 years or so, academic psychology has undergone what many take to be a profound and exciting revolution, variously termed "the mind's new science" (Gardner, 1987), "a major shift in metatheory" (Baars, 1986), or simply "the cognitive revolution" (Gardner, 1987; Baars, 1986). The traditional behaviorist paradigm has been almost wholly abandoned; talk of the mental is no longer scorned as "subjective" and "unscientific"; there is a revival of interest in "the mind." Indeed, many see it as representing the gaining of a maturity that psychology has lacked until now. This is because, as Baars (1986) says in his characterization of this revolution: "It is only theory that gives a sense of unity to physics, chemistry, and the other mature sciences. If the major thesis of the book is correct, we may be moving today toward the first major shared theory for psychological science" (p. 395). Although psychology may not yet have a finally agreed theory of mind, there is now, seemingly, widespread agreement on the sphere of metatheory from which theoretical constructs *should* be drawn. Everyone seems to have something to say about "the mind" in terms of "information processing," "representation," "computational processes," "memory," "retrieval," "encoding and decoding," "stages or levels," "knowledge structures and organization," "decision strategies," "language structure,"and so on.

However, I want to be harshly critical of precisely the hope that "in a matter of decades we shall have a *dominant* theory of human mental processes" (Baars, 1986, p. 415). Baars claims that "the cognitive view of the organism may well shake our conception of ourselves, whether rightly or wrongly, but so far, compared with the previous way of looking at people, the new perspective has increased rather than decreased human self-respect (e.g., Boden, 1977)." To me, the opposite seems the case: To the extent that cognitivism constitutes "a first theoretical integration," "a unifying theory," it makes it even more difficult for alternative voices, speaking words from other vocabularies, to be heard within the arena of psychological debate about ourselves. While the self-respect of those within the field may have increased, their respect for those outside it—that is, the rest of us—has markedly decreased. Others have little chance of being heard and listened to seriously at all. The (unconscious) arrogance of workers in this field needs experiencing to be believed.[1] Thus, what this "revolution" means—the institution by cognitive scientists of an already decided and limited vocabulary for our psychological discussions—is that instead of the rest of us being able to negotiate between ourselves what we mean when we communicate—thus, to talk of our psychological being in our own words—we begin to find that we must talk of ourselves in *their words*, in terms of *their* already-agreed criteria for reaching agreements.

And this, I will argue, leads us to ignore the importance (among many, many other aspects of our lives) of our contacts with each other as embodied living beings, and thus also to neglect those moments in our lives in which we sense our ethical duties to each other, those moments when we cannot refrain from reacting bodily to those around us. As Wittgenstein remarks here: "It is a help to remember that it is a primitive[2] reaction . . . to pay attention to other people's pain-behavior, as one does *not* pay attention to one's own pain behavior" (1980, no.540). Breaking that contact, separating ourselves from our bodily responses to each other's conduct in our communications, can—as Bauman (1989) shows in describing the procedures devised in the bureaucratic implementation of the Holocaust—make the recognition of evil difficult to detect. Simply put, the issue is a matter of the extent to which, and in what context, we are or should be influenced in our lives by calculations or, by the free play of embodied voices, speaking in dialogue with each other.

Language-Games and Forms of Life

To set the scene for exploration of these issues—that is, to grasp the importance of our embodied presence to each other—let me begin by pointing out the obvious: that no one learns such orderly disciplines as cognitive science or computer science *de novo* as a child, prior to learning one's mother tongue, or before learning to write and to do other much more simple things like naming, counting, and calculating with numbers. As a special discipline or practice, it is a vocation, something into which one, only as an already developed member of a social group, can be "called." "One forgets," says Wittgenstein, "that a great deal of stage-setting in the language is presupposed if the mere act of naming is to make sense" (1953, no.257). To answer such a "call," one must already be someone to whom it makes sense to issue commands and instructions, to point things out, to name them, and to make assertions about them, etc. If one cannot respond to such talk (or read such writing), it makes no sense to those within the academic group to attempt to "instruct" one as a novice into the discipline's "mysteries," into the things its practitioners claim to be revealing as hidden in or behind the everyday life appearances with which we are familiar. Thus, central to all such callings are the special speakings (and writings) used: injunctions (authoritative commands such as "look at this"), namings ("this is a triangle"), imperatives ("count"), assertions ("2 times 2 equals 4"), accounts ("this is justified because . . . "), descriptions ("this unknown circumstance is of *this* kind"), attributions ("so and so is good, so and so is bad"), and so on. (See, e.g., Rotman, 1993.)

Some of these speakings are used in the more informal, supplementary, and epiphenomenal talk surrounding the more privileged domain of formal or professional talk within which an academic discipline usually situates itself, while others only have their use within that central domain (Rotman, 1993; Shotter, 1993a). In other words, those within cognitive science see it (similar to the way mathematicians see mathematics) as being solely *about* a supposedly special extralinguistic domain of events, discovered as existing in a "reality" beyond and underlying the "mere appearances" of our everyday life affairs. We can see it another way: as an imaginary reality created and sustained by a special kind of language-game (to use Wittgenstein's 1953 term), as a special kind of writing that *some* of us have constructed from within our already existing, everyday linguistic involvements with each other—from within, again to use Wittgensteinian terms[3], the already existing "bustle" or "hurly-burly" of our everyday, embodied forms of life. Indefinite and chaotic though this background activity may be, living embedded within it, we have developed into the kind of ordinary, everyday persons we are, embodying most of the kinds of reactions and responses, competencies and abilities, required to "go on" within it. And we make use of these abilities spontaneously and unthinkingly in everything else we do—even in the intelligible formulation and testing of our scientific theories. "What has to be accepted, the given, is—so one could say—*forms of life*," he says (1953, p. 226e). Only from within a form of life does our use of words make sense.

Wittgenstein (1953) introduced his notion of a language-game in an effort to break the stranglehold of that view of language in which it functions solely representationally, in which words (in themselves) stand for things (in themselves) outside a human form of life. For Wittgenstein, the point of focusing on language-games is that "it disperses the fog to study the phenomena of language in primitive kinds of application in which one can command a clear view of the aim and functioning of the words" (1953, no.5). They show how words might function in a reactive or immediate way to direct and control people's behavior. He begins by discussing a complete primitive language consisting of just the four words, "block," "pillar," "slab," and "beam." "In the practice of the use of [this] language one party calls out the words, the other acts on them . . . . I shall call the whole, consisting of language and the actions into which it is woven, the 'language-game'" (no.7). Associated with a language-game is also a form of life, in the sense that those who live in it (1) have certain limited identities and subjectivities in relation to each other, for example, a "builder" and his "assistant"; (2) occupy a certain "world," for example, of "blocks," etc.; and (3) have certain goals, motives, aims, or inclinations in life, for example, "to build," and its associated sub-goals and sub-motives. Thus, "to imagine a language is to imagine a form of life," he says (no.19). And it is not difficult, of course, to imagine many other such limited language-games with their associated forms of life—

cognitivism, with its limited, computational, information processing vocabulary, being just such a case in point.[4]

What I want to explore is the power of words to "call" us into such limited language-games. What is involved is not the power of patterns of already spoken words to represent already existing states of affairs[5], but another function of our words altogether: the power of our words in their speaking to "move" the others around us, and, in so doing, to create a sense of a reality within which both we and they are involved—what we might call the rhetorical-responsive function of words. Central to the stance I will take is a focus on that flow of living contact between speakers and their interlocutors, in which they cannot fail to react or respond to each other's utterances. It is their bodily "movement" in their speaking that is important. For simply presenting each other with already completed shapes or forms (representations) need evoke no response at all. It is in that flow of activity that they must try to negotiate how, practically, to coordinate what they do, with the activities of the others around them. They must try to "go on" living with them. This contact is central, I claim, for it is in people's use of words (in their reacting and responding to each other in everyday, practical circumstances), that they give a shape to the movements occurring between them, to create between them a sense of their speakings as being "about" something. Indeed, as Wittgenstein sees it, it is in such original circumstances as these, in which our speech is an extension or refinement of our gestures, pointings, or other bodily activities, that we first learn the different meanings, the different uses, of our words (and of our other forms of behavior). Hence his claim that: "Only of a *living* human being and what resembles (behaves like) a living human being can one say: it has sensations; it sees; is blind; hears; is deaf; is conscious or unconscious" (1953, no.281, my italics).

To reinforce this emphasis on the living nature of the reactions or responses involved here, let me quote G.H. Mead who seems to put the matter well when he says: "The mechanism of meaning is present in the social act before the emergence of consciousness or awareness of meaning occurs. The act or adjustive response of the second organism gives to the gesture of the first organism the meaning it has" (1934, pp. 77–78). As second organisms ourselves, the "meaning" we give to the speech of a first-person other in a particular context is developed, so to speak, in how we react or respond to them practically. And we "show" what this meaning is for us, in how we react or respond to such speech—where what can go wrong here is not so much that we find another's speech false, but instead misleading or senseless, that is, we cannot follow it through and it leads us into confusing or conflicting reactions.

Wittgenstein (1953) called his investigations into how we create anticipations and expectations in our practical or living use of words in different contexts "grammatical investigations." But the nature of such investigations is not easy to grasp. For, to the extent that we must shape our use of words (our responses) to fit into a context already partially shaped by previous speakers, their use is a joint outcome influenced both by us and by others. We cannot simply determine it ourselves. It is not "mine," nor "yours," but "ours." Or, to put it another way, it belongs to the circumstances of our talk. "It is *particular circumstances* which justify me in saying I can go on," he says (1953, no.154). "Our talk gets its meaning from the rest of our proceedings" (1969, no.229). " 'Obeying a rule' is a practice. And to *think* one is obeying a rule is not to obey a rule" (no. 202). Thus, all we can determine in our contemplative studies of our use of words is the "possibilities" (1953, no.90) they make available to us, for how we might react or respond to them, that is, what we might expect or to try to do, and whether, in certain contexts of their use, they might be misleading.

In this respect, of course, Wittgenstein was particularly concerned to bring to our attention the fact that in our use of words, it is only too easy for us to "make" or "shape" a sense of reality that does not in fact exist. For, to repeat, due to the joint nature of our responsive use of words, we can create in our talk the sense of an external reality or circumstances full of certain "things" or "entities" (as I think *is* the case in cognitive science) without any awareness of having done so. In such circumstances, we can tend mistakenly to "predicate of the thing what lies in [our] method of representing it" (Wittgenstein, 1953, no.104).

The Cognitive Revolution and Its Technocratic/Bureaucratic Ideology

What I want to do, then, is explore how in our speakings with each other we can construct and sustain in existence, by appropriate forms of talk, such limited forms of life with their associated "sense of reality." For I want to explore cognitivism, cognitive science, and the cognitive revolution, as instituting within psychological debate just such a limited language-game—limited (just like the language of the builder and his assistant) by the practicalities of the form of life with which it is associated. Currently, it seems to me, those practicalities are of a technocratic/managerial/bureaucratic kind, aimed at the goals of efficiency, predictability, quantification, and control. But are these the aims of cognitive science? Is this the form of life imagined in its language-games? For such a limitation does not seem apparent from within the cognitive revolution itself.

Returning to Baars' account of the cognitive revolution, we can note his remark that, "after reading the 17 interviews in this book [with the cognitive revolution's leaders], one is left with the strong impression that the psychological community is marked more by diversity than consensus" (p. 406). Indeed, he finds it difficult to say in what, precisely, the cognitive revolution consists. And, in fact, one finds few recent textbooks that contain a clear definition of it. Most now seem to begin with the assumption that readers already will know what is being discussed. As Eysenck (1984) also remarks, "a part of the reason for the growing army marching behind the banner of cognitive psychology is the increased vagueness with which the term is used" (p.1). And, like Baars, he also remarks on the diversity it allows. Thus, if the diversity of which Baars and Eysenck speak is real, and an agreed theory of mind does not yet exist, in what does the cognitive revolution actually consist? Further, if its basic ideas are so vague, how do all those involved in it create a shared sense of being involved in the same enterprise, and see those—like myself—who criticize it as "living on another planet" (as a cognitive scientist said to me recently)? Something more like an ideology than a theory seems to be involved in holding it together as a unity (Sampson, 1981, 1993). Implicit in it is what Wittgenstein would call a "grammatical picture," one that represents the seemingly fundamental (metaphysical) nature of our reality. What is that "picture" in cognitive science?

In attempting to outline its nature, it will be relevant to examine the amalgam of general notions—drawn unexamined from common sense, philosophy, communication and information theory, and computer science—within which cognitivism has its life. It apparently takes for granted, for instance, the current commonsense conception of the mind as a container of things mediating between the self of the person and reality at large—where these three realms are treated as being ontologically distinct and physically separated from each other. Language too, it seems, is a distinct, self-contained entity with an existence of its own, separate from the reality it is said to represent, and is made up of a collection of independent elements externally linked to each other by a set of rules, that is, a code. In this view (although some now consider it limited in the light of parallel distributed processing [PDP]), it is the function of language to convey information about states of affairs symbolically (or at least, in some other univocal, deterministic way). It thus overcomes this set of separations, the lack of any living (embodied) contact between the inner and the outer, the subjective and the objective, between persons themselves, and persons and their surroundings.

This isolation of everything from everything else typifies cognitivism. For the lack of direct, living contacts inclines it toward a "spectator at a distance" view of knowledge, and necessitates its central focus on inner mental representations to "depict," "picture," or otherwise "re-present" states of affairs in a person's environment. Gardner (1987) captures the import of all these concerns in his account of the mind's "new science."[6] "First of all," he says,

> there is the belief that, in talking about human cognitive activities, it is necessary to speak about mental representations and to posit a level of analysis wholly separate from the biological or neurological, on the one hand, and the sociological or cultural, on the other. . . . [Indeed, there is] a deliberate decision to de-emphasize certain factors which may be important for cognitive functioning but whose inclusion at this point would necessarily complicate the cognitive-scientific enterprise. These factors include the influence of affective factors or emotions, the contribution of historical and cultural factors, and the role of the background context in which particular actions or thoughts occur. (p. 6)

The only realm he fails to mention is one already so de-emphasized in modern life that theorists often fail to notice it at all: the moral realm. Thus, as a complex of general ideas, the cognitive revolution or cognitivism in psychology can be summed up as follows: Behind or underlying what individual people do or say are supposedly hidden orderly processes taking place "in their minds." These processes work in terms of codes or symbols with already decided ways of representing external reality. They thus can be of a computational, calculational, or information processing kind, in which new meanings or mental representations are made by rearranging components of old meanings in an orderly way. And with enough research into the nature of such processes, we will be able to explain all human conduct. Lachman, Lachman, and Butterfield (1979) put it thus:

> Computers take symbolic input, make decisions about the recoded input, and give back symbolic output. By analogy, that is most of what cognitive psychology is all about. (p. 99)

Central to such procedures are codes, rules, or other conventions linking symbols to their already decided meanings. Assigning pre-established meanings to the words and other signs that make up the "information" received by a person eradicates the (inefficient) living moment of free interaction, in which people negotiate meanings between themselves. In other words, such a view has its application in a world in which self-contained, isolated, atomistic individuals—without any ethical feeling for others, without a background context, without a history, without a culture—live their lives, by reasoning in an orderly, instrumental, problem-solving fashion about, how, quantitatively, it is best for them to live.

It would be difficult to find a more obvious parallel to Weber's (1964) account of the embodiment of a formal, instrumental rationality in the structuring of a modern bureaucracy. Cognitivism achieves the main aims of bureaucratic organization— efficiency, predictability, quantification, and control—first, by the substitution of nondecision-making human functionaries for decision-making persons, and then by substituting nonhuman for human functionaries. In other words, its central claim is that all human behavior can be mechanized, that there is nothing distinctively human that machines cannot be programmed also to accomplish. As human beings, it seems that we can be excluded from the world as we know it without noticeable consequences! For in a world of already decided meanings, there is no need for us.

Indeed, this is precisely the argument Dennett uses (1978) to account for what he calls "the remarkably fruitful research strategy" of the artificial intelligence (AI) programmer. In a world consisting solely of a set of basic objects and their different configurations, all "intelligent" activities may be executed by a hierarchy of functionaries. The programmer's

> first and highest level of design breaks the computer down . . . into a committee or army of intelligent homunculi with purposes, information and strategies. Each homunculus in turn is analyzed into smaller homunculi, but, more important, into less clever homunculi. When the level is reached where the homunculi are no more than adders and subtractors, by the time they need only the intelligence to pick the larger of two numbers when directed to, they have been reduced to functionaries "who can be replaced by a machine" . . . [and] if the program works, then we can be certain that all homunculi have been discharged from the theory. (pp. 80–81)

Such a hierarchy of functionaries, he claims,[7] is what all intelligent activities *are*. Indeed, in this respect, it is worth exploring further the character of the general technocratic/bureaucratic ambience within which cognitivism as a whole is embedded— and the way in which it is concerned to render "invisible," or altogether eradicate, the living moment of interaction of which I spoke above.

A further vivid example of this can be found in an account by Broadbent (1980)—one of the British innovators of the cognitive revolution. He argues (Broadbent, 1958)—that in the research of "an applied experimental psychologist . . . one should not start with a model of man . . . [but] from practical problems" (p. 117). In his account, he first describes, in the abstract, the information processing model that, he claims, has arisen from *his* practical concerns. He then adds:

> If you have found the abstract description . . . difficult to follow, you can (*at some risk*) think of it this way. Imagine a man sitting in an office. On one side of him is his 'in-baskets,' into which people keep putting pieces of paper. . . . On the other side of the

man are his 'out-baskets,' into which he puts papers which are going to leave the office. . . . If the man wants to keep information more permanently, and without blocking the use of his baskets, then he uses a filing cabinet which is placed behind him. . . . [And] in addition the man has in front of him a desk on which he can put papers which he is using at the moment. (p.125; my emphasis).

The techno-bureaucratic form of activity, on which his model of man is modeled, could not be more plain.

But why Broadbent feels that it is risky to personify the information processing involved in talk of this kind is even more revealing. "The reason it is risky," he says, "is that we are trying to model man, and yet there is a man in the model; if we are trying to analyze human function, we must not appeal to the thing we are trying to analyze. [But] the processor in the middle [of the model] need not be a complete man with all his properties of foresight, memory, reasoning or purpose. It can be a totally passive robot which merely obeys instructions found in its four stores [in-baskets; out-baskets; filing cabinet; and desk], at the moment when it reads them" (p.125). In other words, Broadbent is not concerned with whether people are actually being treated as passive robots, but with whether that which makes them unpredictable, unquantifiable, uncontrollable, and thus inefficient, has been fully eliminated. For only then (paradoxically) will we be in a position to claim to have theories dealing with the whole human being. And to the extent that Broadbent feels he has all but succeeded in this then, like other cognitive psychologists, he claims that "at some point during the last decade or so we reached a major threshold in the development of psychology" (p.126). Circumstances now are such that, "For the first time in the history of the subject, we now have a rough and ready framework for the working of the human system as a whole" (p.126).

Minsky too, in his *The Society of Mind* (1986), sees bureaucratic[8] heirarchies as the solution to the understanding of intelligent, human activities. This book assumes," he says (p. 322), "that any brain, machine, or other thing that has a mind must be composed of smaller things that cannot think at all." Thus, in discussing how a simple agent called *Builder*[9] executes its task of building something, he points out: "As an agent, *Builder* does no physical work but merely turns on [the agents] *Begin*, *Add*, and *End*. Similarly, *Add* just orders *Find*, *Put*, and *Get* to do their jobs. Then these divide into agents like *Move* and *Grasp*. It seems that it will never stop—this breaking down into smaller things. Eventually, it all must end with agents that do actual work, but there are many steps before we get to all the little muscle-agents that actually move the arms and hands and finger joints. Thus *Builder* is like a high-level executive, far removed from those subordinates who actually produce the final product. Does that mean that *Builder's* administrative work is unimportant? Not at all. These lower level agents need to be controlled. It's much the same in human affairs" (p.34).

For human affairs in general, then, the single, unified theory for cognitive science would seem to offer hierarchies of more and more mindless functionaries, knowing less and less about what they are doing and why. But what of our human freedom to decide, not only what to do, but also what kind of people to be? For, as Minsky (1986) asserts, "everything that happens in our [whose?] universe is either completely determined by what's already happened in the past or else depends, in part, on random chance. Everything, including that which happens in our brains, depends upon these and only these: (i) A set of fixed deterministic laws; (ii) a purely random set of accidents. There is no room in the middle on either side for any third alternative" (p.306). Does that mean that we should "put aside the ancient myth of voluntary choice? No. We *can't* do that: too much of what we think and do revolves around these old beliefs. . . . No matter that the physical world provides no room for freedom of will: that concept is essential to our models of the mental realm" (p.307). So again, the moment of interaction—in which those involved can decide their meanings for themselves—is eradicated here in favor of any "instructions" they might receive in an already decided code, supposedly arranged to provide an optimum means toward an already given end. But precisely in this elimination of those moments, in which the people involved decide the circumstances of their lives, they are dehumanized.

Why do the theories proposed by different cognitive scientists all seem to be the same? Because, I want to suggest, they draw their sense from the same "grammatical picture," from the image of a "community" (or system) of functionaries or mechanisms that in themselves are mindless. The nearest living example is the image of the modern bureaucracy or corporation. This seems to be the image implicit in theorists' ways of talking that works to draw their attention to what they feel it is important to model or explain. They seem bewitched by it, unable to talk in other terms. Wittgenstein described his own bewitchment by a related "picture" (the "picture theory of meaning") thus:

> A *picture* held us captive. And we could not get outside it, for it lay in our language and language seemed to repeat it to us inexorably." (1953, no.115)

Far from worrying about how they might escape their entrapment in it, cognitive scientists celebrate it. It does, however, raise some difficulties.

Relational Ethics and the Holocaust

I claimed that only in people's living contact with each other, through the "movements" embodied in their responsive speakings do they get a sense of each other as living human beings. It is in their embodied reactions to such movements that people get a sense of each other's *being*, of each other's identities, of who and what they are (or could be) for each other within the current circumstances of their talk. Indeed, we must note here the importance of Wittgenstein's (1953, 1980, vols. 1 & 2) emphasis on people's *right* to make first-person claims (declarations or avowals) about themselves ("I feel afraid," etc.), as well as to exhibit "expressions" (*Äusserungen*) of pain, surprise, fear, etc. It is not just a matter of us, in our pain-behavior and so on, informing each other how the world seems to us, but something much more. For in our responsive talk we are concerned, actively, not only properly to address those to whom we talk, but also to be answerable for our own, unique, momentary sense of our "position" in existence—in the hope that they too will be answerable for theirs. "Being sure that someone is in pain, doubting whether he is, and so on, are so many natural, instinctive, kinds of behavior towards other human beings, and our language is merely an *auxiliary* to, and further *extension* of, this relation" (Wittgenstein (1981), no.545). In other words, it is in our responsive bodily presence to each other that we exhibit the forms of respect for each other's unique being that is called *ethical*: the forms of respect beyond mere convention, in which (like mothers with their babies) we respond to the hints and tendencies in the behavior of others that enable them to realize themselves.

In only being allowed to act as functionaries in relation to each other, in being only presented with, and in being only allowed to present, finalized (already coded or conventionalized) formal meanings, we are denied access to just such living forms of life. The flow of living contact with each other, in which our concern for each other's (and our own) *being* makes its appearance, is eradicated. If this is true of cognitivism, if the "true theoretical language, one that can naturally represent the elements of psychological reality" (Baars, 1986, p.399) is of such a kind that it provides no place for such moments, then it needs to be vigorously contested. With its obsession with separating and isolating everything from everything else, and its determination to obliterate all living (embodied) contacts between us, it can estrange us from ourselves. The promised theoretical integration (and exclusion of competing voices) can lead to the exclusion of just those moments and spheres of activity in our lives in which we treat each other ethically. (See Shotter, 1984; 1993a and b; 1995.) The consequences of this need not be evil, but they can be. For, as such moments are eliminated, this results in fewer occasions on which our care for each other can be aroused. Nowhere were the consequences of this more appalling than in they way it was used in managing the "industrialized" murder of Jewish people in Nazi Germany.

Bauman (1989) discusses many of the "procedural" problems that arose in administrating the "process": "The most shattering lessons deriving from the analysis of 'the twisted road to Auschwitz' is that—in the last resort—*the choice of physical extermination as the right means to the task of* Entferung [the achievement of a *judenfrei* Germany] *was a product of routine bureaucratic procedures*: means-end calculus, budget balancing, universal rule application. . . . The 'Final Solution' did not clash at any stage with the rational pursuit of efficient, optimal goal-implementation. On the contrary, *it arose out of a genuinely rational concern, and was generated by bureaucracy true to its form and purpose*" (p.17). Central to its implementation was, as Bauman terms it, "the social production of moral indifference: for, as I have already indicated, it was not an intensification of hatred that was important, but a lessening of care. In the famous phrase of Hannah Arendt, the most difficult problem that the initiators of the *Endlösung* encountered (and solved with astounding success, as it were) was 'how to overcome the animal pity by which all normal men are affected in the presence of physical suffering' (Arendt, 1964, p.106)" (pp.19–20). Successive "improvements" in the technology of killing were required. At the *Einsatzgruppen* stage, the rounded-up victims were killed at point-blank range in front of machine guns. But with this method, it was difficult to sustain the morale of the perpetrators, and another was sought. This led to the invention, first of mobile, and then to stationary gas chambers: here, the role of the killers was reduced to that of "sanitation officers." As Bauman details, the problem was solved by the production of a bureaucratized separation between victims and their murders,where the precise point of the bureaucratization was the social production of a distance, a "separation" between people, so that those involved could simply say, in Eichman's infamous phrase, "I was only doing my job"!

Wittgenstein and "the Moment of Interaction"

Once produced, such bureaucratic activities involve functionaries who have little further need to think at all. However, as I pointed out above, even if one were interested, he or she in any case would find it difficult to become aware of how one's own actions in fact related to the social outcomes produced. For we are not ourselves, as individuals, wholly responsible for the outcomes produced in our *joint* activities. But, if this is so, how might we study our contribution to such activities? Here, I think, it is relevant to reflect on Wittgenstein's methods for drawing our attention to the nature of our activities in such circumstances. For no matter how distanced we might be from each other, no matter how formal our communications, there still is a point at which we face the task of understanding how, by the *use* of words (written or spoken), we can be "moved" in some way. We need to grasp how in the *living moment* of their use, our words can exert an influence on us in the social activities into which they are interwoven.

Currently, as I made clear above, we tend not to pay attention to this "moment of interaction." "It all goes by so quick, and [we] should like to see it as it were laid open to view" (Wittgenstein, 1953, no.435). But instead of attempting to develop a set of "tools" appropriate to that task—the task of "seeing" (*describing* precisely) what we are doing, in what we are doing, in our different particular uses of words— we fall victim, Wittgenstein suggests, to a temptation to search beyond or behind the phenomena for something else that will explain it. Thus, here, he says (1953, no. 436): "It is easy to get into that dead-end in philosophy, where one believes that the difficulty of the task consists in our having to describe phenomena that are hard to get hold of, the present experience that slips quickly by, or something of the kind, where we find ordinary language too crude, and it looks as if we were having to do, not with the phenomena of every-day, but with ones that 'easily elude us, and, in their coming to be and passing away, produce those [crude] others as an average effect.' (Augustine)." Hence, when we reflect on our everyday talk, say, of a plan or an expectation, we feel that there must be a *something* beyond them that they are a plan or an expectation *of*. And only if we could get a picture of *that*, could we understand why we talk about our plans and expectations as we do: for then we would know what they *represent*.

However, Wittgenstein suggests (1953, no.435), "If it is asked: 'How do sentences manage to represent?'—the answer might be: 'Don't you know? You certainly see it, when you use them.' For nothing is concealed." In other words, while it is true that everything goes by so quickly that we find it difficult to attend to all the influences "shaping" the nature of what we say, nonetheless, our talk is shaped by no one else but ourselves. Thus, says Wittgenstein (1953, no.373), "Grammar tells us what kind of object anything is." We can come to a grasp of the influences at work, for, "nothing is hidden" (no.435). We shape the words we use according to circumstances; and it is in its shaping that its use is revealed. "Every sign *by itself* seems dead. *What* gives it life?—In use it is *alive*" (no.432).

Thus, rather than for theoretically "picturing" states of affairs, Wittgenstein uses words practically, as instruments, to draw our attention to features in the momentary circumstances of our own use of words, features that until now we have failed to notice. For example, very near the beginning of his *Philosophical Investigations* (Wittgenstein, 1953), he asks the following question: "Suppose someone points to a vase and says: 'Look at the marvelous blue—the shape isn't the point.'—Or: 'Look at the marvelous shape—the color doesn't matter.' Without doubt you will do something *different* when you act upon these two invitations. But do you always do the *same* when you direct your attention to the color?" (no.33.) And we are very easily tempted to answer: "Yes, of course we must do the same thing. Paying attention to just a specific color (or to just a particular shape) in such circumstances is paying attention to the *same* color (or shape), isn't it?" But, by arraying a set of everyday example situations, he shows us that in seeming to attend only to a vase's color (or shape), we still attend to it in many different ways, for our attending to something is uniquely a matter of the speech communicational situation in which we are involved. Our seeming to attend only to the color or to the shape of the vase is *an outcome* of our talk in the situation. It is a matter of the different ways in which we relate what we speak of *as* "our attending" to the rest of our proceedings in the circumstance in question.

To illustrate the range of such differences, Wittgenstein uses a number of examples (not all of which I will repeat here) in which we attend in our talk seemingly just to the color blue: "Is this blue the same as the blue over there? Do you see any difference?"—You are mixing paint and you say, "It's hard to get the blue of this sky." "It's turning fine, you can already see the blue sky again," and so on. His point is to "show" us that we do in fact attend to the color blue in different ways in different contexts, according to the different ways in which we draw attention to it *in our speech*. For our attending to something in our circumstances is interwoven with responding to it in ways that *must* make linguistic sense to (certain) others around us.

Wittgenstein's examples remind us of how easily we are inclined to forget the parts played by our talk in such circumstances. We tend, so to speak, to look *through* it, to "see" our attending as taking place simply within us individually, as if independently of our communicative involvement with others. However, our individual ways of responding to our circumstances are embedded in our ways of communicating with each other: "The *speaking* of a language is part of an activity, or a form of life," he says (1953, no.23). And, one could add, many of our activities are embedded in, interwoven with, or "orchestrated" through our language. Thus, talking as we do is not just one of our many activities in *the* world. On the contrary, we and our worlds have their being for us within our linguistic activities. Wittgenstein reminds us of our deep relation to others around us, through their speech and communicational nature. That is, rather than material, mechanical, causal, or physical, our relations to our surroundings are linguistic, since it is in our speaking that we link what is other than ourselves to our selves.

However, given how we tend to think about ourselves—as having our cognitive capacities independently of language and our relations to others (tendencies that Wittgenstein shows are created in us by language)—our ability to make judgments of similarity seems to raise a problem for us. For judging that colors or shapes are in fact the *same* colors or shapes makes us feel that there must be a similarity, somewhere, between all the cases in which we talk of attending to "the blue," or talk of attending to "the shape." How else could we talk as we do? Thus, "We do here what we do in a host of similar cases: because we cannot specify any *one* bodily action we call pointing to the shape (as opposed to the color), we say that a *spiritual* activity corresponds to these words. Where our language suggests a body there is none: there, we should like to say, is a *spirit*," he remarks (1953, no. 36). In practice, according to the different kinds of involvements we might have, we talk of attending to both the color and to the shape of a vase in different ways in different circumstances. But, in theory, we feel there *must* be a similarity. Thus, we invent or imagine a mysterious inner place that must exist in us somewhere, where what we seemingly "know" to be the same, *is* the same. We call this place "the mind." It is "in our minds," we say, that the judgment of similarity must take place. When we become acquainted with some of the problems involved, to do with viewing colors and shapes in different light levels, with different contrasts, angles of view, and so on, we think seeing the same color, or the same shape, must be the result of some very complex and special kinds of inner mental *processes*. "And psychology is now the theory of this inner thing" (1980, I, no.692). But *is* this what is involved?

Perhaps. But in discussing the possible existence of such inner processes, Wittgenstein considers, among other examples, a person being said to *understand* the principle or formula for continuing a series of numbers. Again his strategy is similar to that in the example above: he lists a seemingly indefinite number of different things that a person might be said to be doing in claiming to "understand" the principle or formula of the series, and then shows that there is nothing in common among them. If you still feel tempted to say that they *must* have something in common, then you must say (as indeed cognitive scientists do say) that the common process must be one that is *hidden* somewhere within or behind the person's statements. But, says Wittgenstein, "How can the process of understanding have been hidden, when I said 'Now I understand' *because* I understood?! And if I say it is hidden—then how do I know what I have to look for? I am in a muddle . . . . If there has to be anything 'behind the utterance of the formula' it is *particular circumstances*, which justify me in saying I can go on—when the formula occurs to me" (1953, nos. 153,154). Irrespective of the nature of the supposed inner processes involved, our main task in "going on'" in social life is being able to justify our doings and sayings to the others around us.

Thus, as he sees it, it is not the special workings of a mysterious inner mechanism that produces in us, individually, whether we like it or not, correct judgments of colors or shapes as being the "same" as each other, or of number series as containing this rather than that principle or formula. But it is we who make such judgments, in terms of criteria we propose and develop in discussions between ourselves: "An 'inner process' stands in need of outward criteria," he says (1953, no.580). The "correctness" of a way of going on, a way of talking, perceiving, thinking, or acting is not something we simply allow to be imposed on us by our pre-established natures (whatever they may be). No matter how much we may worry over them, debate them, or even fight over them, we settle them (at least for certain practical purposes) *between* us, in the course of our talk with each other. The "agreement" between us "is not agreement in opinions but in form of life" (1953, no.241). That is, we agree not just that a stated claim about our world is indeed true of it, but we agree on the *way* that it is true of it. Thus, the giving of grounds comes to an end "not [in] an ungrounded proposition: [but in] an ungrounded way of acting" (Wittgenstein, 1969, no.110); "it is not a certain kind of *seeing* on our part; it is our *acting*, which lies at the bottom of the language-game" (1969, no.204). For language-games cannot themselves, of course, be based on grounds. They are what makes the formulation, testing, and doubting of claims possible. A language-game "is not reasonable (or unreasonable). It is there—like our life." (Wittgenstein, 1969, no.559). Without it and others like it, we would have no culture, no history, no society, no intelligible linguistic relations with the other people around us. To repeat, language-games are a part of who and what we are, and of the world in which we live.

Conclusions

As professional academic theorists, we clearly inhabit a very different form of life from that of many of the people around us. In our disciplinary "reality," it does seem to us that we exist as independent individuals, in an "external world" of already existing objects (as in the classic image of the contemplative subject). In it, using the disciplinary resources our field provides, our task is to attempt to produce *true* representations ("pictures") of states of affairs within it, and thus to explain (predict, control, and manipulate) the happenings occurring in and between the supposed "objects" it contains. But if we are mistaken about the nature of the relation between language and "the world"; if the primary function of our speech is not the assertion or denial of already existing facts, but the making and sustaining of living relationships; then we can be seriously self-deceived as to what it is that we are doing when we claim to be producing "the truth" about the "objects" in our institutional worlds. Just as in a good piece of science fiction writing, we can be "bewitched" by the power of the language used into thinking that the institutional objects of our academic talk are the "true" versions of things, somehow there but hidden from us in our daily lives. It is this that concerns Wittgenstein. When a professional academic uses a word for his or her own idiosyncratic, theoretical purposes, claiming to have discovered its "true meaning"—for example, what "knowledge," for instance, *really* is—then Wittgenstein suggests that one always needs to ask himself: "Is the word ever actually used this way in the language-game which is its original home?—What *we* [Wittgensteinians] do is to bring words back from their metaphysical to their everyday use" (1953, no.116).

If we take this aim seriously, we arrive at a very different image of the function of language than that which we usually have taken for granted, and which informs cognitivism. Instead of assuming that we understand each other's speech simply by grasping the inner ideas or representations they supposedly have put into their words, that picture of how we understand each other becomes the exception rather than the rule. In Wittgenstein's view, most of the time we do not fully understand one other; we mostly just respond and react to other people. If shared understandings occur at all, they only do so by people testing and checking each other's talk, by questioning and challenging it, reformulating and elaborating it, and so on. For, in practice, shared understandings are developed or negotiated between participants over a period of time, in the course of their responding and reacting to each other. In responding to each other in an attempt, pratially, to "go on" with the others around them, people construct one or another kind of social relationship or form of life. It is the character of these conversationally developed and developing relations, these forms of life, and the events that occur within them, that they are of much greater importance than any shared ideas to which they might (or might not) give rise. For, it is from within the context of these actively constructed forms of life that what is talked about gets its meaning. And "the theoretical integration" promised by the cognitive revolution needs to face this question: What is the form of life constructed in such a way of talking? If my account above is correct, it is a form of life that obliterates morality and ethics.

The cognitive revolution in scientific psychology needs to be reassessed. It ignores the important part played by our everyday forms of life in making our lives as professional academics possible. That is, cognitive scientists seem to assume that they themselves already are in possession, so to speak, of the correct facts of our lives. As they see it, there is no problem with the character of our surroundings (they are full of well-defined objects); nor with our language (it works in terms of well-articulated structures of representations); nor with the proper way in which we should arrive at conclusions in processes of reasoning (by calculation); nor with our "minds" (they also work in a calculational or computational manner). But what is overlooked in such claims as these is the fact that, in our everyday affairs, all such conduct is *socially judged*. In other words, no matter what people may do, say, or claim, their actions or statements are judged by others, as good or bad, fitting or inappropriate, and so on. In claiming that the only task is the theoretical one of providing what will count as computational explanations of how, precisely, the mental processes hidden "in our minds' function to bring about our conduct, they fail to respect the social nature of the science to which they claim to subscribe. For even when all alone, thinking or writing, "an 'inner process' stands in need of outward criteria" (1953, no.580), or else we are simply in the position of "whatever is going to seem right to me is right" (no.258). Thus, no more than cognitive scientists, can we determine in our contemplative or theoretical studies of our use of words their single, correct, or right use. But we can explore the "possibilities" (1953, no.90) they make available to us, for how we might react or respond to them, that is, what we might expect or try to do, and whether, in certain contexts of their use, they might be misleading.

As a result of such studies, Wittgenstein confronts us with difficulties that cannot be solved by advancing any theories (scientific or philosophical), or by making any empirical discoveries. "They are solved, rather, by looking into the workings of our language, and that in such a way as to make us recognize those workings: *in spite of* an urge to misunderstand them. The problems are solved, not by giving new information, but by arranging what we have always known" (1953, no.109). For the problems in question are not to do with whether a claimed representation of a state of affairs in the world is true of it, but with whether we can judge what a person says as something we can follow, without finding it confusing, misleading, or simply nonsense. Without that opportunity to judge actions in the course of responding to them, without that moment of living contact between oneself and another, one's opportunity to choose whether to "go on" is lost—and along with it, one's opportunity to construct a shared form of life jointly with other people. That possibility—of our constructing a form of life between ourselves, one that is not yours or mine, but ours—is what evaporates if we must speak in an already decided and finalized code. And this should trouble us in our enthusiasm for "a unifying theory" in cognitive psychology.

The problems confronted by Wittgenstein's philosophy emerge from vague but strong feelings of dis-ease such as these. To repeat: such troubles cannot by resolved simply by finding out more about our world. "They are deep disquietudes; their roots are as deep in us as the forms of our language and their significance as great as the import of our language," he says (no.111). To the extent that our different ways of talking draw our attention to things in different ways, his concern is with whether our current ways of talking lead us to see things aright. Do they lead us to heed the world around us, and others, with the right kind or quality of attention? Should we be merely content to observe what is already there, seemingly "in" the behavior of the individuals around us; or, should we, say, attempt to create new forms of life, by exploring the results of new forms of talk? As Wittgenstein saw it, our current ways of talking not only may be leading us to overlook possibilities in our circumstances that we perhaps ought also to be considering. But there may be something deeply wrong *in us all*, in seeing, thinking about, and talking of, matters in the ways that we do. In particular, we may be deceiving ourselves as to what we take the value of these ways to be. For instance, our current bureaucratic/managerial/technocratic ideals may contain corrupting tendencies difficult to detect from within them. In short, rather than technical problems, Wittgenstein confronts us with certain ethical worries—issues that basically have to do with what he sees as the "poverty" and "darkness" of our times. (Wittgenstein, 1953, p.x). Rather than being concerned with how we, as individuals, best might exploit this or that possibility we see as already existing in the circumstances around us, the problems in question have to do with assessing whether *that* is what we, as members of a social group, ought to be doing at all.

Notes

1. It can be experienced, for instance, in Baars's (1986) claim that "within a matter of decades we will have a dominant theory of human mental processes," and by then, "the notions of 'representation' and 'information processing' are likely to be so obvious to the coming generation of students . . . that they will not be able to understand what all the controversy in the first century of psychology was about" (p.415). Presumably, the worries that others of us have, about the point of our more poetic, nonrepresentational forms of talk— those determinative of our being in relation to the others around us—will be long gone.

2. By the word "primitive" here, Wittgenstein means "that this sort of behavior is pre-linguistic: that a language-game is based on it, that it is the proptotype of a way of thinking and not the result of thought" (1981, no.541).

3. Wittgenstein (1980, II, nos. 625, 626, 629).

4. Wittgenstein himself did not envisage his notion of language-games being put to this particular use—that is, being seen as models. "Our clear and simple language-games are not preparatory studies for a future regularization of language. . . . The language-games are rather set up as *objects of comparison* which are meant to throw light upon the facts of lan-

guage by way not only of similarities, but also of dissimilarities" (Wittgenstein, 1953, no.130). But, as I see it, the hegemonic claims currently being made in cognitive psychology, the coming *dominance* of its terms, are tantamount to a "regularization" of language.

5. This is what elsewhere I have called the *representational-referential* function of words (Shotter, 1993a and b).

6. Gardner (1987) recruits Giambattista Vico to his cause, grouping him with Descartes, Locke, and Kant, and suggesting—because he (Garner) also wishes to talk of a New Science—that "Vico even christened a New Science (*Scienza Nova* (sic)) to deal with these and related matters" (p.4). But even a cursory glance at the *Scienza Nuova* is enough to convince one that it is a work aimed at very different issues than the concerns of those with whom he is united.

7. Dennett's views are not easy to divine unequivocally. For instance, in one of his more recent writings, he realizes that talk of representations is controversial, and describes himself as "one of the most extreme skeptics about mental representations" (1984, p.30). Nevertheless, whether or not he is a skeptic about representations, he in no way seems to be a skeptic in still talking of "systems," "patterns," and "mechanisms" as central to any understanding of reasoning.

8. He begins the section about the agent *Builder*, from which I quote, with the *Webster's Unabridged Dictionary* definition of a bureaucracy: "the administration of government through departments and subdivisions managed by sets of officials following an inflexible routine."

9. The relation to Wittgenstein's (1953) "primitive language-game," apparently, is accidental.

References

Arendt, H. 1964. *Eichman in Jerusalem*. New York: Viking Press.

Baars, B.J. 1986. *The Cognitive Revolution in Psychology*. New York and London: Guildford Press.

Bauman, Z. 1989. *Modernity and the Holocaust*. Ithaca, NY: Cornell University Press.

Boden, M. 1977. *Artificial Intelligence and Natural Man*. Sussex: Harvester.

Broadbent, D. 1958. *Perception and Communication*. London: Pergamon.

Broadbent, D. 1980. "The Minimization of Models," in A.J. Chpaman and D.M. Jones (eds.), *Models of Man*. London: British Psychological Society.

Dennett, D. 1978. *Brainstorms: Philosophical Essays on Mind and Psychology*. Sussex: Harvester.

Dennett, D. 1984. *Elbow Room: The Varieties of Free Will Worth Wanting*. Cambridge: MIT Press.

Eysenck, M. 1984. *A Handbook of Cognitive Psychology*. Hillsdale, NJ: Erlbaum.

Gardner, H. 1987. *The Mind's New Science: A History of the Cognitive Revolution*. New York: HarperCollins.

Lachman, R., Lachman, J.L., and Butterfield, E.C. 1979. *Cognitive Psychology and Information Processing: An Introduction*. Hillsdale, NJ: Erlbaum.

Marx, K. and Engels, F. 1970. *The German Ideology*. London: Lawrence and Wishart.

Minsky, M. 1986. *The Society of Mind*. New York: Simon and Schuster.

Rotman, B. 1993. *Ad Infinitum: Taking God out of Mathematics and Putting the Body Back in*. Stanford: Stanford University Press.

Sampson, E.E. 1981. "Cognitive Psychology as Ideology." *American Psychologist*, 36, pp. 730–43.

Sampson, E.E. 1993. *Celebrating the Other: A Dialogic Account of Human Nature*. Boulder, CO: Westview Press.

Shotter, J. 1984. *Social Accountability and Selfhood*. Oxford: Basil Blackwell.

Shotter, J. 1993a. *Cultural Politics of Everyday Life: Social Constructionism, Rhetoric, and Knowing of the Third Kind*. Milton Keynes: Open University Press.

Shotter, J. 1993b. *Conversational Realities: Constructing Life Through Language*. London: Sage.

Shotter, J. 1995. "In Conversation: Joint Action, Shared Intentionality, and Ethics." *Theory and Psychology*, vol. 5, pp. 49–73.

Weber, M. 1964. *The Theory of Social and Economic Organizations*. (Trans. A.M. Henderson and Talcott Parsons). New York: Free Press.

Wittgenstein, L. 1953. *Philosophical Investigations*. Oxford: Basil Blackwell.

Wittgenstein, L. 1969. *On Certainty*. Oxford: Basil Blackwell.

Wittgenstein, L. 1980. *Remarks on the Philosophy of Psychology*, vols. 1 and 2. Oxford: Basil Blackwell.

23

Rom Harré

"Berkeleyan" Arguments and the Ontology of Cognitive Science

The Ontologies of the Old and the New AI/CS[1]
Models in the Context of Realism

In recent developments in theoretical psychology we have gone from positivism (behaviorism)—assuming no direct public access to the mind, to realism (cognitive science)—allowing indirect public access to the mind via the hypothetico-deductive method. There seems to be a possibility for obtaining indirect access to private mental phenomena via the general principle that successful tests of the logical consequences of hypotheses about models of unknown processes gives us confidence in the verisimilitude of such models. This way of penetrating the bounds of sense is characteristic of such sciences as chemistry, and is commonly thought of as ontologically significant. This signficance is assumed by nearly all working scientists. We know from the paradox of Clavius, sometimes called "the underdetermination of theories by any available empirical data," that hypothetico-deductive access is weak in the absence of an ontologically plausible interpretation of the logically most efficacious model. In interpreting cognitive science in the framework of scientific realism, we must play close attention to ontological issues. Ontologies make themselves felt overtly as the supertypes of those type-hierarchies that we use to control our construction of our models' unobservable processes, so that these models are ontologically plausible (Aronson, Harré, & Way, 1994).

The study of ontological questions is indispensable to the development of a science, not only as the prime instrument for the control of model building, but also in formulating new directions of research with respect to the type-hierarchies through which one can nicely express the structured content of a scientific discourse. Throughout this chapter I will assume that the most revealing way to enter into a discussion of the ontology of a psychological science is to employ the Aronson-Way treatment of the ontology of physical science, based on the use of a type-hierarchy representation of the relation between a theory's models and the reality that, under realism, they must be taken to represent (Way, 1991; Aronson, 1990).

The questions to be posed are: Is the ontology of mind presupposed by the first cognitive revolution, the ancestor of the representationalist version of contemporary AI/CS, defensible? And, if not, is the ontology presupposed by the second cognitive revolution, the shift to discourse as the "substance" of cognition, of the same "hidden mechanism" type that is characteristic of those sciences that count as realist in the Aronson-Way sense? Is the best model for the theory a subtype in an ontological type-hierarchy that includes among its other subtypes some aspect of the real world? There are "hidden mechanisms" in the global ontology of the discursivist psychology of the second cognitive revolution: but they are not mental, except by courtesy of a subtle semantics. In particular, they are not hidden behind the overt phenomena of thought and action. They are "hidden" because they are immanent in the phenomena of cognition.

As I understand the matter, there are currently two ontologies on offer in AI/CS:

1. The ontology of many who support a "representationalist" position comprises certain atomistic states of individual human beings. Each is a representation and the ontology of AI/CS consists of such representations.
2. The ontology of many who support a "connectionist" position comprises the relational networks that obtain between meaningful signs, and their meaning consists in their relations with other signs. If there is any use for the notion of representation in this ontology, it is that representations are immanent in the network.

I will be expounding and defending the view that AI/CS should be founded on two type-hierarchies: a hierarchy of interpersonal relational structures, the units of which are acts (i.e., signs in use), and a hierarchy of personal skills, whose existence is a necessary condition for the existence of the relational structures. While meanings are immanent in the networks, mental models are immanent in the hierarchies of personal skills and the "rules" with which their exercise can be evaluated. Developmentally, according to the Vygtoskian position that I will take for granted in this chapter, the former pre-exists the latter. Since the concept of "skill" is a relational concept, linking personal grounding to achievement valued interpersonally, the skills hierarchy is also a hierarchy of relations.

As far as I am concerned, the target in critical discussions of the ontology of psychology is Cartesianism in the philosophy of mind, which exerts a subtle but baleful influence on the ontologies of psychological science. Cartesianism as an ontology has two pillars:

1. It is dualistic; that is, mind and body are held to differ as substances. This is the familiar Cartesian doctrine of the Rylean "category mistake" argument.
2. It is "atomistic," in that the basic cognitive (mental) entities are individuatable states of the mind-stuff, and as such are states of particular persons.

Though there is much talk of "minds" in the writings of the AI/CS crowd, few, when pressed, would subscribe to (1). A version of (1) does, however, have some currency in the ontology presupposed in the writings of Fodor (1975, 1988, etc.) though he does not seem to subscribe to (2). Although I shall devote considerable space to developing a general argument aimed at refuting all versions of Fodor's position, the major question to be addressed in this chapter is the viability of (2). The Berkeley style argument I will use against Fodorian dualism involves an analysis of cognitive acts that will also serve as the basis of an argument against the second pillar of Cartesianism.

Von Eckardt's (1993) "Domain-Specifying Assumptions"

The assumptions just mentioned would specify any psychology other than behaviorism; in fact they do rather well for discursive psychology. What von Eckart calls "basic general properties of cognition" are, for her and for discursivists, actually features of cognitive capacities. Capacities are cognitive, according to von Eckardt if (and only if?):

a. "Each capacity is intentional ('states have content')."
b. "Capacities are pragmatically evaluatable"—that is, they can be exercised with varying degrees of success, since they are under normative constraint.
c. When successfully exercised, each of the evaluable capacities (in ordinary language, skills) has a certain coherence or cogency.
d. Most capacities are reliable—that is, more often successful than unsuccessful, that is, productive (i.e., can be manifested in a variety of ways).

Von Eckardt defines the cognitive science project in four questions:

Q1. For the normal, typical human adult (why not person?) what precisely is the human capacity to. . . . ?
Q2. In virtue of what does a normal, typical adult have the capacity to . . . (such that this capacity is intentional, pragmatically evaluable, coherent, reliable, and productive)?
Q3. How does a normal, typical adult typically exercise his or her capacity to . . . ?
Q4. How does the capacity to . . . of the normal, typical adult interact with the rest of his or her capacities?

This is a perfect characterization of discursive psychology, and shows that it is to count as a cognitive science. For me, it is *the* cognitive science. For discursive psychology, the primary cognitive reality is the array of public and interpersonal linguistic and practical acts, a necessary condition for the existence of which is a repertoire of natural expressive acts, among which are Wittgenstein's moans of pain. The task of psychology is two-fold: on the one hand to identify and describe the arrays of nested acts, and on the other to identify and describe interlinked hierarchies of skills, by which people perform all manner of acts in the carrying out of all sorts of tasks— manual, cognitive, perceptual, and so on. This is not behaviorism, even though people have to learn the majority of these skills, because skill is a normative concept, requiring on the one hand a semi-permanent state of the body of the skilled person and, on the other, some system of norms that may be and usually are extrapersonal. Perhaps they would have to be extrapersonal. Wittgenstein thought they would have to be based on public exemplars, since they could not depend on the existence of this or that group of other persons. The relational structure of a rule system is not necessarily, though it is usually, social.

Cartesianism as Displayed in the Writings of
Proponents of the First Cognitive Revolution

First let me offer some quotations that seem to propose a dualistic Cartesian ontology, the first feature I distinguished above:

1. "The conjecture (or 'hypothesis') that human beings are thinking things (i.e. that human beings have minds)" (Fetzer, 1991, p.2).
2. "Computational approach . . . describe[s] the mind and thinking with concepts drawn from artificial intelligence and related forms of computer modelling" [of human performances] or [of the procedures by which the performances are accomplished]? (Boden, 1989, p.27).
3. "It is characteristic of intentional propositions that their truth depends on psychological truths about the subject" (Boden, 1989, p.75).
4. "Seen from this viewpoint the mind is a system for manipulating symbols" (Boden, 1989, p.27).
5. "Humans [a.k.a. persons] can try out many new ways of thinking to see what they will find" (Boden, 1989, p.117).

We now have a further shift away from a brain ontology to one in which people as active agents *use* new ways of thinking.

Let us look at some quotations that seem to offer a brain-based ontology, some from the same book by the same author, as some of the above:

6. "It makes sense for him [the psychologist] to suggest that the brain is an information processing system that somehow models the environment" (Boden, 1989, p.72).
7. "Computational processes in . . . computers may be significantly similar to the processes going on in human brains" (Boden, 1989, p.126).

This last remark is interesting, since how we are inclined to read it will depend on which ontology we favor. Is the mind just the brain, in the sense that it is the brain and only the brain that is the organ of thought? If so, why not say so? Who is doing the manipulating of symbols? Is it individuals who might use their brain-minds, that is, this material system, for manipulating symbols for some purpose? What is doing the computational jobs, the brain or the mind or the person? Boden's position is eclectic to the point of incoherence.

Again we find ontological incoherence in this quotation from von Eckardt (1993):

8. "The general theory of imaging describes the various representational structures and processes that supposedly underlie the human ability to image. . . . It does not, however, actually give a blow-by-blow account of how we exercise our imagining capacities or how we carry out particular imagining tasks. . . . Kosslyn [adds] various specific models for these tasks. . . . [H]umans can construct images on the basis of descriptions of how parts are to be arranged" (von Eckardt, 1993).

Here we have the ontological problem appearing in an acute form. Von Eckardt might mean that *people* can become skillful users of these "modules for imaging," just as they might become skilled users of CorelDRAW or of tennis rackets. Sometimes the tool one is using happens to be inside the user's head rather than in a plastic box on the table or in the cupboard of sports equipment. It is the use that is the psychological topic, not the tool alone.

But why should we care about this ontological nitpicking so long as we are coming up with successful predictions? To answer that question we would need to turn our attention to a debate of great antiquity between those who take the positivist view of the scientific enterprise and those who defend scientific realism, that is, those who believe that the ultimate aim of science is to give as accurate an account as possible of the world in all its depth and variety.

Diagnosing the Debate

Why am I dissatisfied? What is the nature of the intuition that speaks of some deep incoherence in the unanalyzed AI/CS position, as it is presented in this bouquet of quotations? It comes out clearly in reflection on the ontological sections of Sperry's address (1993, p.879), which runs as follows:

> The new position is mentalistic, holding that behavior is mentally and subjectively driven. This, however, does not mean that it is dualistic. In the new synthesis, mental states, as dynamic emergent properties of brain activity, become inseparably interfused and tied to the brain activity of which they are an emergent property. Consciousness, on this view, cannot exist apart from the functioning brain.

Even at a superficial glance, the defects in this characterization are only too obvious. The most problematic dualism of all still survives, namely, that between subjective (inner) and the objective (outer). Further, what can it mean to talk of an emergent property "interfused with and tied to"? What a cocktail of metaphors! It is at once a dualism and a monism. Of course, in a sense that is correct. But the duality is not between a substance and its properties, emergent or native, but between a tool and its uses.

There are two outstanding difficulties with Sperry's ontology:

1. It reintroduces consciousness as the prime domain of the cognitive, without any general account of this culturally dependent and elusive notion. Nor does Sperry offer an account of what it would be to hypothesize that something cognitive was occurring, but the person "in" ("to"?) whom it was occurring was not conscious of it, as AI/CS seems often, rightly and necessarily to do. Is the realm of the cognitive broader than the realm of the mental? (Of course I shall be querying the need to introduce the realm of the mental at all!)
2. Consciousness is proposed ontologically as an emergent property of the struture of the brain. So what's the problem?:
   (a) Any number of properties of that chunk of organized chemistry are emergent! Following from (1) we have the difficulty that arises when we are supposed to be picking out which are mental by the 'consciousness' criterion. It makes the realm of the cognitive far too narrow. There would be no room in this ontology for cognitive processes of which we are unaware and, more important for cognitive processes immanent in interpersonal interactions, that is, in relational networks of the relevant kind of actions between people.
   (b) We are certainly presented with a dualism in the Sperry summary, a dualism to be found in the writings of many authors, especially, for instance, Boden; but it is not a substantialist dualism.

Sperry's simplistic account allows us to see very clearly that this ontology takes up one of the leading tenets of Cartesianism. (Reminder: Cartesianism as a whole package includes two principles: (1) the idea that the mental consists of a collection of occurrent states and (2) that these are the states of a mental substance.) Although repudiating the second pillar of Cartesianism, it celebrates the first. Thus, not only does it run into the problem of what is to be the status of unconscious cognitive states (mental?); but by virtue of (1) it brings the debate into the territory of a familiar and ancient controversy between the occurrentists and dispositionalists—the debate between Descartes on the one hand, and Leibniz, Priestley, and Ryle on the other. And this is itself a special case of the grander debate between atomists and relationalists.

Wrong and Right Ways of Defining the Cognitive Science Program

THE WRONG WAY: "The notion that linguistic form and the conceptual structure that is behind it is a reflection of the way states and events are perceived or otherwise experienced is one of the basic tenets of cognitive linguistics" (Ganser, 1993).

THE RIGHT WAY: Conceptual structures are immanent in linguistic forms.

This ontological distinction, which we could roughly characterize as that between a static and a dynamic formulation, is crucial to giving AI/CS a sound ontological foundation. There are no linguistic forms *and* conceptual structures. There are only linguistic forms in which, with suitable analytical tools, we can discern conceptual structures.

Sharpening the Question to be Discussed

The discussion in this chapter will not include in any but the sketchiest way:

1. Critical exploration of the "information processing" metaphor in relation to the task of constructing the relationist ontology for AI/CS theorizing;
2. A discussion of the problem of how "active use of skills" is to be dealt with—that is, where active powers are to be located, and to whom or what they are to be ascribed;
3. Problems with the concept of "representation"—which really focuses on the "resemblance fallacy" from another point of view. That is, knowledge representation systems must not include things we are supposed to know!
4. A discussion of what might be meant by the phrase "mental states."

It will include:

5. Arguments for the relational or network account of meaning—that is, of cognitive acts;
6. Arguments against Fodor's type of ontology—as a special case of the "mind behind the mind" or "resemblance" fallacy. Here I will recruit Berkeleyan style arguments.

Neither argument can of course settle the question of the best ontology for AI/CS—that is, what type-hierarchy should be used in devising models for the formal structures developed by AI/CS techniques. But both offer pointers where to go.

A Sketch of the Discursive Program

People's discursive skills are realized in patterns of mutual activity that I will call generically "conversations," of which everyday conversations are a central exemplar. I will always identify the generic concept by the use of quotation marks. Patterns of mutual activity belong in the domain of psychology just insofar as they are sufficiently similar in relevant respects to that exemplar. "Conversational" interactions are to be taken as the archetypes of cognitive processes, both etiologically (Vygotsky) and conceptually (Wittgenstein). We could take this as the animating principle of the discursive turn. Just how similar "sufficient similarity" is, and which aspects are relevant, cannot be laid down in advance. Tennis clearly belongs in the conversational domain. The reason is that it is skillful—that is, intentional and normatively constrained. It is a matter for debate whether and in what respects sexual intercourse or passive smoking also belong in this domain.

Two basic principles animate psychology conceived as the study of "conversation":

1. People use all manner of material "tools," including sounds, images, hands, brains, and even other people to accomplish their "conversational" projects. We should see ourselves as *homo habilis*.
2. Material stuff, as used for these purposes, has different properties from material stuff interacting otherwise. In particular, some of the relational properties of a system of material stuff when it is being used for discursive purposes are not found among the relational properties of material stuff interacting otherwise.

I conclude that, insofar as AI/CS has a characteristic subject matter, in short an ontology that gives meaning to its theoretical structures, the semantic realm, even though it is always sustained materially, is ontologically distinct. In this chapter I am concerned with providing an explication and defense only of the second principle. I will argue that the ontology of what AI/CS models represent is exhausted by this domain. Contrary to the ontological presuppositions of the authors quoted above, I hope to show that there is no ground for the overt (Fodor) or covert (Boden) reintroduction of a Cartesian realm of mental activity—that is, a realm that is neither neurophysiological nor "conversational" in the extended sense, and that is "atomistic" rather than relational. Ridding ourselves of the "inner/outer" metaphor and replacing it with "public/private" as the literal distinction required to delineate the domain of the mental allows the concept of "conversation" to be generalized to both domains. And, in accordance with the theory expounded in this chapter, the concept of "the cognitive" generalizes with it. Since the ontology of the conversational realm is relational, that generalization, if successfully carried through, would defeat the second pillar of the Cartesian point of view.

I would like to argue from proposition 2 above that the dichotomy "discursive/neurophysiological" is exhaustive. But that is far too large a task for this chapter. I will content myself with trying to bring out the fallacy in the ontology espoused by the arch neo-Cartesian, Fodor, and with at least some considerations to motivate putting a relationist ontology in its place.

As a model for such an argument, I will offer an interpretation of Berkeley's way of demonstrating the emptiness of the Lockean thesis that in addition to what we do or could observe, and to the potent entities that produce these observations, there is a third ontological realm, the substance of which is Lockean "matter"—a kind of stuff that partakes of some of the properties of each of the other two realms. Its qualities as primary are supposed to resemble the ideas of primary qualities displayed in what can be observed. Its active powers are supposed to resemble some of the qualities of "spirits"—namely, their power to cause ideas in human beings. It is important to make clear at this point that I am not taking any of the content of Berkeley's argument over to the realm of the psychological. Intead, I am borrowing the structure and style of the argument for my purposes.

Berkeley's Arguments

It is easy to slip into interpreting Berkeley's overall argument as similar to or even the same as van Fraassen's ontological positivism (van Fraassen does not try to defend any version of semantic positivism, that the meanings of theoretical terms are exhausted by the observation statements that theories incorporating them entail, together with descriptions of conditions of application of the theory). But this would be a serious error of exegesis, since Berkeley's argument calls for the admission of real, active powers. We are tempted to fall into this error partly because we are uncomfortable with the kind of active powers he offers us as an alternative to Lockean active matter, namely, spirits!

Berkeley's case for powers (and, for our purposes, we need not be "picky" about which ones) is rooted in the following question: How can qualities that are manifestly inactive in the domain of what can be observed, miraculously acquire causal potency when located in the domain of the unobserved, as the corpuscular real essences of material things?

The three-step argument is complicated and subtle:

1. What occurrent properties are there? The answer is only properties that are manifest in human experience (and I would add in the reactions and states of instruments and apparatus that human beings have constructed). Since this could not exhaust reality (how do we account for the existence of the unobserved unless we realized that reality is more complex?), we must add causal powers to our ontology to account for how what we can observe at one time can exist unobserved at another. This is what I take to be the point of Berkeley's extension of "*esse est percipi*" to "*esse est percipere*"—that is, from "is perceived" to "would be perceived" to mean, which can be traced from the *Principles* to *Siris*.

2. The Newtonian criteria, on which Locke depends for a principled demarcation of the ideas of primary qualities (viz., universality in all we experience and invariance in all experiential contexts, which are supposed to select those properties that are qualities of both what is observed and what is unobserved) are defective as a device for selecting ontologically distinct categories of properties—that is, in the terminology of the time, for distinguishing primary from secondary qualities.

3. Berkeley introduces a deep criticism of the use of observable properties as the supertype for the type-hierarchy, on the basis of which all models of reality are to be constructed. He does this by ironizing the Lockean resemblance doctrine as follows (in Berkeley's terminology): According to Locke, primary qualities (the true nature of

material reality) resemble our ideas of primary qualities—that is, they resemble a certain subset of the qualities of things as we observe them, a subset identified by Newton's criteria. We must follow that thesis through. As ideas, primary qualities are causally inert (even according to Locke). Therefore, as real qualities of bodies, they also must be inert, by the principle of complete resemblance. And so the corpuscular structure—that is, the set of real primary qualities constitutive of a material thing—cannot have the powers to act on other material bodies, nor to excite ideas in us, that Locke requires of them.

Shorn of its religious connotations and translated into the domain of physics, Berkeley's "spirits" would appear as charges and their associated fields. These beings have active powers, but in no way resemble the things of the common world of ordinary observation. Translated into the domain of psychology, they would appear as active agents, in no way resembling the discursive phenomena they engender. I will not address the question of whether we should assign the necessary activity to the brain and nervous system, or to the human person, though I believe that the only coherent account involves choosing the person. (See Coulter, this volume.)

The upshot of the Berkeleyan argument is not only that the way we arrive at the concept of material substance is flawed, but that the concept is otiose. One can see already that in certain respects ontologies like Fodor's are etiologically similar to the primary quality/real essence ontology espoused by Locke for physics. The mind-stuff is characterized by properties that purport to resemble the properties of cognitive activities as we perform them overtly; and the Fodorian mechanism ought to be passive in just the sense that Berkeley argues that Locke's real essences would be passive, if they existed. In this chapter it is not the activity/passivity issue I am concerned with, but the issue of the choice of type-hierarchy to construct models of cognitive processes. Various aspects of the activity/passivity issue, and its relation to the homunculus fallacy, are extensively addressed in Hyman (1991).

"Conversation" as an Ontological Domain

To reformulate Berkeley's arguments for my purposes, I need to show that a large class of "mental states and processes" cannot be individuated as occurrent and intrinsic properties of individual human beings, but only as immanent in normatively ordered, dynamic relational structures among them. Or, in other words, I need to prove that the domain of the mental include both public and private exercises of cognitive skill, on all sorts of diverse "materials." The "supervenience" argument, that the mental has a material substrate, in the many forms that it has taken over the centuries, and as it has resurfaced in Sperry's ontology, presupposes some version of Cartesianism, since it sets out to defend the irreducibility of mental states and processes, despite their realization in material substrate. The discursive turn in psychology and philosophy of mind shifts the problem to the defense of the irreducibility of conversational acts to their material substrate in sounds, marks, bodily movements, and so on.

If all cognition is "conversational" in character, then a defense of the irreducibility of ordinary conversational acts is a defense of the irreducibility of at least some mental states, processes, etc., but not as elements of a Cartesian domain—that is, not as entities of some special mental substance, nor as emergent properties of *individual* brains. I will choose one case only, that of the cognitive activity of speech-act performances, and try to prove the relational theory for just that case. Other cases will have to be tackled either by showing some relevant and essential similarity to this case, or tackled *ad hoc*. I do not think that the demonstration of the point for a conversation seen in terms of speech-acts proves the point for all possible cognitive processes; but it is at least a minor victory for relationism.

The argument I propose to develop for the autonomy of the conversational domain is based on three seemingly simple premises:

1. The relations between acts, the elementary units of the conversational domain, are semantic; and in these relations rules, conventions, and customs are immanent.
2. The relations between the elementary units of the material substrate of the conversational domain, are causal (in which biological, chemical, and physical mechanisms, some of which are known indirectly by model building and hypothetico-deductive inference, are operative).

In general, rules are not causes of semantic regularities (Wittgenstein, 1953). There are no semantic mechanisms—in particular no hidden mechanisms. The question of what brings a semantic act into being is answered by pointing to the persons who do so, the projects they are engaged in, and their skillful conforming to local rules and conventions. For the Berkleyan style argument, we will need to show that acts cannot engender acts.

These apparently simple premises presuppose that much arduous philosophical work in clarifying concepts and revealing and curing conceptual confusions already has been done. In particular, it presupposes that a Wittgensteinian account of the role of rules in the maintenance of social and semantic order has been established, where these rules are expressions of local norms rather than causes of discursive and practical regularities.

3. Conversational acts exist only as products of joint actions by two or more actual or notional conversants.

The argument for this principle depends on drawing once again the distinction between actions as intended performances and the acts or, more correctly, act-forces that the performance of actions in concrete contexts and as performed by specific persons achieve. Even if there is only one category of sign uses, whose meaning cannot be completely accounted for by anything that is true of the individual person who performs the action, this would be enough to defeat the basic tenet of the representationalist AI/CS ontology—namely, that cognition is wholly a matter of states of individuals. In most cases, social acts are jointly produced by two or more people. For instance, in the simple case of a speech-act, an utterance is an achievement of a certain kind of speech-act, only if it is taken up as such by those to whom it is intentionally addressed. And that uptake is in many cases no more than the performance of the appropriate and relevant reciprocal action. A speech-act then consists in a relation between two or more actions. This is not the same point as the familiar Dreyfus argument that much that is required to complete the sense of a statement is encoded in the environment. The point I am making has to do with the interaction between people, rather than between a person and an environment.

In most of his best known writings Searle appeared to be a strong advocate of the view that speech-acts as instances of speech-act types could be wholly categorized as such by attending to the intentions, beliefs, and so on of the speaker. It was that view that was strongly criticized by Muhlhausler and myself (1991). But Searle seems have had a change of doctrine, inclining him more toward the interactive view. His new views are worth quoting: "The dialogic sequence of initial utterance and subsequent response is internally related in the sense that the aim of the first speech-act is only achieved if it elicits an appropriate speech-act in response" (Searle et al., 1992, p.10). The thorough-going relationist would add that this concession seems to entail that the speech-act does not exist as such, unless the relational condition is satisfied. We might dispute as to whether that response need be overt. I am inclined to argue that the overt case is etiologically prior, and it assumes special importance when one comes to look at larger chunks of public cognitive processes than a mere two-turn exchange. If we add the action/act distinction to our means of expression, we can readily distinguish what was intended by speaking from what was achieved in speaking, reserving the term "action" for the former and "act" for the latter. Each speaker performs a speech action, the second speaker's action, taken as an act, serves to create the act force of the first speaker's speech action. It follows that a speech-act is internally related to other speech-acts in the sequence that makes up a conversation, in that the former would not be what it is (have the illocutionary force and perlocutionary effects it has, as such and such a type of speech-act) were the reciprocal and complementary speech-act not to have occurred. Stephen Straight has pointed out to me that there are many cases in which a speech-act is properly said to have occurred even when the addressee has not provided the public confirming act. It seems to me that one can account for such cases either by supposing a private confirming act, or by looking for the confirming act as immanent in the way in which the subsequent act sequence unfolds.

Thus, in at least one important case, cognitive phenomena exist only in a relational network among individual human beings and what they do publicly and privately, rather than as independent states or performances of those people. It is worth emphasizing once again that this argument rests on the claim that the relation between speech-acts is internal or constitutive, in that none would be what it is, unless others were what they were.

Expression as the Primitive Cognitive Act

According to Searle, neither Gricean maxims nor conventions that seem to be immanent in turn-taking practices are able to account for moment-by-moment conversational order, since we need to invoke momentary expectations and presuppositions. At this point, both Searle (1992) and Shotter (1995) make an interesting claim, which I shall not pursue in this chapter, but that seems to be a matter of central importance for the foundational questions of cognitive science. To quote Searle (1992, p.147), "[Again while] the intentionality of the individual speech-act . . . is in general representational, conversation as a whole is characteristically not representational." Why? Because while the concepts of knowledge and belief, suitably disinfected of the entitative status imposed on them by both defenders and critics of "folk psychology," may be a useful local way of referring to the grounds of discursive (cognitive) skills in an individual, conversation is grounded in "practicalities," such as primitive ways of expressing and responding to bodily and environmental conditions. (An example would be Wittgenstein's analysis of how pain talk can be intelligible both to the one who speaks and the person who hears and responds appropriately; or that we shiver in a cool breeze, or sneeze if pollen is about. One of my friends always says "Allergy!" when he sneezes, transforming the primitive reaction into an intelligible but non-intentional response.) Knowledge and belief are not required (they may be acquired), but rather some elementary skill.

I will make use of the sign/symbol distinction to further the analysis. The usage I adopt from semiotics is close to that advocated by Newell (1980). By a sign I mean any material entity that satisfies the familiar Saussurean criterion for having linguistic value, and that is used by a (notional) community of persons for "conversational" purposes. By a symbol I mean a sign that is used to stand for or represent something else.

One of Wittgenstein's major contributions to linguistic theory is the demonstration that not all the signs of a language are used as symbols. Or perhaps this would be better put as follows: Not all uses of signs are as symbols. I will elaborate this below.

This discussion supplements the theory of the ineliminably relational quality of "conversational" acts, finally conceded by Searle. For instance, the representational theory of cognition, as it appears in Boden's writings, requires (I think) that all sign use is use as symbols. So, for any sign use, we can ask what the sign stands for. In expressive uses of signs, the sign does not stand for anything, whether that use is to express "how it is with me" (pain) or to perform an illocutionary act like promising. There is no internal state of the human organism or its conscious experience that gives or could give meaning to the sign as so used. Its meaning must lie in the relational network with other signs in use, expressing how it is with other human organisms. So also, to my giving one of the signs of distress such as saying "I've got a terrible headache," there must be the complementary sense-completing act of your expression of sympathy. Were you to say "Stop being such a wimp," the illocutionary force of my remark in the context now created would not be that of a legitimate complaint. Since I am not reporting my state to you in avowing it, the meaning of the expression cannot be the state that it expresses. Its meaning is illocutionary, not descriptive.

This point goes very deep indeed. Wittgenstein's distinguishing the expressive from other uses of language shows not only that in many psychologically important contexts sign uses are not symbolic of anything, but also that non-symbolic uses provide the grounds for the possibility of language. The existence of natural and trained expressions of how it is with me makes possible the subsequent development of further uses of language—in the end, even the symbolic.

Will has put the point neatly. "[This is] . . . a wider view, the more basic units of which are not propositions, statements, sentences, linguistic or quasi-linguistic entities of any sort, but ways of behaving, proceeding in thought and action" (Will, 1988, p. 112). But these just are Wittgenstein's primitive expressions, which may not be linguistic per se, but are necessary precursors of language, and properly partake of some of its essential features.

Recruiting Berkeley Against the Invention of a "Mind Behind the Mind"

Here we need to return to the Berkeleyan style of argument. A fundamental principle that plays a decisive role in Berkeley's criticism of Locke's ontology, and in the development of his own, is the thesis that ideas are passive, that the qualities of the things we observe do not stand in causal relations to one another. The fire, he remarks, is not the cause of my pain but the mark that forewarns me of it. The cause of "ideas" must be looked for elsewhere. The fact of their orderliness and of their seeming to evolve when no human being is observing the world, speaks for a non-human cause. Whether this position can be sustained for an ontology of material stuff is neither here nor there for my purposes. I prefer fields to God for this explanatory purpose. But this modern prejudice is irrelevant here since Berkeley's argument is only serving as a model for a style of argument. Transferring the principle of passivity to discursive psychology, we have the thesis that cognitive acts (of which speech-acts are a species) do not stand in causal relations to one another. My apology does not cause your gracious act of forgiveness; rather it provides the occasion for *you* (or whatever entity it is to which we assign causal powers) to perform your act, which has to be produced according to rule and either may be felicitous or infelicitous. This is not unlike Berkeley's "mark" account of the properties of physical things as we observe them. Again, I am not concerned with where and to what substance causal power is assigned, but with the fact that it is not to be found in the phenomena.

The Berkeley-style argument against Fodor's "language behind the language" thesis runs as follows: Since discursive acts (language sustained or otherwise produced) do not cause each other, a hidden realm of linguistic acts cannot be supposed to cause the elements that appear in the overt world of discursive acts.

Generalizing the Berkeleyan-style argument, we can say that it makes no scientific sense to put an unobserved discursive process as *explanans* behind an observed discursive process as *explanandum*. The unobserved process is as much in need of explanation as what has been observed. As Wittgenstein put it, "Having got rid of *the mind* behind what people think and do, let us not put the brain behind it." All that is "behind," in an explanatory sense, any cognitive activity is a set of skills. A special case is the hopelessness of trying to explain the meaning of a word by reference to some mental picture. In order for that picture to be relevant as the entity that gives meaning to the word, we must know how to take it. That is, we must know what it means. So we haven't solved the problem of meaning by adding mental entities to our ontology; we have just displaced it from the public to the private domain.

Explaining regularities by reference to rules (explicit or implicit) and norms, conventions, and customs is not like explaining regularities by reference to causal mechanisms. Rule-type explanations require the concepts of skill and discursive and/or practical project to tie the person acting to what he or she thinks, says, or does. A skill is exercised; and sometimes one may use some mechanism in exercising it. I may use the neurochemical mechanism in my brain that stores memory traces, in my display of personal interest so as to flatter supporters in a political campaign by remembering their names, just as I may use my hand in that same campaign when "pressing the flesh."

Setting Up an Ontology for AI/CS as the Working Tool for Carrying on Systematic Studies in Discursive Psychology

We remember that the von Eckardt criteria for a cognitive science are satisfied by discursive psychology. My claim for artificial intelligence/cognitive science as having a central part to play in a developing discursive psychology rather than as an adjunct to some neo-Cartesian ontology puts me under an obligation to set out more carefully the ontology presupposed by discursive psychology, and to show how an AI/CS program would fit into that scheme. To contrast an ontology of mental states and processes with an ontology of speech acts and "conversations" is not enough. Or, alternatively, to argue that to be in a mental state is to be engaged in discursive activity, and that mental processes are best interpreted as discursive activities, needs further argumentative support. I will not undertake that analysis here.

In what way could cognitive acts be said to be a distinct ontological category? Let us look first at the idea that an ontological category is defined by reference to the distinguishing characteristics of a substance. We recall that this is how Descartes argued for his two substance ontology. Not content with pointing out that the properties of cognitive entities were, across the board, different from the properties of material entities, he concluded that this entailed that there were two distinct substances of which each set, taken generically, would be the defining properties. But the move from distinguished sets of properties to distinct sustaining substances is not mandatory. To put the matter in seventeenth-century terms, it seems perfectly reasonable to claim that distinct specific instantiations of the same set of property-types should constitute the real essences of two beings that, with respect to their nominal essences, are radically disjoint. Since this cannot be ruled out *a priori*, it seems to me that we may find it just the right move to make in the unraveling of the ontological assumptions that underlie some science or other, for instance, the science of psychology as the investigation of discursive activities and the skills necessary to their performance.

To see how this insight could be applied in general psychology, we again appeal to the distinction between behavior, action, and act. A behavior is identified by anatomical/physiological criteria; an action is an intended behavior identified by an actor's intentions; and an act is the social force (or meaning) of the performance of an action as it is taken up in a structure of the relevant acts of others. A simple case is a farewell wave, and its acknowledgment by the reciprocal gesture of the one who is leaving. The ontological thesis of discursive psychology is that while behaviors as material events might be related causally, acts as discursive events are not. They are Berkeleyan passive. They are related semantically ("grammatically" in the larger, Wittgensteinian sense of that word). We would write down the principles of orderliness that we might discern (as psychologists) in the flow of discursive acts as rules, conventions, or customs, depending on what we supposed the modality of the sequential structure might be. A rule format might be used to express an apodeictic modality, laying down which reciprocal act must be performed for sociality (intelligibility and warrantability) to be preserved. But rules, conventions, and customs are not causes. They fix what is correct, not what must happen. Grammatical orderliness is rule-orderliness, not cause-orderliness.

In one sense the substance of material events and discursive acts is the same— namely, molecular and, ultimately, electromagnetic. But the network that links cognitive acts into structures is distinct from that which links material events in the same space-time region into material structures. If, as seems reasonable, we take the relations between acts to be internal—that is, the relations in which they stand are, in some measure, constitutive of them as acts, as I argued above in the context of Searle's apostasy—then the genera and species of cognitive acts will (indeed must) constitute an autonomous domain. In other words, they belong in type-hierarchies that are distinct from the type-hierarchies of the physical or biological sciences. The work of AI/CS is to construct formal grammars of that domain. The ontology of AI/CS in this role is then nothing else but the genera and species of that domain— that is, acts constituted by an interpersonal network of semantic (grammatical) relations, backed up by a multiple hierarchy of discursive and manual skills. I will call this the "first role, since to call it primary would be to beg the question of the priority of this to other possible roles.

One could express the point about the public nature of cognition in different rhetorics. That is, one could say that the illocutionary force of a speech act is not complete in itself, but only as part of a multiperson sequence of speech acts; or that the content of a mental act is not expressible in the Turing states of any one human individual considered as a finite automaton, but exists only in the network of relations between such beings.

What of the individual mind? In the ontology of discursive psychology, what it means to say an individual human being "has a mind" is no more than that a certain human being is a person competent in the performance and management of a certain minimal class of discursive (cognitive acts) with others, and ultimately reflexively in conversation with oneself, coupled with competence in some minimal class of material practices. The domain of the cognitive, of the mental, is essentially a public domain, a network of relational structures between the actions of individuals, so that they jointly bring acts into being. As Vygotsky was among the first to point out, the privatization of part of that network *is* an individual mind. The ontology of mind is then the ontology of "conversation"; and that is an ontology of internal or constitutive relations between skillfully produced cognitive acts. I believe the construction and testing of formal grammars of those "conversations" is a proper role of AI/CS.

Conclusion to the Argument, So Far

The Fodorian ontology commits the fallacy of resemblance—namely, that the primary qualities characterizing what produces what we observe (experience) must resemble the ideas of primary qualities, or certain qualities of what we observe. As applied to the philosophy of psychology, the fallacy appears as the assumption that what produces sequences of discursive acts must itself be discursive. Or, to put the matter more abstractly, models of mental functioning must be built on type-hierarchies of types of mental functioning, as if models for explaining the behavior of gases only could draw on such concepts as "pressure," "volume," "temperature," and "viscosity" as used in phenomenological gas physics. In the context of psychology, "Berkeley" comes down hard on any account that purports to explain the patterns of discursive activity by putting another domain of discursive activity behind the discursive activity we can observe. The feature of AI/CS that makes it such a powerful research tool is that it puts an abstract model behind overt cognition, about whose interpretation (the concrete models for which) we can debate.

We can now say something about the status of a central element in the ontology of AI/CS—namely, mental models, or representations that people have of all sorts of aspects of their worlds. Of course, mental models figure among the many instruments or tools we have on hand for various projects, like finding our way home. One of the tasks of the AI/CS project is to reveal them. According to the principles of discursive psychology, as I am setting them out here, mental models cannot exist as mental entities *behind*, or transcendent to, the cognitive and material practices to which they are relevant, together with the open sets of rules, conventions, and customs that define the necessary skills. Mental models must be immanent in these practices, just as meanings are immanent in the networks of act-linking actions, which we skillfully accomplish. By creating representations of mental models, as entity-like or process-like constructs in the AI/CS manner we, as psychologists, present in explicit form that which exists only immanently in the practices that engender the cognitive realm.[2]

Just as with any other instrument, when we use mental images to express one of our immanent mental models to ourselves, we have to know how to use whatever it is we have. The point was well made by Wittgenstein when he asked, "What makes a picture of him a picture of *him*? Nothing in the picture will determine this but only how I or someone else takes it, or uses it." Again the point is clear in his example of the stick figure on an incline, and all the different cognitive acts that it can be used for. *Homo habilis* (the human tool user) is my bet for the best general picture of the nature of human kind (who is *homo sapiens* only because some of the tools he or she uses are discursive). Therefore psychology, for me, is the study of the appropriate tools and their manner of use for the accomplishment of cognitive tasks—that is, the skills required to employ them correctly. One of the main tools is language. The discursive psychologist accepts the way language is used for cognitive tasks as a model for all other cognitive tool use.

Some Problems That Remain

This still leaves a question: Is there any further mileage to be got from AI/CS in asking whether the way its models implement the hypothetical rule systems, the grammars, casts any light on the way groups of people implement their cognitive—that is, discursive and manual—skills? It is particularly important to keep distinct the problems of finding powerful expressions for "grammars," and of discovering how "grammars" are implemented in action.

One way of entering into the debate is to try to make sense of the technical concept of "information." This is a not an actualist, but a dis-positional concept. That is, while the claim that a physical system embodies information requires that that system be in some structural state, a structure is information-bearing to the degree that it can play a role in some human form of life—in particular, in the use of skills in the realization of a project. In this sense, a molecular structure in the brain can be information rich, as can an inscription in an unknown language, or a fossil skeleton before it has been excavated, and so on. This also means that semantic content exists in relational structures and not in atomistic units. And it also allows for a dis-positionalist, connectionist account of the concept of "mental state."

I only can give a sketchy account of the second role; its discussion can be independent of discussion of the first role. For my money, the information-processing interpretation of AI/CS theorizing is a working metaphor (perhaps indispensable methodologically, like fluid metaphors in electrodynamics) for formulating concrete hypotheses about brain architecture and function. However, I believe the interpretation into neurophysiology never could be complete, since the discursive activity it purports to explain is located in the network of relations between cognitively active users of discursive tools. This follows from the ontology of skills, and the thesis of the immanence of meanings, rules, and mental models. It boils down to the following, which links the discussion back to von Eckardt's desiderata for a cognitive science:

1. A cognitive tool is related to its use by convention.
2. Its use is subject to normative constraint.

It follows from (1) that any cognitive tool can be put to a new use, and from (2) that the uses of cognitive tools are skillful uses. For me, the tools and techniques of AI/CS play an indispensable role in developing discursive psychology, in just the way that differential calculus played an essential role in the development of physics. Many important physical intuitions never could have been either expressed or developed without that tool. Interestingly, it still is an open question whether the ontology presupposed by the calculus (namely, the order type of spatial and temporal manifolds) is acceptable as a basis for a realist reading of physics.

Finally, consider one of the most persistent and important problems that becomes clear from using the Berkeleyan argument style as a model. It is the question of where the active powers to perform skillfully are to be located and to what manner of entity they should be assigned. Hyman (1991) offers a wide variety of arguments that the assignment of active cognitive powers must be to persons and not to their brains or any other part of the physical entity in which persons are embodied. This position is very much tied to the "tool-use" account of human cognition, with the brain being just one of the various bodily parts it is possible to use for this or that purpose. Thus, we use the brain for calculating, as we use the hand for writing down the results of our calculations. However, it would be better to consider this issue in another analysis.

Notes

1. Throughout this chapter I will use the abbreviation AI/CS for artificial intelligence-cognitive science—that is, for the point of view that computer modeling is or should be the main tool of research into cognition. My interest in the ontological problems this program raises has been stimulated by discussions with Jerry Aronson, Eileen Way, William de Vries, and Paul Macnamara, among others.

2. I am grateful to Jerry Aronson for his perceptive critical comments that enabled me to formulate the immanence thesis more clearly.

References

Aronson, J.L. 1990. "Verisimilitude and Type-Hierarchies." *Philosophical Topics,* 12, pp. 17–33.

Aronson, J.L., Harré, R. and Way, E. C. 1994. *Realism Rescued.* London: Duckworth.

Boden, M. 1989. *Artificial Intelligence in Psychology: Interdisciplinary Essays.* Cambridge: MIT Press.

Eckardt, B. von. 1993. *What Is Cognitive Science?* Cambridge: MIT Press.

Fetzer, J.H. 1991. *Philosophy and Cognitive Science.* New York: Paragon.

Fodor, J.A. 1975. *The Language of Thought.* New York: Thomas Crowell.

Fodor, J.A. 1988. *Psychosemantics.* Cambridge: MIT Press.

Ganser, M. 1993. "The Structure Grounding Problem," in *Proceedings of the Fifth Annual Conference of the Cognitive Science Society*, p. 151. Hillsdale, NJ: Earlbaum.

Hyman, J. (ed.) 1991. *Investigating Psychology: Sciences of the Mind after Wittgenstein.* London and New York: Routledge.

Muhlhausler, P. and Harré, R. 1991. *Pronouns and People.* Oxford: Basil Blackwell.

Newell, A. 1980. "Physical Symbol Systems." *Cognitive Science,* 4, pp. 135–83.

Searle, J.R. et al. 1992. *(On) Searle on Conversation,* H.Parret and J. Vershueren (eds.), Amsterdam and Philadelphia: Benjamin.

Shotter, J. 1995. "In Conversation: Joint Action, Shared Intentionality, and Ethics," *Theory and Psychology*, vol. 5, no. 1, pp. 49–73.

Sperry, R. 1993. "The Impact and Promise of the Cognitive Revolution." *American Psychologist,* 48/8 vol. 5, pp. 878–85.

Way, E.C.1991. *Knowledge, Representation and Metaphor.* Dordrecht: Kluwer.

Will, F.L. 1988. *Beyond Deduction: Ampliative Aspects of Philosophical Reflection.* Chapman Hall, New York: Routledge.

PART SIX

Christina Erneling

HISTORICAL APPROACHES

How wide should we cast our net in attempting to understand mental life? The chapters in the preceding parts suggested that a substantive widening was necessary if we want to understand the complexity and diversity of mental phenomena. The two chapters in this part add yet another dimension: time. Both Donald and Johnson, from different perspectives, argue that we need to understand how the human mind has developed through past centuries, both in response to morphological changes and particular social and cultural inventions. They claim that the "working model" of the mind currently employed by cognitive science—the Western literate, logical, and scientific mind—is misleading because it is insensitive to how cognition and other aspects of mental life have been formed by such historical factors.

Donald focuses on the evolution of language. He, like Chomsky and Green together with Vervaeke in Part Two, takes language to be a unique human adaptation. Presupposing and interacting with changes to what was essentially a primate brain, language emerged in a social context as an ability to repeat, refine, and remember already existing action patterns and non-linguistic skills, which then took on a communicative and representative function. Thus, to understand a phenomenon like language, we need to look beyond the internal structure of present-day humans, studied by Chomsky and other cognitive scientists, to both non-mental skills and the historical contexts that allowed these skills to develop and take on new functions. Historical investigations not only tell us how skills of this sort developed, but also what they are.

Donald's chapter also brings out examples of the mind's cultural and dynamic character. Mental processes are not static; they change over time, where the mind is considered from both a phylogenetic and an ontogenetic point of view. These changes are not only, or even primarily, the result of genetic or biological changes, but are due to changes in human culture and social practices. Culture, as it were, invades the mind and alters its internal structure. Donald illustrates this by discussing the skill of reading. This skill is a relatively recent cultural invention; yet it changes specific subsystems of the brains of all those people who learn to read and write. No one would claim that reading is something genetically coded. And therefore the fact that there are specific and identifiable brain structures linked to these particular mental skills points to the cultural shaping of the mind in this and in other cases.

While Donald primarily discusses the evolution of the human mind in terms of changes made to the primate brain, Johnson sees the evolution of the human brain as a presupposition for having a mind, but not a sufficient condition. The mind is the result of an evolved brain plus a specific intellectual invention arising from particular historical events, which resulted in new cultural traditions.

Johnson claims that a scientifically adequate understanding of the human mind and how it works requires a recognition of how human thought patterns have changed over time. It is wrong to suppose that folk psychology—the view that humans have mental states like beliefs and desires capable of influencing their actions, which is the basis for our everyday thinking about each other, and cognitive science's "working model" of the mind—reflects the mind's true, timeless, and unchanging structure. Rather, mind came into being, according to Johnson, at some time between 1100 and 750 B.C. with the shift from magical or mythic thought to logical and early scientific thinking. (See also Johnson, 1987.) In fact, scientific thought patterns and so-called folk psychology are so closely connected that we cannot have the one without the other. This, he argues, is a case against the view of those cognitive scientists and philosophers who suppose that folk psychology is a quasi-scientific theory that is bound to disappear with the advancement of science, especially neurophysiology. The elimination of folk psychology would be the elimination of scientific thinking as well, according to Johnson. Thus, he believes the philosophers called eliminative materialists are correct in claiming that folk psychology is not essential to humans, and not a natural characteristic of the human mind, but they are wrong in predicting that this way of thinking will be eliminated in the course of scientific progress, since science presupposes and is built on the foundation of folk psychology.

The chapters in this part, like previous chapters, are critical of present-day cognitive science. But all of them, in different ways, are true to and attempt to develop one important component of the cognitive revolution of the 1950s—namely, its aim to integrate all the disciples that deal with human mental life.

References

Donald, M. 1991. *Origins of the Modern Mind*. Cambridge: Harvard Univeristy Press.
Johnson, D. 1987. "The Greek Origins of Belief." *American Philosophical Quarterly*, 24, 4, pp. 319–27.

24

Merlin Donald

The Mind Considered from a Historical Perspective

Human Cognitive Phylogenesis and the Possibility of Continuing Cognitive Evolution

Thirty (some would say, forty) years ago, the proponents of the so-called Cognitive Revolution set out an agenda that grew out of a unique historical context. Their brainchild, cognitive science, is a product of the postindustrial present and, like many intellectual movements, it has used analytic tools and ruling metaphors that reflect its own immediate cultural environment. Thus, it is not surprising that our working models of cognition tend to typify literate Western (mostly English-speaking) adults, and tend also to resemble the computers that we have built into the fabric of our culture.

In principle, one could and should challenge the generality of any set of models with such a limited reach. The human mind has a long evolutionary history, and we have good reason to believe that it is still changing in significant ways. Certainly, the human mind has not always been the way it is in our society. The cognitive science mainstream could be accused of focusing excessively on the most recent cultural acquisitions of humans: paradigmatic thinking, precise denotation, causal formalisms, and so on. This emphasis on literacy-dependent logical and quasi-scientific thought reflects the roots of cognitive science in intellectual traditions deriving from logical positivism, and doesn't necessarily invalidate it as science. But none of the biases that characterize the field would encourage us to accept *a priori* that it adequately encompasses (or even acknowledges) the range of its subject-matter, namely, the whole of human cognition. Street language, the metaphorical (and very ancient) logic of mythic traditions, magic, custom, and ritual, and the whole range of non-verbal modalities of expression that still make up the greater part of human interpersonal communication are generally swept aside by cognitive scientists, while literacy-based skills are hugely overstressed in our theories. But in science, as in the courtroom, evidence of bias requires us to take a critical stance when we are evaluating the generality of our own theories.

The field of cognitive science has been split between those for whom the Cognitive Revolution has apparently revolved around the inherent computability of mind, and those for whom it has revolved around a set of experimental paradigms aimed at modeling the various "modules" and "mechanisms" of the human mind. Both camps have traditionaly shared a common fascination with mechanistic models, and the invisible (and, as Hebb observed, largely conceptual) brain-machines that support the putative mechanisms proposed by those models. Both camps tend also to strip away the richness of culture and context, in the belief that the "atoms" of mental processing and the "generators" of language can be revealed more precisely by paring away context, isolating general principles, and using a theoretical approach that is perhaps more typical of mathematics or classical mechanics than of biology (in this, AI, cognitive psychology, and even Chomskian linguistics have remained curiously similar to the behaviorism they were meant to supplant).

This predominantly mechanistic/positivistic approach has produced some very useful theoretical progress, and a great deal of interesting new data (the recent book by Posner & Raichle especially comes to mind); but a lot of AI research and much of cognitive psychology remain locked into a few paradigms that are too narrow to do justice to the enormous range of phenomena they are meant to explain. Maybe this is because they are overly concerned with rigorous reductive mechanisms, and not enough with old-fashioned scientific integration. A host of other scientific and scholarly disciplines have focused on the human mind and its representations, and it is worth considering (once again) the possibility that these other disciplines might contribute something fundamental to cognitive science. I am not suggesting that this should take the form of a reduction of one level to another, for instance the reduction of psychology to physiology or genetics; rather, I am suggesting a process of theory-construction that is essentially integrative, not reductive. Chomsky (1993) recently pointed out that strict reductionism often leads us down the garden path into a maze of contradictions, and in any case has rarely been successful, even in physics or chemistry. The more typical process of scientific understanding has been one of constant *integration* across disciplinary lines, rather than the "reduction" of one discipline to the terms of another. Such integration involves a process of successive approximation in which both parties to the integrative process must revise their theories, and importantly, as Chomsky observed, it has often been the more "fundamental" science that has been revised during this process of integration.

In the spirit of integrating knowledge gathered from various disciplines relevant to the study of human cognition, I would suggest that one dimension above all stands out as a potential integrative device central to all of our subject-matter: time, that is, the historical dimension. Mental processes are not static; they change over time, in both the short and the long run. And like many other functions, the structures underlying mental processes would be more easily visualized and modeled if they were viewed in the time dimension. "Mind" is an aspect of evolving life, and evidence of mind (loosely defined) abounds in many organisms. Like all aspects of life, it has a phylogeny; and humanity's version of mind has, in addition, a cultural, as well as a personal history. The description of human phylogeny and cultural history should not leave out cognition; surely it is the most interesting and important part of the story. But ironically, most efforts to reconstruct human emergence have done exactly that: they have left out cognition.

The phenomena we call "mental" emerge only at the highest organismic level; that is, several levels of complexity above inert matter. And, as with many complex functions, there may be more than one evolutionary solution to the construction of "mentalities." Thus, just as there is no universal solution to the problem of thermoregulation, or sight, or navigation, there need not be a universal, once-and-for-all solution to functions such as language and communication. There might be many conceptual and actual solutions to the problem of memory storage, event-perception, or problem-solving; or even representation and communication. Thus, our understanding of cognitive evolution might be better served by moving away from abstract models of hypothetical and supposedly species-general cognitive mechanisms like learning, spatial cognition, and memory; and instead take aim at species-specific theories whose primary purpose is to understand specific cognitive adaptations tailored to particular environments. Human language is one such adaptation, very particular to humans, and there are other cognitive trademarks special to humanity that probably also depend on a unique biological solution to a specific cognitive challenge. Lineage is an important consideration in such a context, because evolution proceeds in a conservative manner, necessarily building on pre-existing adaptations. The lineage of a species has a determining influence on the types of cognitive structures that are available to be subjected to selection pressure, and this should provide a major clue to the nature of structures that emerge later in the evolutionary chronology. In the case of humanity, our lineage is primate, and therefore the cognitive characteristics of primates are the starting-point, the basic working material out of which our kind of mind evolved.

Recently, the subject of cognitive evolution has been rediscovered with a vengeance (e.g., Bickerton, 1990; Bradshaw & Rogers, 1993; Calvin, 1990; Corballis, 1991; Donald, 1991; Dennett, 1991; Greenfield, 1991; Lieberman, 1991; Pinker & Bloom, 1990; Pinker, 1994). Although there is certainly no consensus in these, and many other, efforts to reconstruct human cognitive origins, new efforts at tracking our mental origins are throwing light on cognitive science itself. It may be objected that the problem of human mental origins is unsolvable since it cannot be addressed by direct observation, but this objection also holds for most of the other really interesting problems facing scientists—the origins of the universe, the ultimate nature of matter, etc. We cannot pretend that our species has no documentable history, or that we know nothing about human origins. We know at least as much about the origins of humans as we do about those of any species, and perhaps more. The problem has been that when we have studied our origins in the past, we have not typically looked at them in the context of cognitive theory. Human cognitive evolution has hitherto been addressed obliquely, mostly from the theoretical viewpoint of ethology and sociobiology (Lumsden & Wilson, 1983; Barkow, Cosmides, & Tooby, 1992). But ethology and sociobiology have not approached the problem of cognition as their primary concern; their main focus is on selection and biological fitness, not cognitive structure or mechanism. They have traditionally left open the question of representational mind and its changing nature.

Any serious effort we make toward understanding the evolutionary origins of human cognition should profit everyone, and force us to develop an integrative view of human cognition. In the study of other species, we have already moved in this direction. Animal behaviorists long ago realized the limitations of Skinner-box paradigms and moved toward naturalistic observation, systematic ethology, and a tighter fit between laboratory data and evolutionary theory. Some cognitive psychologists, ably represented by Bruner and Neisser in this book, have heralded a move away from black-box laboratory paradigms toward the complexities of the real world; and this constitutes an important albeit small step in the direction of improving the ecological validity of the field. But cognitive research also needs the broader perspective provided by the study of human prehistory and evolutionary biology, and the more comprehensive database offered by the direct study of human culture. This does not necessarily mean that cognitive science should change its basic approach to model-building. But it should be modeling a wider sample of human cognitive reality than it is, and should be using a natural, ecologically valid database to narrow down the enormous range of alternatives open to it. This approach is already found to a limited degree in some cross-disciplinary fields like neurolinguistics, cognitive anthropology, and neuropsychology.

The Roots of Language in Action

Having said this, is there any single theme that is central to human cognitive evolution, or a property of mind that can be held up as the quintessential human innovation, the one that allowed humans to create the technologies and cultural structures that surround us today? The most obvious candidate for such special status is language; and indeed the origin of language is often the only issue discussed in evolutionary theories coming out of language-oriented disciplines like linguistics and philosophical psychology. However, I would ask my colleagues in these fields to try to see the problem of language from a different perspective.

Human language must have come into existence during some specifiable period of time, in a particular chronology. It could not exist in isolation from its nonlinguistic predecessors, and cannot have evolved in some sort of Creationist vacuum. The strong form of a discontinuity position is simply untenable in an evolutionary framework. But the simplest form of a continuity position also runs into trouble. It can no longer be argued that human language lacks uniqueness—for instance, that it is no more than a variation on some universal quasi-computational "language of thought" found in many species. The evolutionary theorist cannot avoid either horn of this dilemma. In effect, language is unique, and apparently discontinuous with what preceded it, and yet logically it must be continuous with the primate mentality from which it has sprung.

One possible solution is that language did not spring fully armed from the minds of primates in a single step. Rather, there were earlier cognitive adaptations that set the stage for it, on both a cognitive and a social level. Human language undoubtedly has special features, but these features came into existence in the context of a unique pongid cognitive environment; and the human brain is, for all intents and purposes, a variant of the primate brain. Since chimpanzees are genetically much closer to us than they are to most other primates, we must share a great deal, in the cognitive realm, with chimpanzees. There is no credible alternative to the conclusion that the human evolutionary journey started with a primate mind, living something like a chimpanzee life style. And this simple insight has implications for the kinds of language origin theories we can accept as feasible.

For one thing, it implies that the vast majority of the structures and systems supporting human cognition preceded language in the evolutionary chronology, and that modern human language emerged as the end-product of a series of changes to the basic primate mind that we must still possess under the surface. Put bluntly, language was an evolutionary latecomer, an add-on, albeit an add-on with a revolutionary impact. It follows that if our cognitive models of the human mind are to be accurate, they should build carefully from primate to human, and *give precedence to the primate part of the equation*, which is precisely the part left out of most cognitive science. An adequate model of human representation has to start with a primate (read: nonlinguistic) infrastructure that should logically account for most of the built-in features of human cognition. Primates have highly complex and subtle event-perceptions that enable them to understand a great deal more of human culture than any other species, and they achieve this without language.

Assuming this kind of starting point, there is the question of mechanism. How did the earliest humans dig themselves out, figuratively speaking? What new mechanisms would a basically primate brain have needed in order to support the spontaneous appearance of language in the wild? There is no possibility of giving this complex topic a fair hearing in this short chapter, and there are many possible answers, as can be seen from the variety of hypotheses in the current literature. There are, however, a few basic points that need to be kept in mind when thinking about language origins. First, any significant change to the primate mind would have immediate social and cultural consequences; no cognitive adaptation can be convincingly argued without working out this aspect of its emergence. Second, language emerged in a social environment, and is inherently social even in modern humans. Humans cannot learn to speak in communicative isolation, the way, for instance, they can learn to walk or see or use their hands; language requires a public forum for its development. Third, the only public forum for communication is action; and specifically, the actions, or motor activity, of a species must show a great deal of *morphological variability* to support language as it is manifest in humans. Fourth, other primates, despite their capacity for understanding, are very limited and stereotyped in action, relative to humans. It follows that before any public forum for the invention of language could have emerged, primates had to generate much more morphological variability in their action-patterns.

This variability could not be random, however. To support a skill like language, the variable morphology of action must be (1) rehearsable and subject to practice and purposive refinement; (2) rapidly communicable to other members of the species by some means; (3) replicable by other members of the species; and finally (4) driven by some sort of representational agenda. The fourth item doesn't have to be immediately engaged, however; it is entirely conceivable that this kind of morphological variability did not initially emerge in a communicative framework. It is also worth keeping in mind that the well-documented morphological skills of humans extend well beyond speech and language. It is true that infants babble and human languages seem virtually infinite in their variability; but this is generally true of human action. Humans generate this same apparently infinite morphological variety in their play, their athletic skills, their crafts and customs, and in their nonverbal communications and representations. The variability of action is every bit as unique to humans as human language, and logically prior, since it is the precondition of any voluntary communication system.

This suggests that a good starting point for any evolutionary theory of language origins would be in the realm of action. In what part of the system would the morphological skills of humans have originated? My suggestion in this regard, is that, first and foremost, the primate brain needed some changes to its *memory* representation mechanisms. In particular, it needed some means for achieving *voluntary access* to memory. In order to refine and extend action, actions must be rehearsable and autocuable. This is a form of recall; one might argue that the capacity for purposive rehearsal is the most basic form of voluntary recall. It is self-triggered, or autocued recall. But in this primordial case, the self-trigger cannot be a lexical address; rather, it is an image that is tied into an action-schema. The autocued recall of action-schemas is a skill that is probably unique to humans, although there might be a case for a limited degree of this capacity in pygmy chimpanzees. Such recall does not depend on language; it is present in very young human children, who spend much of their time repeating and refining action-patterns. Humans seem to experiment with, and reflect on their own action from an early age, and alter it accordingly. A child might spend an afternoon practicing standing on one foot, for no apparent reason; or throwing stones, or turning somersaults. Or, in a more social vein, spend a lot of time re-enacting episodes. The capacity to rehearse action, and to generate novel actions through iterative refinement, may not appear representational. But the repeated act is in effect representing itself. And the point is, once a capacity for self-initiated rehearsal has been established, it sets the groundwork for similar public representations.

The representational use of autocuable action would not necessarily lead directly to language. It would lead to a general increase in socially diffused skills— toolmaking, for example—and in the range and variability of social communication systems and gestures. The complex of actions that would be triggered by this capacity may be called "mimetic" and finds its cultural expression in what I have labeled as "mimetic culture." Mimetic skill involves much more than imitation; but it must have started as an extension of primate imitative ability. It probably began as an extension of primate imitative skill into the realm of intentional expression, and remains the basis of many modern forms of human expression that are principally based on body-metaphor: dance, expressive games, ritual, much social custom, acting, athletic skills, and toolmaking. Although this capacity was a necessary cognitive bridge to the later evolution of language, it lacks certain key linguistic properties. Mimesis remains an *analog* representational strategy; driven largely by imagery, which refers by means of perceptual resemblance. Mimesis does not use denotation or grammars, or construct lexicons; however, it might generate iconic and metaphoric gesture.

Mimetic representations are evident in human children before they acquire language competence. Their representations include re-enactments, explicit imitations, pantomime, rehearsal and intentional repetition, and reciprocal mimetic exchanges such as those that control joint attention, mutual emotional reactions, and mimetic games. They continue to be important in adults, taking the form of highly variable social customs, athletic skills, and group expressive patterns (such as mass demonstrations of aggression or rejection). Mimesis is the predominant skill underlying dance, simple forms of song, acting, and the transmission of nonverbal arts and crafts. It can be present in the absence of language in certain neurological cases, and in certain of the congenitally deaf (Donald, 1991). An interesting dissociation of mimetic skill, language, and social understanding can be seen in certain well-documented autistic savants, whose artistic skills developed before they had any measurable language skills (Sacks, 1995).

Despite their nonlinguistic nature, mimetic representations constitute a sort of "proto-symbolic" representational system that combines the intentional use of repetition and imitation, gesture, facial expression, gaze, body language, mutual attention-getting, and group reactions and displays. Nonverbal self-reminding is another important spinoff of mimetic capacity. Mimetic capacity is the underlying support system of language; it made hominid society more complex, increasing the need for a more powerful expressive device. At the same time, it allowed hominids to generate the morphological variation in action that, given the right cognitive incentive, would lead to lexical invention. But, without an underlying capacity to generate a wide range of retrievable morphological variation, there would have been no production system in existence to respond to whatever selection pressures favored the appearance of large lexicons and more complex communication systems.

In the field of human memory research, voluntary recall is sometimes known as "explicit" memory (Graf & Schacter, 1985), although the term is usually reserved for linguistically driven explicit recall. But a capacity for explicitly recalling action-schemas through imagery is surely a more fundamental form of explicit recall (explicit procedural memory?) without which the underlying morphology of language would be unacquirable and inaccessible. The rehearsal and refinement of action must have preceded spontaneous lexical invention in the wild, since lexical entries are built from elementary action-patterns, or articulatory gestures that must be imitated, rehearsed, and refined. The acquisition of basic morphophonology is still an essentially mimetic phase in language acquisition, which usually occurs before linguistic reference is acquired. And this skill must be supra-modal, since the morphological components of language are not restricted to vocalization (Poizner, Klima & Bellugi, 1987).

This is evident when we consider the nature of the most elementary component of any language, the lexical entry. Lexical entries are built around addresses, which usually take the form of phonetic production systems, which may be used to "point" to semantic content, and vice versa. This system of counterposing address and semantic content creates a tension that gives the language user great power over the (formerly unaddressable) contents of semantic memory systems, in the sense of clustering and cross-referencing material under a form of self-referential control. But the *sine qua non* of the lexical entry system remains, at the ground floor, an autocuable, rehearsable, refinable bit of morphophonology, and a public system of nonverbal communication, both of which reside at the mimetic level.

The Question of Continuing, Culturally Driven Cognitive Evolution

The evolutionary wedge, the bridge from primate dependency on environmental cuing to the relative autonomy of human memory access, probably originated in a revolution on the level of action and its expressive product, action-metaphor. Needless to say, the limited power of explicit recall found in mimesis was greatly extended when the second level of explicit access—oral language—evolved. I will not try to discuss that issue here, but I am sure I do not have to convince linguists and philosophers that the transition from mimetic to linguistic representation required another very major evolutionary step, one in which a capacity for constructing narrative descriptions encompassed (without superceding) a pre-existing capacity for mimetic invention and retrieval. The net result was a second level of retrievable representations, and an extension of the reach of explicit memory.

Is cognitive evolution continuing in any significant sense? The continuing story of human cognitive evolution looks very much like an extension of the earlier theme: a continuous expansion in the degree of endogenous control over the contents of memory. Of course it involves a great deal more than this; memory storage media and retrieval strategies are only aspects of the human capacities for representation; but they are essential aspects. And once the human capacities for mimetic and linguistic representation were established, the main scenario of cognitive evolution shifted into the realm of new external memory media. The externalization of memory was initially very gradual, with the invention of the first permanent external symbols. But then it accelerated, and the numbers of external representational devices now available has altered how humans use their biologically given cognitive resources, what they can know, where that knowledge is stored, and what kinds of codes are needed to decipher what is stored. Both mimetic and linguistic internal representations can now be externally driven, formatted, recombined, and retrieved by means of a tremendous number of new external memory media. In fact, the whole hierarchy of biological retrieval mechanisms and subsystems acquired over the past two million years has gradually been wired into a fast-moving external memory environment, with results that are difficult to predict. One very good reason for re-examining the longer time frame of cognitive change is to achieve a clearer view of one of the most dramatic outcomes of human evolution, namely the modern, fully literate, electronically plugged-in mind.

This can be regarded as a hybrid system that has acquired various "layers" of representational skill over a long evolutionary period (Donald, 1991). It is not easy to develop a clear picture of the modular structure of such a complex cognitive system because of the interconnectedness and interpenetrability of its highest representational levels. When we study literate English-speaking adults living in a technologically advanced society, we are looking at a subtype that is not any more typical of the whole human species, than, say, the members of a hunter-gatherer group. What would our science look like if it had been based on a very different type of culture? The truth is, we don't know, but it would profit us greatly to find out, because the human cognitive system, *down to the level of its internal modular organization*, is affected not only by its genetic inheritance, but also by its own peculiar cultural history. The idea of culturally imposed cognitive modularity (or quasi-modularity—we don't have to be too rigidly Fodorian here) is not a vague or obscure concept; the existence of one form of this can be seen clearly in the case of reading and dyslexia. Reading can break down in a variety of ways after brain injury, and this suggests that several semi-autonomous brain systems may be involved in reading. Moreover, these systems can break down without the loss of other visual or linguistic functions, suggesting that, to some extent, they are dedicated systems. This implies the existence of specialized subsystems in the brain dedicated largely to reading, and the literature on dyslexia is filled with speculations about the underlying modular structure of these purported subsystems (Morton, 1980; Coltheart, 1985). Usually their putative modular structure is rooted in brain anatomy, whether in a highly localized manner (McCarthy & Warrington, 1990) or in a more distributed, somewhat less localized system (Hinton & Shallice, 1991). These approaches are quite credible on empirical grounds, because there appear to be some neurological regularities in the way literacy skills are represented in the brain.

But no one has seriously proposed that humans could have evolved a specialized reading module in the brain; literacy is far too recent for that to have happened, and besides, members of human groups that have never developed an indigenous writing system can acquire literacy skills in a single generation. This suggests that the brain's reading systems must be "kludges" or culturally jury-rigged arrangements that employ existing neuronal capacities to create the neural complexes that support such an esoteric skill. Above all, neuronal plasticity must be the key to humanity's flexibility in acquiring such radically new cognitive adaptations. Recent work on brain plasticity shows that extensive skill-training can have a major impact on the way the brain allocates its available resources (Merzenich, 1987). Such plasticity would be a prerequisite for cognitive survival in a culture that changes as fast as our own society does. Thus, the way such a recently acquired skill as reading sets itself up in the brain is probably a by-product of neocortical plasticity, amplified by and interacting with the rapidly changing human representational environment (Donald, 1991, chaps. 8, 9).

The plasticity principle applies to a wide range of cognitive subsystems that support not only alphabetic reading, but symbolic literacy in the broader sense. These subsystems are all evident in selective cases of brain injury; together literacy-related functional subsystems of the brain must support several types of lexicons, phoneme-to-grapheme mapping rules, specialized grammars, scanning conventions, novel nonverbal and symbol-based thinking skills, and new classes of semantic content (Donald, 1991; Shallice, 1988; McCarthy & Warrington, 1990). These skills, and the social institutions that support them, form the infrastructure of symbolically literate cultures.

There is also an external infrastructure of symbol-based cultures that takes the form of a variety of external memory devices—writing and counting systems, large artifacts such as books and reports, indexing and classification systems, electronic retrieval and a variety of other symbolic storage devices. Individual minds carry the burden of serving as a link between the external infrastructure of representation, and the real world knowledge that only brains seem (so far) capable of acquiring. But the changing external structure continually imposes new patterning on developing brains as cultures change. In principle, at least for cognitive changes as massive as the shift to literacy, the modular structure of both mind and brain must have been influenced by culture.

Once granted, this general principle, that culture might directly influence and "install" cognitive architecture (both inside and outside the biologically defined individual), should change the way cognitive scientists do business. We have always assumed that culture affects cognition only in ways that we would consider irrelevant or unimportant—for instance, in determining the particular language or version of history we learn, or in specifying the specific customs, experiences, stereotypes, and attitudes we might acquire. Such cultural variations were regarded as trivial because they have no implications for the basic researcher, who is presumably more interested in cognitive universals. However, the idea that certain major cultural shifts can "invade" the brain and impose major structural, change down to the level of fundamental representational architecture, suggests that cognitive fundamentals are not always universals in the biological sense. Some cognitive fundamentals may be largely cultural products, and thus culture cannot be safely ignored by basic researchers in cognitive science. Note that the word "culture" also encompasses our current technological changes in electronic communications media, which are undoubtedly having a major impact on individual cognition—another issue that needs further systematic study by cognitive scientists.

Thus, the externalization of memory, which came quite recently, brought with it new mechanisms of storage that had radically different properties from those of internal biological memory (Donald, 1991). Perhaps even more important, however, were the novel representational and retrieval paths that have been created by various new electronic media. We might gain some control over what is happening if we study the external memory environment and its effect on us. We should do this optimistically, just as we did when we decided to focus on the physical environment; with the hope that things might turn out a little better as a result of our efforts. This should be a priority for the near future. If cognitive scientists do not look into the cognitive implications of current technological change, I cannot imagine who will.

References

Barkow, J.H., Cosmides, L and Tooby, J. 1992. *The Adapted Mind*. New York: Oxford University Press.

Bickerton, D. 1990. *Language and Species*. Chicago: University of Chicago Press.

Bradshaw, J. & Rogers, L. 1993. *The Evolution of Lateral Asymmetries, Language, Tool Use and Intellect*. New York: Academic Press.

Calvin, W. 1990. *The Ascent of Mind*. New York: Bantam.

Chomsky, N. 1993. *Language and Thought*. Moyer Bell.

Coltheart, M. 1985. "Right Hemisphere Reading Revisited." *Behavioral and Brain Sciences*, 8, pp. 363–365.

Corballis, M.C. 1991. *The Lopsided Ape: Evolution of the Generative Mind*. Oxford: Oxford University Press.

Dennett, D.C. 1991. *Consciousness Explained*. Boston: Little, Brown.

Donald, M. 1991. *Origins of the Modern Mind*. Cambridge: Harvard University Press.

Donald, M. 1993a. "What We Were, What We Are Becoming: Cognitive Evolution." *Social Research* 60, pp. 143–170.

Donald, M. 1993b. "Précis of Origins of the Modern Mind." *Behavioral and Brain Sciences,* 16, pp. 737–748.

Donald, M. 1993c. "On the Evolution of Representational Capacities." *Behavioral and Brain Sciences,* 16, pp. 775–791.

Graf, P. and Schacter, D. 1985. "Implicit and Explicit Memory for New Associations in Normal and Amnesic Subjects." *Journal of Experimental Psychology: Learning, Memory and Cognition,* 11, pp. 501–518.

Greenfield, P. M. 1991. "Language, Tools and the Brain: The Ontogeny and Phylogeny of Hierarchically-Organized Sequential Behavior." *Behavioral and Brain Sciences,* 14, pp. 531–95.

Hinton, J. and Shallice, T. 1991. "Lesioning an Attractor Network: Investigations of Acquired Dyslexia." *Psychological Review,* 98, pp. 74–95.

Karmiloff-Smith, A. 1992. *Beyond Modularity.* Cambridge: MIT Press.

Levelt, W. 1989. *Speaking.* Cambridge: MIT Press.

Lieberman, P. 1991. *Uniquely Human: The Evolution of Speech, Thought and Selfless Behavior.* Cambridge: Harvard University Press.

Lumsden, C. and Wilson, E.O. 1983. *Promethean Fire.* Cambridge: Harvard University Press.

McCarthy, R. and Warrington, E. 1990. *Cognitive Neuropsychology.* New York: Academic Press.

Merzenich, M.M. 1987. "Dynamic Neocortical Processes and the Origns of Higher Brain Functions," in J.-P. Changeux and M. Konishi (eds.), *The Neural and Molecular Basis of Learning.* New York: Wiley.

Morton, J. 1980. "The Logogen Model and Orthographic Structure," in U. Frith (ed.), *Cognitive Approaches to Spelling.* New York: Academic Press.

Nelson, K. 1990. "Event Knowledge and the Development of Language Functions," in J. Miller (ed.), *Research on Child Language Disorders*, pp. 125–41. Boston: Little, Brown.

Olton, D.S. 1984. "Comparative Analysis of Episodic Memory." *Behavioral and Brain Sciences,* 7, pp. 250–251.

Pinker, S, and Bloom, P. 1990. "Natural Language and Natural Selection." *Behavioral and Brain Sciences,* 13, pp. 707–784.

Pinker, S. 1994. *The Language Instinct: How the Mind Creates Language.* New York: Morrow.

Poizner, H. Klima, E.S., and Bellugi, U. 1987. *What the Hands Reveal about the Brain.* Cambridge: MIT Press.

Posner, M. and Raichle, M. 1994. *Images of Mind.* New York: Scientific American Library.

Sacks, O. 1995. *An Anthropologist on Mars.* New York: A. Knopf.

Savage-Rumbaugh, E.S., Murphy, J. Sevcik, R.A., Braake, K.E., Williams, S. L., and Rumbaugh, D. 1993. *Language Comprehension in Ape and Child. Monographs of the Society for Research in Child Development.*

Shallice, T. 1988. *From Neuropsychology to Mental Structure.* Cambridge: Cambridge University Press.

Tomasello, M., Kruger, A.C., and Ratner, H.H. 1993. "Cultural Learning." *Behavioral and Brain Sciences,* 16, pp. 495–552.

25

David Martel Johnson

Taking the Past Seriously

How History Shows that Eliminativists' Account of Folk Psychology Is Partly Right and Partly Wrong

Not many years ago, it was common for people to suppose that any appeal to empirical considerations in the context of a philosophical argument was necessarily irrelevant, lacking force, and vulgar. For example, Bertrand Russell defended himself against such a charge of "empirical vulgarity" in his discussion of Kant's epistemology (1945, p.716). But today this no longer is the case. On the contrary, most philosophers now claim, in the style of Karl Popper, they have as much right as any other investigators to employ all available facts and methods in their search for truth. (See Passmore, 1985, chapt. 1, esp. pp.10–11.) In fact, quite a few affirm the even more extreme Quinian view that "There is no first (i.e., pre-scientific) philosophy," and, because of this, "Philosophy of science is philosophy enough." The relatively new discipline of cognitive science provides a clear example of this idea (also accepted by the earliest Greek philosophers) that philosophy and the natural sciences are identical, or at least intimately blended and mutually supportive. More particularly, cognitive scientists maintain—consonant with Quine's ideal of "naturalized epistemology"—that certain facts drawn from observational sciences like biology, physiology, and psychology not only are relevant to a proper understanding of mind, but play a crucially important role in this understanding.[1]

However, even though philosophers in general and those interested in cognitive science in particular are again willing to consider empirical research relevant to their concerns, there still seems to be one area where the once fashionable charge of empirical vulgarity retains its full vigor. What I am talking about is the currently popular view that discoveries made by historians could not possibly be helpful for answering (scientifically influenced) philosophical questions, because the discipline of history is essentially unscientific.[2] For example, cognitive science is supposed to be a synoptic enterprise that combines several narrower fields that once were considered unrelated. What sub-fields? According to Howard Gardner, the principal parts of cognitive science are philosophy, psychology, linguistics, artificial intelligence, neuroscience, and anthropology. (See his schematic representation on page 37 of 1985.) More recently, Francis Crick (1994, p. 272) has proposed an even narrower list—namely, linguistics, cognitive psychology, and Artificial Intelligence. Why does neither of these commentators also mention history? Is it not plausible to suppose that the record of humans' past struggles to think in helpful and illuminating terms might have something to teach us about the nature and functioning of mind?

People often conceive of cognitive science as specifying a middle analytical level of mental representations, "wholly separate from the biological or neurological, on the one hand, and from the sociological or cultural on the other" (Gardner, 1985, p.6); and some might suggest that this already explains why the great majority of cognitive scientists ignore history. Nevertheless, I think there is a still more basic reason. That is, people like Gardner and Crick extrapolate from a certain interpretation of "fundamental" sciences like physics and chemistry to the idea that the proper subject matter of all the sciences—and of philosophical thinking insofar as it relates and appeals to science—is things that never change (at least in essential respects). It follows from this Aristotle-like view of scientific knowledge that while historical investigation might be capable of satisfying certain antiquarian interests, it cannot tell us anything important about the world itself. For example, suppose— in accordance with the view we are discussing—that all members of our species, present, past, and future, had the same basic organization of their minds. In that case, there never would be any reason to think about minds historically. Or, roughly the same idea expressed in different terms, commentators might claim that accounts of historical events are scientifically worthless for approximately the same reasons as poetry. In other words, historical events are products of various accidental and non-repeatable factors; therefore it always is necessary to specify them in terms relative to certain thoughts, values, etc., rather than as instances of the objective and stable laws and patterns characteristic of scientific thought.[3] For example, Merlin Donald (1996) reminds us that human minds not only are (partly) formed by physical and physiological factors common to the whole species, but also by the experiences and cultural achievements of single persons—for example, literacy and the first language someone happens to learn. But someone might challenge this claim by insisting that all the latter influences are merely quirks associated with particular individuals and groups, and therefore cannot fall under any of the abstract, repeatable natural kinds that it is the job of science to discover and investigate.

Still more narrowly, commentators might argue that the reason science cannot make theoretical use of historical discoveries is that over the long run historical events always cancel one another in such a way as to leave everything else more or less the same. For example, they might say that so-called human progress cannot really accomplish anything, since it has no ultimate effect on things themselves. We find this idea expressed in the familiar saying, "To read history is to read about ourselves," and in the claim some anthropologists make that, "One of our best relics of early man is modern man." (See Howell et al., 1965, p.172)[4]

I suggest that the view just described is wrong; and because of this, cognitive scientists are not justified in taking radically different attitudes toward experimental science on one side and the study of history on the other. The crucial point is that not all historically determined changes associated with human thought have been ephemeral, self-canceling, and thus necessarily irrelevant to science. My argument to show this turns on just one example—namely, the fact, attested to by most historians of humanity's earliest periods, that the thinking of human beings has passed through two main successive phases. Following a suggestion of Henri and H.A. Frankfort in 1949, it is convenient to describe this progression by saying that an earlier phase of "mythopoeic" thought was later replaced by "logical and objective" thinking. Virtually all historians also agree that the shift from the one to the other happened sometime between 1100 and 750 B.C.E., in the area of the eastern Mediterranean,[5, 6] and that the change in question apparently has been permanent. (These data are mentioned, for example, by Chester G. Starr in 1965/1991, p. 186.) This means that even if the human race survives millions of years into the future, it is reasonable to expect that the mentality of all those future humans will continue to be characterized by logical thinking. To summarize, my principal thesis is that cognitive scientists will not succeed in developing a scientifically adequate understanding of the human mind and how it works, unless and until they come to see this particular historical shift not just as centrally important to mind, but as constitutive of it.

Can this point really have a bearing on usual philosophical debates about the mind? Consider the following example. Eliminative materialists give the name "folk psychology" to unreflective ways of thinking and talking about ourselves and our actions, to mark the fact that they believe these conceptions amount to a (quasi-) scientific theory about human beings' inner workings. Here are two representative principles of this theory, suggested by Paul Churchland (1988, p.59):

1. "Persons who feel a sharp pain tend to wince."
2. "Persons who want that P, and believe that Q would be sufficient to bring about P, and have no conflicting wants or preferred strategies, will try to bring it about that Q."

Churchland and other eliminativists think future scientific progress eventually will discredit folk psychology, and replace it with other, more informative and empirically justified accounts of the same underlying subject matter. Moreover, since "mind" seems to be the most central concept of folk psychology, this also amounts to the prediction that science will show there are no such things as minds. That is, eliminativists claim scientists will establish that the notion of "mind" is just as illegitimate and misleading as "phlogiston," "demon," "unicorn," "caloric fluid," etc.—and for roughly the same reasons.

Eliminativists admit that folk psychology has been associated with human beings for a long time. Thus, Paul Churchland says readers of Sophocles' plays should find it obvious that this author conceived of the thought and behavior of human beings in approximately the same way as the vast majority of people still do today. Nevertheless, folk psychology's lengthy career is not a reason for thinking that it is true. *Au contraire*, its having remained essentially unchanged during the 2,500 years since Sophocles shows that it "is a stagnant or degenerating research program, and has been for millennia" (1981, p.75). And this in turn is grounds for believing folk psychology is false. Presumably, Churchland and other eliminativists would say that even if humans have thought about themselves and their behavior in terms of folk psychology from the very beginning of the species until the present moment, this still amounts to nothing more than an historical accident, because humans always have had the power to criticize, reject, and replace this "naive" way of thinking—even if they did not begin to do so self-consciously until the advent of modern neuroscience.

By contrast, certain critics of eliminativism (e.g., Horgan & Woodward in 1985, *passim*, esp. pp.197–199) say the whole idea of replacing folk psychology with something else is nonsense, since folk psychology is an essential—and therefore unchangeable—part of human nature. These theorists maintain that there never were, will, or could be human beings who did not conceive of themselves according to folk psychology's rules and categories, because any creature who failed to do this would not count as a (normal, adult) human.

At first this may strike one as a conflict between incompatible timeless intuitions, which only could be settled empirically by waiting to see in what direction experimental and theoretical science proceeded in the future.[7] But, over and above narrowly scientific considerations, I think (equally empirical) historical facts are also relevant here. More especially, it follows from the hackneyed historical points mentioned before—right now, without waiting to observe future science— that eliminativists and their critics are each partly right and partly wrong. That is, these facts imply that the earliest members of the species *homo sapiens sapiens* did not either recognize or employ folk psychology; and therefore eliminativists are right to say that folk psychology is not a permanent part of human nature. On the other hand, if it really is true that folk psychology did not appear at the same time as—or earlier than—the beginning of our species, when, where, and for what reasons did it originate? Again, the answer to this question implied by the work of historians is that folk psychology first took shape in association with the cultural achievement we call the invention of (proto-) scientific thought, about 1100 to 750 B.C.E., in the area of the eastern Mediterranean. This means critics of eliminativism are also right when they claim that scientific research never will replace folk psychology with a better alternative, because science presupposes the same distinctions between truth and falsity, and between the relevant and the irrelevant, which only became generally known and important with the advent of folk psychology. In other words, it is just as nonsensical to look forward to a time when scientific theories will supplant and eliminate folk psychology, as to expect that someday sprinters will be able to run without their legs. The one depends on, and cannot occur in the absence of, the other.[8]

Some recent philosophers suppose the most important question to ask about mental states is simply whether—and, if so, how—they exist. But a still more basic question is whether purported subjects of the states in question think and behave in ways that are capable of "supporting" them. For example, Daniel Dennett locates himself at the approximate middle of a scale of different views on the existence or nonexistence of mental states, when he distinguishes

> Fodor's industrial strength Realism (he writes it with a capital 'R'); Davidson's regular strength realism; my mild realism; Richard Rorty's milder-than-mild irrealism, according to which the pattern is only in the eyes of the beholder; and Paul Churchland's eliminative materialism, which denies the reality of beliefs altogether. (1991, p. 30)

Dennett foresees no great problems about locating and describing the thought and behavior appropriate to subjects' having mental states like beliefs and desires, since he thinks a very wide range of thinking and acting would suffice to show that this was the case. More particularly, his mild realism dictates that, under the right circumstances, almost any activity that depends on stored information could be the expression of a belief, and any almost bit of goal-seeking behavior could express a desire. The single crucial point is that one or several observers should find it natural, and perhaps also effective, to adopt an "intentional stance" toward the behavior in question. It follows from this that even primitive, non-human animals like jellyfish, snakes, and squids can be literal believers and have literal desires. And the same must be true as well even of certain artificially constituted machines like chess-playing computers.

Nevertheless, I maintain that Dennett's view is wrong, and that this is proved by the fact that there was a time when humans (much less jellyfish and computers) did not have any beliefs or desires at all, because they had not yet learned to think and act in the ways required for the existence of these mental states. Consider the following point. It is incoherent to speak of someone or something as a literal possessor of beliefs and desires (and therefore as participating in folk psychology), unless the thoughts and actions of the creature or object in question are set against a background of relatively sharp and consistent distinctions between (1) truth as opposed to falsity, and (2) relevant evidence as opposed to irrelevancies.

Steven Stich described the case (1983, p.55) of elderly Mrs. T., who suffered from a degenerative condition similar to Alzheimer's disease. Stich reasonably concluded that even though Mrs. T. asserted repeatedly and vehemently that President McKinley had been assassinated, these words did not count as an expression of something she believed, because of the fact that she no longer could affirm and deny certain other propositions connected with this first one in appropriate and consistent ways. For example, if someone asked, "Is McKinley dead?" she failed to produce the expected answer, "Yes." Analogously, even relatively superficial knowledge of ancient and (human) pre-history is enough to show that early members of our species could not have had literal beliefs and desires, because these people thought in ways that involved quite different assumptions from the ones that operate in the thinking of normal people today.

Anthropologists[9] of our century have uncovered a series of clues about the intellectual activity that stood behind cave paintings and etchings produced by Cro-Magnon people in Europe (early *homo sapiens sapiens*) over 40,000 years ago. One such clue is the fact that those who made these paintings almost never decorated the place where they lived, but saved their artistic skills instead for special underground chambers that are (or were in Cro-Magnon times) extremely inaccessible. For example, even modern spelunkers using electric lights, ropes, clamps, helmets, boots, etc. are not able to reach some of these places until after having clambered and crawled laboriously through rough, narrow, and dangerous passages for up to a full hour. Second, the vast majority of the pictures on walls of the caves are of animals belonging to species that the early humans hunted; many of the pictures show the animals as wounded and bleeding, or with spears sticking into them, or caught in traps. Apparently, the pictures did not have a merely aesthetic function, but also were employed ritualistically because of certain practical effects they were assumed to have. In particular, the pictures' creators probably conceived of them as playing an important role in bringing about good fortune in hunting. (Some anthropologists say the same thing is indicated by the fact that there seem to have been "lucky" places on cave walls, where artists painted one picture over another many times.) Finally, a third clue is that very few pictures of the humans who hunted the animals occur in the caves—and those that do tend to be simple stick figures.[10] Again, the natural explanation of this is to say the painters wanted to avoid exposing themselves and fellow hunters to the strong, possibly fatal magical influences that they supposed their pictures already had unleashed on prey creatures. (See Howell et al., 1965, chap. 7, esp. p.148.)

In today's world, the great majority of artists have no trouble distinguishing their own creations from real, extra-artistic things. And the few unfortunate individuals who sometimes cannot do this—for example, the comedian Jonathan Winters—find themselves forced to undergo medical treatment as a result. Nevertheless, it is not safe to assume that the cultural world of paleolithic times operated along approximately the same lines as ours, or took the same things for granted.[11] In particular, even if various things the Cro-Magnons did and said had a superficial resemblance to our expressions of beliefs and desires, this is not a good reason—remembering Stich's Mrs. T.—for supposing that their speech and behavior must literally have been such expressions.[12] For example, because these artists conceived of their creations not as passive descriptions, records, or celebrations of hunting success, but as powerful (partial) producers of this success, it probably was not a sign of madness, but rather of ordinary "common sense" for them not to recognize a sharp distinction between what they painted and the world outside their paintings. For instance, whoever drew the image of a mammoth struggling in a trap reproduced in Howell et al. (1965, p.148—also see pp.134–135) did not consider this production as either separate in space and time from, or causally irrelevant to, the living mammoth he and his associates expected to meet in the next few days' or weeks' hunting. Further, it is likely that this same "magical mentality" extended beyond religious and ritual contexts to more ordinary situations as well—for example, deciding what clothes to make and wear, trying to guess the species of a distantly glimpsed prey animal, or choosing a mate. Thus, a Cro-Magnon woman might assume without question that the exact pattern assumed by a set of bones which a medicine man threw on the ground somehow was involved in—or even partly constitutive of—her future happiness or grief with a prospective husband.

Should we describe cases like these as more or less ordinary beliefs in nonphysical causes; or should we say instead that there were no beliefs (or desires) present here at all? I favor the second, more radical answer because the background of these peoples' thoughts and actions led them to draw "incorrect implications" for belief.[13] More specifically, it is plausible to say that beliefs must be capable of being true or false. But if an imaginary time-traveler from our century asked a paleolithic painter—concerning the set of marks the artist just had daubed and sprayed on a cave wall—"Is this a mammoth?" he could not obtain a clear or unambiguous answer. Again, much the same would apply if the traveler asked a Cro-Magnon woman peering at small objects that a witch doctor just had thrown on the ground, "Is this a set of bones?" Furthermore, if beliefs and desires are the essential kernel of folk psychology generally, this also implies that these ancient humans were not participants in folk psychology itself.[14] In other words, although it is obvious that the artists-hunters-magicians we have been discussing were capable of surviving and thriving in their environments, the intellectual means they employed to do so were different from the means that virtually any person would deploy in similar situations today. To deny this (as, e.g., Dennett does) is to be guilty of the ethnocentric fallacy of projecting oneself and one's distinctive style of thinking into an alien context for *a priori* rather than empirically justified reasons.

Paul Churchland claims the main effect of early humans' magical view of the world on their broad intellectual categories was that it led them to apply folk psychology more inclusively than we do now. He says,

> [t]he presumed domain of [folk psychology] used to be much larger than it is now. In primitive cultures, the behavior of most of the elements of nature were understood in intentional terms. The wind could know anger, the moon jealousy, the river generosity, the sea fury, and so forth. These were not metaphors. Sacrifices were made and auguries undertaken to placate or divine the changing passions of the gods. Despite its sterility, this animistic approach to nature has dominated our history, and it is only in the last two or three thousand years that we have restricted [folk psychology's] literal application to the domain of the higher animals. (1981, p.74)

But the idea of a river's believing it had an obligation to be generous, or the wind's desiring to rip down some person's animal skin shelter, is either nonsense or merely poetic imagery, because—and here I agree with Putnam (1975)—beliefs and desires are not self-contained, punctiform states of single subjects, but are essentially connected with many other objects and situations as well. Thus, a river could not believe something, or the wind desire something else, unless the whole world were organized along quite different lines from those that scientific investigation tells us are actually the case.

What is the upshot of this discussion? Eliminativists have made an important contribution to our self-understanding by correctly pointing out that folk psychology—that is, the sort of logically consistent, relevance-based thinking employed by most people today—is *not* an indispensable part of the human condition. In fact, far from being essential, folk psychology is not even "natural" to humans, as proved by the fact that the members of our species did not think in these terms during the vastly greater amount of time the species has existed. Nevertheless, in another respect eliminativists are mistaken, because they give a topsy-turvy account of the actual import of folk psychology. Rather than an impediment that scientific knowledge eventually will expose and discharge, it is an irreplaceable foundation for this knowledge.

A still broader moral is that one should not think of our civilization (i.e., the culture shared by virtually all people now living) as beginning at the point in history when one or more individuals discovered a new set of answers to certain eternal questions available to humans just by virtue of self-examination. Instead of answers, it makes more sense to suppose these first representatives of the so-called Western tradition discovered (more correctly, invented) a new set of problems. In fact, a quick way of summarizing the content of this chapter is to say that two of these newly uncovered problems—which people had not contemplated in a serious or systematic way before then—were (1) "What is the mind?" and (2) "How does it work?"[15]

Notes

1. Are they really in a position to make such a claim? Francis Crick believes most are not. He expresses this idea by quoting the following two cynical comments from psychologist Stuart Sutherland: (1) "Cognitive scientists rarely pay much attention to the nervous system." And (2) ". . . the expression ['cognitive science'] allows workers who are not scientists to claim that they are" (1994, p.272).

2. Cf. Nagel (1995, p.7):

> In the United States, philosophers don't have to know much about history or anything about literature, but they are expected to know some science, to have at least an amateur's grasp of the contributions of Newton, Maxwell, Darwin, Einstein, Heisenberg, Cantor, Gödel, and Turing, as well as of some representatives of the human sciences like Chomsky, Sperry, and Arrow—all of which provide data for philosophical reflection.

3. See Bruner (this volume):

> Modern historiography teaches one big truth that should sober us. The Past (with a capital letter) is a construction: how one constructs it depends upon your perspective toward the past, and equally upon the kind of future you are trying to legitimize.

4. Pre-World War I historian, Wilhelm Windelband, remarks that

> History has remained an object of indifference to most philosophical systems, and has emerged as an object of philosophical investigation relatively late and in isolated cases (1901/1958, p.5).

Notice that Windelband only speaks of history as an object of philosophical investigation, and not as part of the means philosophers use to investigate other things.

5. This is a theme affirmed both by "classical" as well as more recent authors. See, for instance, Zeller (1883/1980, pp.2–3, 9, 19); Goperz (1901/1964, pp.3–42); Burnet (1914/1978, pp.4–10); Cornford (1932/1979, p.5); Snell (1948/1960), *passim*, esp. chaps. 9 and 10); Kirk, Raven, and Schofield (1957/1983, p.99); Wheelwright (1960, Introduction, esp. p.1); Lloyd (1970, Chapter 1, esp. pp.1 and 8); Hornung (1971/1982, p. 249); Guthrie (1975, pp.15–16); Brumbaugh (1981, pp.5–10 and chap. 1); Ring (1987, p.9).

6. Some claim that a person might as well eat anything he or she happens to enjoy, since professional nutritionists—the supposed "experts"—disagree among themselves. But careful investigation shows that nutritionists' disagreements are almost always about superficial matters of detail, but that they are virtually unanimous concerning the basic principles of their discipline. I maintain that historians are similar to nutritionists in this respect.

7. For example, this is essentially what Churchland claims (1988, p.61).

8. I think Bronowski's opening statement in 1958 is literally true—namely, "The most remarkable discovery made by scientists is science itself."

9. Readers may think I am being unfair to Howard Gardner because, as noted earlier (p. 365), he lists anthropology among the sub-disciplines of cognitive science. My reply is that the remainder of Gardner's book does not show him to be particularly interested in the historical dimensions of anthropology—presumably because he does not think such historical matters have a high degree of scientific respectability.

10. There are always exceptions. For example, there was a detailed portrait of a human face on a limestone plaque left in a cave near Vienne in the upper Rhone Valley. A photograph of this object appears on page 8 of Prideaux et al. (1973).

11. The Frankforts sometimes propose to account for the non-logical thinking characteristic of our ancient ancestors by comparing it with present-day imaginative, creative, or esthetic thought. This is misleading if it inclines one to suppose that ancient people had some sort of

innate disposition to distinguish the literal and metaphorical in the same way as we do now. It is a familiar fact that cultural factors can influence brain development irreversibly, and thereby transform inborn capacities. Thus, even if every Cro-Magnon started life with exactly the same innate physiology people have today, this does not mean the patterns of thought available to them in adulthood could not have been profoundly different from ours.

12. A similar point applies to non-human animals. For example, Leiber says,

> I would agree with the claim that Washoe-Delta [a chimpanzee] has no "sense of grammaticality." But when she makes the two hand signs for "give banana" there is absolutely no doubt in my mind that she *means* bananas, that she is *expressing her desire* for a banana—just watch what she does when you show up with something else! (1985, p.37)

Nowadays, however, many philosophers point out that desires and beliefs are interdependent by virtue of playing complementary roles within a larger context. In particular, I maintain that desires presuppose the same background of consistent action and thought, and objective methods of evaluating evidence, that beliefs also presuppose. Thus, if Cro-Magnon humans did not possess such a background, it follows that non-human animals like chimpanzees could not have it either. To speak of non-humans as having beliefs and desires is like describing a dog as "walking"whenever it briefly manages to balance upright on its hind legs—that is, walk *like us*. Literally speaking, one only is entitled to attribute "as-it-were" or, at best, "proto"-beliefs and desires to animals of this sort. (I also mentioned this same parallel in another place. See 1988, pp.288–289.)

13. Cf. Fodor (1990, p.320):

> The line of argument that leads to this conclusion is familiar from the writings of Dennet, Davidson and others. Roughly, it's a condition on entertaining a belief that one be disposed to draw the appropriate inferences, and which inferences it is that one is disposed to draw partially determines which belief it is that one is entertaining.

14. For example, Stich's "map of mind" on page 75 of 1983 is divided in two major parts, reflecting the fact that the mind is fundamentally a "believing-desiring mechanism."

15. I am indebted to my co-editor, Christina Erneling, for insightful comments on an earlier draft of this chapter.

References

Bronowski, J. 1958. "The Creative Process," *Scientific American*, September. (Reprinted in Gingerich, 1952–82, pp. 2–9).

Brumbaugh, R.S. 1981. *The Philosophers of Greece*. Albany: State University of New York Press.

Burnet, J. 1914/1978. *Greek Philosophy: Thales to Plato*. London: Macmillan Press.

Churchland, P. 1981. "Eliminative Materialism and the Propositional Attitudes." *Journal of Philosophy*, 78, pp. 67–90.

Churchland, P. 1988. *Matter and Consciousness: A Contemporary Introduction to the Philosophy of Mind*, rev. ed. Cambridge: MIT Press.

Cornford, F. 1932/1979. *Before and After Socrates*. Cambridge: Cambridge University Press.

Crick, F. 1994. *The Astonishing Hypothesis: The Scientific Search for the Soul*. London: Simon and Schuster.

Dennett, D.C. 1991. "Real Patterns." *Journal of Philosophy*, Vol. 78, pp. 27–51.

Fodor, J.A. 1990. "Psychosemantics or: Where Do Truth Conditions Come From?," in W.G. Lycan, *Mind and Cognition*, pp. 312–37. Oxford: Basil Blackwell.

Frankfort, H.; Frankfort, H.A.; Wilson, J.A.; Jacobsen, T. 1949. *Before Philosophy: The Intellectual Adventure of Ancient Man*. London: Penguin Books.

Gardner, H. 1985. *The Mind's New Science: A History of the Cognitive Revolution*. New York: Basic Books.

Gingerich, Owen (ed.). 1952–82. *Scientific Genius and Creativity* (*Readings from Scientific American*). New York: Freeman.

Gomperz, T. 1901/1964. *Greek Thinkers: A History of Ancient Philosophy*. New York: Humanities Press.

Guthrie, W.K.C. 1975. *The Greek Philosophers from Thales to Aristotle*. New York: Harper Colophon.

Horgan, T. and Woodward, J. 1985. "Folk Psychology Is Here to Stay." *The Philosophical Review*, XCIV, No. 2, pp. 197–226.

Hornung, E. 1971/1982. *Conceptions of God in Ancient Egypt: The One and the Many*. Ithaca, NY: Cornell University Press.

Howell, F. C. and the Editors of *Life*. 1965. *Early Man*, New York: Time Inc.

Johnson, D.M. 1987. "The Greek Origins of Belief." *American Philosophical Quarterly*, Vol. 24, No. 4, pp. 319–27.

Johnson, D.M. 1988. "Brutes Believe Not." *Philosophical Psychology*. Vol. 1, No. 3, pp. 279–94.

Kirk, G.S., Raven, J.E., and Schofield, M. 1957/1983. *The Presocratic Philosophers: A Critical History with a Selection of Texts*. Cambridge: Cambridge University Press.

Leiber, J. 1985. *Can Animals and Machines be Persons?* Indianapolis, IN: Hackett.

Lloyd, G.E.R. 1970. *Early Greek Science: Thales to Aristotle*. London: Chatto and Windus.

Lycan, W.G. 1990. *Mind and Cognition*. Oxford: Basil Blackwell.

Nagel, T. 1995. *Other Minds: Critical Essays 1969–1994*. New York and Oxford: Oxford University Press.

Passmore, J. 1985. *Recent Philosophers: A Supplement to a Hundred Years of Philosophy*. London: Duckworth.

Prideaux, T. and the Editors of Time-Life Books. 1973. *Cro-Magnon Man*. New York: Time-Life Inc.

Putnam, H. 1975. "The Meaning of 'Meaning'," in K. Gunderson (ed.), *Minnesota Studies in the Philosophy of Science*, Vol. 7, *Language, Mind and Knowledge*. Minneapolis: University of Minnesota Press.

Ring, M. 1987. *Beginning with the Pre-Socratics*. Mountain View, CA: Mayfield Publishing Co.

Russell, B.1945. *A History of Western Philosophy: And Its Connection with Political and Social Circumstances from the Earliest Times to the Present*. New York: Simon and Schuster.

Snell, B. 1948/1960. *The Discovery of the Mind*. New York: Harper.

Starr, C.G. 1965/1991. *A History of the Ancient World*. New York: Oxford University Press.

Stich, S. 1983. *From Folk Psychology to Cognitive Science: The Case Against Belief*. Cambridge: MIT Press.

Windelband, W. 1901/1958. *A History of Philosophy, Vol. I: Greek, Roman, Medieval*. New York: Harper.

Wheelwright, P. (ed.), (1960). *The Presocratics*. Indianapolis, IN: Bobbs-Merrill.

Zeller, E. 1883/1980. *Outlines of the History of Greek Philosophy*. New York: Dover.

Christina Erneling

Cognitive Science and the Future of Psychology

Challenges and Opportunities

"For in psychology, there are formal methods and ontological disagreement." I use this paraphrase of Wittgenstein's (1953, p.232) evaluation of psychology as a means of pointing out the fundamental challenge as well as opportunity that the chapters in this volume provide. The primary sort of reassessing that confronts psychology today is not methodological or disciplinary, but metaphysical. Psychology may go on to develop into a more and more fragmented science, and perhaps fall into the danger of disappearing altogether into other disciplines like biology, neurophysiology, genetics, linguistics, sociology, anthropology, or even literature. On the other hand, psychology can address its metaphysical past and future. Many of the chapters in this volume undertake the second task in both a critical and constructive way. For example, some chapters forcefully defend the traditional cognitive science conception of mind and mentation (e.g., Chomsky, Boden); others equally forcefully reject it (e.g., Bruner, Harré, Putnam). This disagreement is perhaps most clearly illustrated by the contrasting approaches to the study of language explored in the chapters in Parts Two and Five. There is still no consensus about what constitutes the cognitive glue or shared conception of the mind that Johnson discussed in his introduction. Of course this is not something new; it has been a recurrent theme in the history of psychology. Yet the question remains and needs to be addressed: What is the nature of psychological phenomena?

In the debate about the cognitive science paradigm and its role in psychology, people commonly ascribe its problems or achievements to a Cartesian conception of mind, while overlooking the importance of Kant. But as Shanker reminds us in his chapter (Part One), Kant's conception of mind has played a central role. In fact, I intend to argue in the following summarizing commentary that crucial concerns of both the history of psychology, and the current debate about the future of cognitive science (as illustrated in this volume), can be illuminated by focusing on the Kantian influence. To be even more specific, I will try to show that psychologists have been trying to refute Kant's claim that psychology is not and cannot become a science. They have done this by trying to show that one can apply experimental and formal methods to psychological phenomena, but without seriously questioning Kant's view of psychology's subject matter—in particular without questioning his conception of psychological phenomena as consisting of individual inner mental processes and their behavioral manifestations. Even if, as Johnson shows in his introduction, each of the various revolutions in psychology—structuralism, behaviorism, and cognitive science—has involved a shift in the conception of mind, none of them has radically questioned the Kantian conception of mind and its associated view of how mind is related to the physical and social environment. Thus, none has been genuinely revolutionary in the sense of replacing one set of basic metaphysical assumptions with another. The focus on Kant in this chapter is not intended to minimize or deny other important influences on psychology, like Hume's associationism or Darwin's evolutionary biology, but is an attempt to highlight certain recurrent themes and problems.

Let me now turn to the Kantian conception of mind and the role it has played in psychology.

While Descartes, Locke, and Hume made mind into a separate sphere of study, it was Kant who separated this new field of study from philosophy. Although, for example, Hume tried to emulate Newton's physics in the study of mind he, like Locke, constantly confused psychology with epistemology. On the other hand, Kant made a sharp distinction between these two types of inquiries. According to him, philosophy's role was the *a priori* investigation of the rational structures or pure rules of understanding that were capable of organizing the mind and synthesizing experience. This Transcendental Ego or pure thought was lacking in content and could not be studied empirically, since it was the precondition for all empirical and scientific study. But humans also had bodies and psychological functions—that is, empirical selves—that could be studied empirically. Thus, it followed that psychology was limited to the empirical aspects of mental life or the empirical self—that is, to subjective experience or the internal sense, and human behavior.

Although Kant thought that psychology was an empirical discipline, he did not give it scientific status, but thought of it only as natural history—as classification and description of psychological phenomena. Psychology was not a science according to Kant, because internal sense or subjective experience could not be mathematically described and analyzed, which he thought was a precondition for science. Furthermore, it is not possible to study internal sense or experience experimentally, because it could not be dissolved into recombinable elements, and introspection altered what it studied. In addition, a person only had access to his or her own experience, which ruled out the possibility of intersubjectivity. Thus, psychology was no more than the systematic ordering of facts (Kant, 1786/1970). However, the study of mental life was not limited to internal sense but encompassed the study of character, moral and other kinds of behavior, different faculties, and many other things, and was part of what Kant called anthropology. This was an interdisciplinary enterprise using many different methods like introspection and the study of human behavior, and exploiting other disciplines like history and literature, as well as commonsense knowledge (Kant, 1798/1974).

Various later developments in psychology, from Herbart's mathematical psychology and Fechner's psychophysics to today's cognitive science, can be seen as attempts (and failures) to develop a science of psychology contrary to Kant's prescription—that is, to give experimental, mathematical, or formal account of mental phenomena, but without questioning his fundamental assumptions about psychology's subject matter, that is, the individual inner experiences and their behavioral manifestations. Structuralism or introspectionist psychology, behaviorism, and, more recently, cognitive science are all examples of such attempts. Each of them explores a different aspect of the Kantian framework: that is, subjective experience, behavior, and the mind's organizing categories.

Let me now briefly explain how structuralism, behaviorism, and cognitive science all fall inside the boundaries of the Kantian conception of mind, and the recurrent problems to which this has led.

Structuralism, which grew out of Wundt's experimental research program, was an attempt to apply experimental methods to Kant's empirical self or inner sense—that is, subjective experience. Wundt was well aware of the difficulties Kant had pointed out about introspection, and tried to overcome these problems by restricting his and his disciples' work to the study of inner sensation and perception. Simple physical stimuli were manipulated in aspects like size, density, and duration, and such things as the subjects' reaction time or their naive or unreflected experience were immediately reported. According to Wundt, other aspects of mental life such as thought, language, emotion, and different kinds of social and collective behavior could not be studied scientifically, but were part of *Volkerpsykologie* along the lines of Kant's anthropology.

Later developments, like Külpe's and Titchner's attempts to expand this experimental approach to include more and more aspects of subjective consciousness, were, as Johnson points out in the introduction, a failure. The first attempt to turn the "Kantian mind"—internal sense or subjective experience—into an object of science failed in a way that would be repeated by subsequent attempts—namely, it either excluded (as in the case of Wundt) or failed to account for central psychological phenomena (e.g., Titchner and Külpe).

Behaviorism, understood as a reaction to structuralism's methods and subject matter, failed in a similar way. Instead of introspection, which had failed to satisfy any of the criteria Kant set up for scientific psychology, psychologists turned to the study of behavior, which was part of Kant's anthropology. Unlike subjective experience, behavior was a public intersubjectively observable entity that could be controlled, subdivided, and measured. Laws, preferably mathematical laws, were proposed to link independent and dependent variables. (Segalowitz and Bernstein—in Part Three—mention Hull's model of rote learning as a paradigmatic case of this.) The behaviorist approach, drawing on diverse sources like associationism and evolutionary biology, was an improvement in method, but at the expense of subject matter. Subjective consciousness was totally dismissed; with time it became apparent that behaviorism could account for less and less of mental life. Skinner's failed attempt to explain language and language acquisition is an excellent illustration of this.

The next step, the cognitive revolution in the 1950s, is interesting in several aspects. It provided a new departure both in methodology and in subject matter, albeit still inside the Kantian framework.

As Shanker shows in his chapter, instead of studying subjective consciousness or behavior, psychologists and others turned to the study of the mind's organizational activity—the basic categories organizing how humans utilize experience, which amounted to a naturalization of Kant's transcendental psychology. In a sense this was a radical attempt to refute Kant, in that cognitive science focused on the Transcendental Ego, which, according to Kant, could not properly be a subject of empirical investigation, even less an object of scientific study. Thus, this amounted to an attempt to apply scientific methods to the only remaining aspect of the Kantian conception of mind (after structuralism and behaviorism had tried and failed to do so with the others).

Cognitive science also systematically adopted a different scientific approach, in the sense of deserting the dominant inductivist conception of science in favor of hypothetical deductive methods (see, e.g., Harré & Gillett, 1994). The mind's basic structure or organization, which is not accessible by introspection nor directly reflected in behavior, is indirectly accessible through a set of hypotheses about unobservable mental processes that can be tested by their consequences. (Agassi in Part Two critically discusses Chomsky's contribution to this change.) This was an important step toward opening the door for the formal and mathematical hypotheses and models of mind that Turing and others introduced. Boden's chapter, as well as the chapters in the connectivist section (Clark, Bechtel, van Gelder, Dror, and Dascal) give examples of further advancements.

The cognitive revolution promised to provide explanatory tools to account for the "higher mental processes" with which Vygotsky and Piaget were concerned— that is, symbolic, or in Bruner's term (see Part Five), the meaning-making activities of humans. It promised "to illuminate all areas of psychology: human and animal, normal and abnormal, cognitive and affective, perceptual and motor, social and individual" (see Boden's chapter in Part One.) It promised to study mental processes inaccessible to public observation by proposing and testing computational and later connectivist models of the mind. It promised to do what structuralism and behaviorism failed to do: namely, to give a scientific—experimental and formalized— account of hidden mental processes. In short, it promised a revival of scientific mentalistic concepts and explanations.

Furthermore, the revolution was interdisciplinary in its intent, proposing a unifying approach for psychologists, linguistics, anthropologists, sociologists, historians, etc. It thus also promised an integration not only of psychological research, but of many different disciplines concerned with mental life.

Have these promises been fulfilled? Is psychology finally in possession of formal methods capable of accounting for the very center of mental life—that is, the mind's own categories and organizing principles? Again, the same kind of criticisms that people applied to the other two attempts have been raised against the new view as well—namely, that cognitive science fails to deal with phenomena like subjective consciousness and meaning, which are located at the very center of mental life. Early critical voices were raised, for examples, in books like Dreyfus's *What Computers Can't Do* (1972). But a more challenging and interesting fact is that people who once were some of the strongest proponents of the cognitive revolution now have repudiated it. This volume presents several such cases, for example, Jerome Bruner and Ulric Neisser in psychology and Hilary Putnam in philosophy. Why have these former supporters rejected it? And what explains the criticisms of others, like Harré, Shotter, and Coulter?

In his book *Acts of Meaning* (1990), Bruner argues that the promise of the cognitive revolution—to bring "mind," that which behaviorism had neglected, back into the human sciences—has not been fulfilled. On the contrary, cognitive science has come to deal with more and more narrow and trivialized aspects of cognition. In his chapter in this volume (Part Five), he focuses less on the negative aspects of computational cognitive science, but claims that it is, "in some ways, even more dismissive of the private sphere" than behaviorism. Furthermore, it cannot deal with how human beings construct or achieve meaning, which is one of the most central aspects of human mental life and culture.

Feldman, like Bruner, is skeptical of formal models of meaning, consciousness, and qualia. She describes the situation as one's running around in circles trying to go from what seems like mind (formal models) to what the mind is, and a preoccupation with methods, so that the tool one uses has become the model, and places limits on what can be studied. Bialystok expresses the same worry when she says that cognitive science has a tendency to confuse metaphor with the very thing it is studying. A Neo-Piagetian like Pascual-Leone claims that cognitive science not only has proved unable to account for notoriously difficult areas like creativity, but also for such things as learning and development.

If we turn from psychologists to philosophers, we find a similar picture. Putnam argues in his chapter that functionalism in the contemporary sense—that is, the idea that there is a synthetic identity or theoretical identification between mental properties and computational properties—has failed to account for central aspects like meaning. Furthermore the computational model of the mind is a failure, he argues, because a complete formalist account of the mind (which this model assumes) is impossible. Harré, Shotter, Coulter, and Stenlund, in this volume as well as in other writings, express fundamental skepticism about this model and give persistent criticisms of it. Inspired by the later Wittgenstein, they claim that cognitive science altogether misses the mark. The individual psychological processes it purports to study are mere stipulations or products of reification, and thus not genuine objects of study. The formal models employed by cognitive scientists fail to account for any important aspect of mental life, since they are mistaken about the very thing they try to model.

Critical voices are also raised among neuroscientists. In their chapter, Segalowitz and Bernstein focus on connectionist models and argue that this more recent way of developing cognitive science promises more than it can deliver. They claim that such models are unable to provide us with an understanding of either how human mental processes actually work, or how the brain works.

The above claim, that cognitive science can only deal with certain limited aspects of mental life, is rejected by some of its defenders (see Boden's chapter in Part One). But, interestingly enough, some of these supporters also accept this same idea. Jerry Fodor's well-known case of methodological solipsism—namely, the idea that scientific study of the mind has to be limited to its formal internal structure—

supplies one such example. (See Fodor, 1981.) In this volume, Chomsky similarly puts limits on the scope of linguistic cognitive science, by restricting its object of study to what he calls I-language, or the innate, individual aspects of linguistic competence. (See Chomsky's chapters in parts I and II.)

So, not only critics of formal and computational models, but also prominent defenders like Chomsky, claim that at best this approach is only capable of accounting for limited aspects of human mental life. Instead of a fulfillment of earlier expectations, it seems to demand an increasingly limitation of psychology's subject matter. This is not unlike what happened to both structuralism and behaviorism. Introspection failed to give an acceptable account beyond very simple perceptual tasks; and behaviorism remained limited to extremely simple behavior.

The second promise—that is, integration of different areas of psychology and different disciplines—does not seem to have been fulfilled either. Neisser points to the fact that there are lots of experimental results, that is, we know much more about certain things now than we did in the 1950s—but there is no common view of the mind that holds all these results together. Thus, cognitive science has not so far been able to provide the desired cognitive glue, nor even to tackle some of the most central aspects of mental life.

To sum up: there is ample evidence, provided by the failure of structuralism and behaviorism, and the limitations of cognitive science, that the Kantian conception of mind—that is, of individual mental processes and their behavioral manifestations— has led the study of mental life up the same blind alley again and again, because it does not provide a way of accounting for many higher mental processes. It is because of this dilemma that the contributions to this volume are so welcome. Together they provide a better picture of the strengths and limitations of contemporary cognitive science (as outlined above), and give anyone interested in human mental life an opportunity to assess the basic metaphysical assumptions that underlie both cognitive science and the alternatives approaches suggested by its critics. On one hand, some of the most creative and productive representatives of the "Kantian paradigm" present their research programs and models (computational, connectivist or dynamic), and argue that we still can gain much understanding from this approach, despite its inherent limitations. On the other hand, we also have some equally creative and productive critics of contemporary cognitive science, who not only present negative evidence, but also propose alternative metaphysical foundations, research programs, and models of mental life. Instead of focusing on the inner structural aspects of the mind, as defenders of cognitive science urge us to do, these critics argue that we should study such things as discourse, actions and interpersonal networks, historical developments, and so on, that are not linked by causes but by norms, narratives, or story lines, and that re-conceptualize how humans interact mentally with the physical world. The sometimes searching and metaphorical language these authors use should not make us overlook the fact that they are proposing substantial new conceptual frameworks and research programs. For example, Neisser and Reed show what an ecological approach entails. Bruner, Harré, Stenlund, Coulter, and Shotter remind us of how much of the social and discursive environment is hidden from us, yet utilized in our mental life, when we are engaged in social and solitary activities of different kinds. Donald and Johnson claim that even though the human mind is largely a cultural and even historical object, it still is possible to study it scientifically.

Thus, the opportunity that faces psychology today is a plurality of metaphysical options, all connected to dynamic research programs. What is new today is the existence of alternatives to the traditional Western conception of psychological phenomena (which I have used Kant to illustrate in this chapter) as limited to the individuals' inner mental processes and their behavioral manifestations. As mentioned above, representatives of these alternative approaches argue that we should reconceptualize the traditional conception of how the mind is related to physical and social environments. Although the question "What is the nature of psychological phenomena?" has not yet received an unanimous or unambiguous answer, the contributions to this volume show that today we have more alternatives than ever to choose from.

The challenge we now face is to make use of this diversity and richness—that is, not to succumb to temptations that leads toward isolation or fragmentation, but to continue the sort of debate, exchange of ideas, and empirical research that the contributions to this volume have begun.

References

Bruner, J. 1990. *Acts of Meaning.* Cambridge: Harvard University Press.
Dreyfus, H. 1972. *What Computers Can't Do.* New York: Harper & Row.
Fodor, J. 1981. "Methodological Solipsism Considered as a Research Strategy in Cognitive Science," in *Representations.* Cambridge: MIT Press.
Harré, R. and Gillett, G. 1994. *The Discursive Mind.* London: Sage.
Kant, I. 1786/1970. *Metaphysical Foundations of Natural Science.* J. Ellington (trans.), Indianapolis, IN: Bobbs-Merrill.
Kant, I. 1798/1974. *Anthropology from a Pragmatic Point of View.* M. Gregor (trans.), The Hague: M. Nijhoff.

Citation Index

Subject Index

abnormal psychology, 61–62
acceleration, 235
act, 348–49
action, 243, 302–15, 348–49
actional mode, 283
action at a distance, 27
action-metaphor, 362
action processes, 82, 97n.6
action-schema, 360
activation patterns, 174–77
Acts of Meaning (Bruner), 111, 276, 283, 311–12, 380
affects, 83
affordances of objects, 256
AI. *See* artificial intelligence
AI/CS. *See* artificial intelligence/cognitive science
algorithms, 234
ambient information, 268–69
American Sign Language, 256
analog representational strategy, 360
analytic pathology, 268
analytic philosophy, 49–50
ancient peoples, 370–71, 373–74n.11
animal learning, 143–45
animal psychology, 59–60
anthropology, 111, 358, 377
anxiety, 62
ape language studies, 156–57, 158
apodeictic modality, 349
arithmetic problems, 197
Aronson-Way treatment, 335–36
artificial intelligence (AI), 45–52, 57–59, 61
    Boden's views, 68, 69, 70, 71, 73
    connectionist, 104, 137
    difficulties of, 238
    evolution of, 137

metasubjective processes and, 76
    programming for, 323
    in realism context, 335–37
    strong vs. weak, 106–7
    traditional, 218
    See also GOFAI
artificial intelligence/cognitive science
    (AI/CS), 335–37, 340, 344, 348–49, 350, 351, 352n.1
Artificial Life (A-Life), 60
assimilation principle, 88
associationism, 263
attention, joint, 254, 256
attention, object of, 282–83
autocuable action, 360
automatic prototype extraction, 172

back-propagation model, 206n.1, 212
basic level categories, 254
behavior, 307–10, 311, 348–49
    insect, 60
    intelligent, 196
behavioral data-language, 305
behavioralism, 304, 311
behaviorism, 4, 6–7, 8, 10n.3, 378
    cognitive revolution and, 110
    cognitivism compared with, 192–93, 276–77, 304, 310
    conceptual framework of, 304–5
    failure of information-based programs, 261
    functionalism as alternative to, 165
    information-processing psychology compared with, 265–66
    learning continuum in, 47
    methodological, 224
    psychological theory and, 47, 49
    subjective, 264–65
    views of mental processes, 262–63